BREAKING THE CHAINS OF AGING

A biochemical drama

Mikhail Shchepinov

"I guess there are never enough books."
— John Steinbeck

"People who like quotes love meaningless generalizations."
— Graham Greene

Book design by Arina Alaferdova and Mikhail Shchepinov

First published in the United Kingdom in 2025 by
The Cloister House Press

ISBN 978-1-913460-97-6

Моему папе, который никогда не терял в меня веры.

For my papa, who never gave up on his straying son.

Perspectives and views

Shchepinov is a true biomedical groundbreaker. His insight, now dating back over 20 years, that the isotope effect might have medical utility was so outlandish that even I, one of the field's prouder heretics, was initially inclined to dismiss it. How glad I am that I put my doubts aside and, in small ways, helped his vision to become the proven concept that it is today. In this book, Shchepinov presents the idea and its development in a form that should be easily digestible by the layperson. Its time has come! – *Aubrey D.N.J. de Grey*

Lipid peroxidation is a fundamental mechanism of oxidative damage that has been studied for over 100 years, by such pioneers as Trevor Slater, Hermann Esterbauer, Al Tappel, John Gutteridge, Jason Morrow, Ned Porter and many others. The recent discovery of ferroptosis has re-awakened interest in iron, lipid peroxidation and glutathione peroxidases although old concepts are frequently presented as novel in the ferroptosis literature. In this amusing, provocative and sometimes iconoclastic book, Misha Shchepinov explores the role of fatty acids, lipid peroxidation and iron in human health and disease, and presents an argument that deuterated fatty acids may be important therapeutic agents, especially for neurodegenerative diseases. It is a good read, I recommend it. – *Barry Halliwell*

Food for thought: suppose there were a food supplement that actually protects brain health? Wouldn't taking it be irresistible? This book describes the biology and chemistry that plays a role in many brain diseases. During human evolution we became dependent on long-chain fatty acids for proper function of nerve cell membranes. Oxidative damage to these molecules is rampant as we age and matters a great deal now that the average span approaches 80. The author of this book Mikhail Shchepinov is the inventor of a supplement, deuterated long chain fatty acids, which shows great promise in slowing the damage and may be protective in many age-related diseases. The book is a charming mixture of science and the authors personal recollections and iconoclastic ideas. – *Charles R. Cantor*

With sharp wit and keen insight, my friend Misha takes us on an extraordinary tour through the biochemistry of aging and neurodegeneration. Oxygen—that double-edged sword driving the fire of life—sustains our metabolism while simultaneously escaping to wreak havoc on our most vulnerable biomolecules: the polyunsaturated lipids. Misha's elegant solution to reinforce these molecules of life offers a brilliant approach to taming one of aging's most insidious culprits, potentially unlocking new pathways in our struggle against neurodegeneration. – *J. Thomas Brenna*

For anyone interested in fats - whether in relation to the body, nutrition, health, longevity, or any other aspect - this is essential reading. The proposed approach to healthy longevity is second to none (as a linguistic aside, *second* in Greek is *deuteros*). I find myself tempted to put it to the test, as my needle is rusting out of existence... – *Kashchei Bessmertnyi*

Contents

Abbreviations

A2E, an eye-specific retinol-ethanolamine conjugate; Aβ, amyloid beta; AD, Alzheimer's disease; ADHD attention deficit hyperactivity disorder; ALDH2, aldehyde dehydrogenase; ALS, amyotropic lateral sclerosis; AMD, age-related macular degeneration; APO(A-E), apolipoprotein; APP, amyloid precursor protein; ARA, arachidonic acid; AREDS, age-related eye disease study; ATP, adenosine triphosphate; AV, anisidine value; BBB, blood brain barrier; BEB, blood eye barrier; BMR, basal metabolic rate; CEP, hydroxyethyl pyrrole; CETP, cholesterol ester transfer protein; CL, cardiolipin; CNS, central nervous system; CoQ, coenzyme Q; COX, cyclooxygenase; CR, chain reaction or caloric restriction; CRP, C-reactive protein; CSF, cerebrospinal fluid; CVD, cardiovascular disease; Cyt, cytochrome; DGLA, dihomo-γ-linolenic acid; DHA, docosahexaenoic acid; DR, diabetic retinopathy; DS,Down syndrome; ELOVL, elongation of very long chain fatty acids; EPA, eicosapentaenoic acid; ER, endoplasmic reticulum; ESR, erythrocyte sedimentation rate; ETC, electron transport chain; FA, fatty acid or Friedreich ataxia; GLA, γ-linolenic acid; GPCR, G protein-coupled receptor; GPX4, glutathione peroxidase 4; GSH, glutathione; HD, Huntington' disease; HDL, high density lipoprotein; 4-HHE, 4-hydroxyhexenal; 4-HNE, 4-hydroxynonenal; HODE, hydroxyoctadecanoic acid; IE, isotope effect; IL, interleukin; INAD, infantile neuroaxonal dystrophy; isoP, isoprostane; KIE, kinetic isotope effect; LC-PUFA, long chain polyunsaturated fatty acid; LDL, low density lipoprotein; LHON, Leber's hereditary optic neuropathy; LIN, linoleic acid; LNN, linolenic acid; LPO, lipid peroxidation; LPS, lipopolysaccharide; LOH, lipid hydroxide; LOOH, lipid peroxide; LOX, lipoxygenase; mitos, mitochondria; MCI, mild cognitive impairment; MDA, malonic dialdehyde; MLS, max lifespan; MPI, membrane peroxidation index; MPTP, 1-methyl-4-phenyl-1,2,3,6-tetrahydropyridine; MUFA, monounsaturated fatty acid; MUI, membrane unsaturation index; NADH, nicotinamide adenine dinucleotide plus hydrogen; NBIA, neurodegeneration with brain iron accumulation; Nf-kB, nuclear factor kappa-light-chain-enhancer, a transcription factor family that controls cytokine production, among other things; NRF2, nuclear factor erythroid 2-related factor 2, a transcription factor; NSAID, non-steroidal anti-inflammatory drug; OLE, oleic acid; OS, outer segments; ox-

LDL, oxidized low density lipoprotein; OxPhos, oxidative phosphorylation; oxPL, oxidised phospholipid; oxPUFA, oxidized PUFA; PCR, polymerase chain reaction; PE, ethanolamine phospholipid; PD, Parkinson's disease; PG, PGD, prostaglandin; PI, peroxidation index; PoV, peroxidation value; (VLC)PUFA, (very long chain) polyunsaturated fatty acid; PL, phospholipid; PLA, phospholipase; PTSD, post-traumatic stress disorder; PLG, plasmalogen; QM, quantum mechanics; RBC, red blood cells; RCS, reactive carbonyl species; ROS, reactive oxygen species; RP, retinitis pigmentosa; RPE, retinal pigment epithelium; SASP, senescence-associated secretory phenotype; SkQ, Skulachev ion; SOD, superoxide dismutase; SPM, specialized pro-rezolving mediator; TAU, tubulin-associated unit; TBARS, thiobarbituric acid reactive substances; TBI, traumatic brain injury; TNF, tumour necrosis factor; TOTOX, total oxidation value; TPP, triphenilphosphonium; TRP (A,C,M,ML,P,V), transient receptor potential channels; TTPA, tocopherol transfer protein; UCP, uncoupler protein; vit, vitamin; VLDL, very low density lipoprotein.

1. Preface

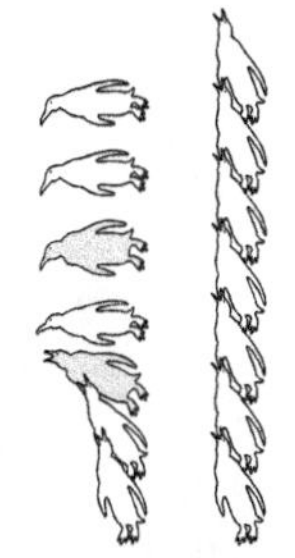

"Knowledge isn't free. You have to pay attention."
— Richard P. Feynman

"No one reads; if someone does read, he doesn't understand; if he understands, he immediately forgets."
— Stanislaw Lem

The tentacles of the main subject of this book - lipid peroxidation (LPO), - reach far and wide. This calls for the examples of LPO to be selected from all walks of life, and death, - resulting in a rather eclectic mix of short stories. It was challenging to do justice to all these diverse lipid oxidation topics in a concise, lapidary fashion, forcing the style to be more **lipidary*** instead. If you find the chemistry in the upcoming chapters too dense with the nitty-gritty, just skip over the formulae until you are comfortably back among words.

---

* The **ANTON** font is reserved throughout this book for mutant bastard words, coined by a half literate non-native speaker.

1.1. PUFA manifesto

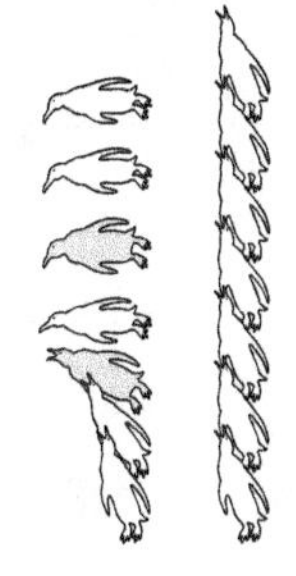

"Nearly everything is really interesting if you go into it deeply enough."
— Richard P. Feynman

"It's simple, if it jiggles, it's fat."
— Arnold Schwarzenegger

Textbooks list proteins, nucleic acids and lipids as the building blocks of life. But how important are these species relative to one another? Imagine spending a day in B, the world's life sciences capital. On a crisp sunny morning you mosey up and down the streets, admiring the views. Tall expensive buildings rise skywards, so high they remind you of reverence-instilling cathedral ceilings, designed to make us lift our heads to look up to Heaven. Those shiny new high-rises provide a great unobstructed view of the skyline, and are filled to the brim with top Scientists, tirelessly striving to push the frontiers of biology beyond that horizon. To tackle the most important challenges that modern medicine is facing, it makes sense to focus on the most important molecule - DNA, and its extended family. The DNA structure is based on just four "letters," these aesthetically appealing blocks of pure logic, combinations of which produce the immense variety of all life forms on Earth. Over the years since the structure of DNA was deciphered, the field has progressed by leaps and bounds. At least 30 people have been awarded Nobel prizes in the RNA area alone (and again, in 2024), culminating in the incredibly fast development of mRNA-based Covid vaccines. The nucleic acids field constantly begets new domains (cloning, Southern blots, antisense drugs, DNA sequencing, DNA synthesis, polymerase chain reaction, oligonucleotide microarrays (DNA chips to an American reader), genotyping, gene therapy and CRISPR to name a few), spawns multiple start-up companies, and attracts the brightest minds, massive funding, prestige and respect. Nucleic acids are the nuclear power source that drives the biotech industry. Trucks and buses shuttle between the buildings, hauling university graduates, wads of research dollars, supplies, materials, TV crews hungry for new discoveries, and, naturally, quite a bit of waste. Findings destined for top-tier journals are discussed at scientific conferences in the nicest locations the world over, with drop dead gorgeous étudiantes, Aspirantinnen and profesoras in attendance. The smartest people keep joining the club, and success begets success.

Gawping in awe at this towering, dazzling citadel of science you keep stepping further and further back to get a more holistic view, eventually retreating far enough to find yourself in an adjacent neighbourhood. Now you are somewhat on the outskirts of the Science capital (with a capital "S"), and being not so close to the epicentre, the buildings look perhaps ever so slightly less glamorous, no longer made of marble, gold and rubies. The grant money and supplies-carrying trucks shuttling between them seem smaller in size and perhaps not moving as fast compared to the DNA district highways. Welcome to the protein sciences area! Still a lot of prestige. Buildings are not as tall, but not by much, not by much. Now wait a minute, what is that grotty eyesore of a building over there that seems to be shut? Oh, that's just a former Department of Protein Chemistry, being refurbished into a new AI facility…

While soaking up the views and making comparisons, you become increasingly aware of some squelching, plop-plop sounds coming off your feet. Finally, it dawns upon you to look down to check it out, and not a moment too soon, as you narrowly avoid tripping over an ajar manhole cover... blimey, what is all this off-putting, jiggling goo that oozes out of the sewer hole, staining the pavement? Looking gross and gloopy, with whiffs of stale rancidity wafting off – perhaps the manhole reaches all the way down to a malfunctioning septic tank, if not straight to hell? It even looks like there is some faint otherworldly light flickering enticingly down there… And of all places, right here, in the dazzling capital of biomedical Science? Horrible. Upset with the indecorous sight of the sludge, you bend down to zoom in, with morbid curiosity, on the cast iron lid. The embossed small letters read:

DEPARTMENT OF LIPID SCIENCES ...

This surreal fable, if preposterous, intends to vent the author's frustration at how the unsung lipids are often deemed less important* , the third-class citizens of biochemistry and research funding alike, not nearly as popular as the other two members of the big three. What a mistreatment for molecules which might have been there first, at the very beginning, helping incipient life to compartmentalize itself, a vital job which the humble lipids continue doing to this day. What an undeserved neglect for the molecules that have also been the cause of more disease and suffering than all other evil agencies to ever afflict the human body. The molecules that have been complicit in killing innumerable billions. What unfair discrimination against a class of compounds essential both for our diets and our membranes. People are just too preoccupied with the central dogma of "DNA to RNA to protein" information transmission. Lipids are relegated to the life sciences backwaters, even though one of the proposed origin-of-life hypotheses posits that once upon a time in the deep ocean, lipids might have made the first crucial step towards life, by creating separate aqueous partitions, the matryoshka doll-type "bags in bags" compartments[(1)] *(Fig.1)*, which allowed various gradients to form across these ancient lipid membrane prototypes, providing the driving force for all kinds of chemical reactions. An astonishing 40% of the total energy output in our bodies is expended maintaining potentials across lipid membranes – that's how important lipids are. Yet current interest in lipids, apart from studying fat as an energy source, largely focuses on helping the body to get rid of them, shedding them or otherwise reducing their load somehow, to decrease the size of the adipose tissue organ – of all biomolecules, lipids are the only class to have their own organ, made of 50 billion adipocytes – or decrease their deposits in our cardiovascular network. There is also a trend, in nutrition but not so much in biochemistry, that quite rightly concerns itself with the worrying abundance of one type of polyunsaturated fatty acid (PUFA), an omega-6 (n-6), at the expense of another, an omega 3 (n-3). This imbalance comes from a massively

---

* Except in atherosclerosis or obesity.

Figure 1. Russian lipid dolls: lipid bilayers form assemblies which fully enclose smaller structures, creating a nested arrangement. This "bags in bags" compartmentalisation is similar to the Matryoshka layout. Bilayers may play a role of relay stations helping transfer oxygen to the smallest compartments.

distorted Western industrial food chain that uses grain (source of n-6) rather than green leaves (mainly n-3) as the source of oils. The normal, one-to-one n-6 to n-3 preindustrial ratio now stands at a shocking twenty to one, wreaking biochemical havoc and creating numerous modern health issues. This critical topic was recently covered comprehensively by Anthony Hulbert of Wollongong University in his popular book[(2)].

As a subgroup of fats, PUFAs make up one the most important, yet one of least addressed, areas of life sciences. Only occasionally does a nonconformist trailblazer appear, trying to challenge the status quo, like the DHA pioneer, nonagenarian Michael Crawford[(3)]. Sadly, these mavericks are few and far between. PUFAs are indispensable for neurons and eyes, mitochondria and inflammation. We are human because we have a brain. As Thomas Edison put it, the chief function of the body is to carry the brain around. (I would modify it ever so slightly, by assigning one more vital function to the body, with the two of them often working in tandem, helping each other. Says Woody Allen about his brain: "It's my second favourite organ.") Our brain is a brain because it is made up of neurons. Our neurons are neurons because they have plasma membranes, which act so to speak like transistors on a chip. All the neuronal DNA-protein innards are in there just to maintain the membranes in working order. And the neuronal plasma membranes are what they are only because they are made of PUFAs. The same can be said about our mitochondria, our eyes, our immune system, and the list goes on. So important are the PUFAs that in case of their dietary deficiency the body goes out of its way to sustain correct levels in key organs by relocating PUFAs from tissues deemed less important, essentially sacrificing those tissues, to maintain the PUFA homeostasis in the brain, just like a pregnant, or breast-feeding mother shares her own brain DHA with her baby. If the resources are depleted, the neurons and other systems start malfunctioning, leading to severe pathologies.

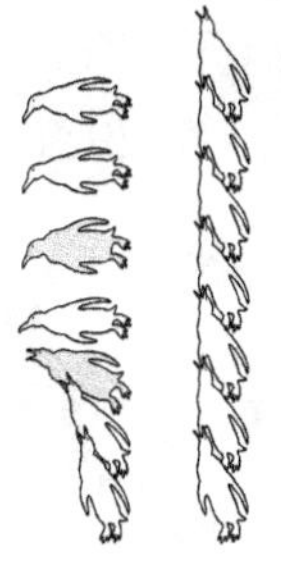

PUFAs are vital for lipid membranes, and key organs and organelles – brain, eye, muscle, skin, mitochondria – cannot function without them. Unfortunately for us, they undergo a unique type of damage - lipid peroxidation (LPO), which proceeds as a chain reaction, so that a single initiating event can lead to multiple PUFA molecules being destroyed, or worse - to be converted into highly toxic end-products. The term "end-products" is a misnomer though, for where LPO ends, pathologies and aging are just getting started. Since LPO is a non-enzymatic process, evolution could not have taken LPO under its direct control. And the evolution's indirect efforts, such as antioxidants and removal of damaged molecules – are not terribly efficient, once the genie is out. It's a bit like using hot air fans and towels to keep the floors dry in a building with a leaky roof: expensive and not fit for purpose. Bad news for us all.

PUFAs are likely key players in neurological and other diseases, particularly age-related, yet even high-profile papers in top journals often neglect to even check if PUFAs are involved. When animal disease models are used in experiments, even if lipids are playing the major role, Experimental sections in published papers regularly show utter disregard for lipids by stating that the mice were raised on "chow" catalogue number so and so. The term "diet" means the components are known and well-defined, while "chow" is more like an unknown unknown, suggesting that the authors have no idea what fatty acids are present in the kibble. Often nor do the manufacturers, for this year a barrel of "hydrogenated fat material" might have been sourced from

Zimbabwe, and the previous year from Argentina, with very different lipid profiles indeed. But when asked they may not necessarily confess, claiming commercial secrecy instead.

Or consider biochemical descriptions of tissues. The available information typically lists protein and fat as a percentage. When pressed, investigators might share some data on phospholipids (PLs), triglycerides and suchlike. Even when the subject of a study is fats, often there is a disproportionate interest in lipids other than PUFAs. One can find many publications providing reams of data detailing the PL types, with all sorts of head groups reported, or perhaps specifics on cholesterol or LDL and HDL ratios, etc. But what about the PUFA composition, PUFA interactions, PUFA transformations? The chain reaction (LPO), the biggest problem afflicting lipid bilayers and lipoprotein particles alike, the chain reaction that sets our bodies smouldering, depends exclusively not on a type of PL head group but on the PUFA composition.

PUFAs, or essential fatty acids, first described as vitamin F (presumably, a tribute to their important role in overseeing the four key physiological activities of all animals, succinctly known as the 4F: feeding, fleeing, fighting, and reproduction) were discovered in 1923. Fast forward a hundred years to a simple question: How does the PUFA composition of mitochondrial membranes in rods or cones compare to the PUFA composition in mitochondrial membranes of the retinal ganglion cells? Or what about PUFA composition of plasma membranes of cones versus rods? Or what mechanisms are responsible for maintaining different PUFA profiles? To this day, these and numerous similar questions, vital for understanding the function of PUFAs, remain unanswered. Never mind possible PUFA-composition variations in 3000 (and counting) different types of brain neurons[(4)], or glia, or mitochondria in different tissues, or inner versus outer mitochondrial membranes. A case in point is an n-3 PUFA known as EPA, a well-known and sought-after fish oil component. Millions take it as a supplement, and it has been approved by the FDA as a cardiovascular drug. It is essential – we (almost) cannot make it and so have to consume it with food, just like all other PUFAs – it is this, and it is that. We have mood swings if it is missing from our diet. With billions of dollars spent worldwide on this commercially successful supplement, it is rather surprising, wouldn't you agree, that no one knows exactly where it all goes once ingested? It is not present in neuronal membranes (although according to Toronto University neuroscientist Richard Bazinet there may be a small – 1% – subclass of glial cells in our brain that may rely on EPA instead of DHA, but even if true that alone would not explain the daily dietary EPA requirement). EPA is not converted into other PUFAs. Perhaps it is converted into some unknown signalling molecules? Is it sitting in some so far undiscovered membranes? Or in lipoproteins? Burned for fuel? Isn't it about time we found out, in the 3rd millennium AD? Yet figuring out the exact geographies and nuances of PUFA tissue and cell distribution would just be the starting point, because the key question is: How do we protect these precious and vital molecules? We may be blissfully ignorant of the hazard, or carelessly neglecting the wellbeing of PUFAs, but Nature knows better than that. Nature is well aware of the clear and present danger of exposing PUFAs to oxidative stress, and spares no resource taking evasive action at various levels. Neural stem-cell (neuroblast) safety is important, but what if the environment puts them at risk by exposing neural stem-cells to oxidative stress? In the workhorse of experimental biology called the fruit fly, PUFAs get temporarily sequestered from membranes of cells exposed to stress, and then packed away into small spheres called lipid droplets and stored in a remote corner, to decrease the chance for reactive oxygen species to detonate the chain reaction[(5)]. When the danger is over, the workfly releases the PUFAs back into lipid membranes.

Can damaged PUFAs be replaced with fresh ones in membranes? The brain works nonstop to minimize the consequences of the LPO that smoulders night and day. While consuming 20% of the body's total energy output, our brain spends a quarter of that (5% of total body energy) continuously re-forming membrane lipids[(6)], by indiscriminately (irrespective of whether PUFA residues are oxidized or not) disconnecting all PUFAs from phospholipid head groups and then re-attaching the good ones back to PLs as part of the eternally running "Lands cycle." A similar approach is practiced by the eye, where 10% of the highly DHA-enriched optical discs of rods and cones are recycled daily by a special retinal pigment epithelium (RPE) cell layer.

Can we just keep taking fresh PUFAs to restock membranes, replenishing the damaged ones which had been removed by this recycling? And as an added bonus, if fish oil is used as a PUFA supplement, then perhaps the distorted omega balance can be fixed at the same time. Would that be two for the price of one? Sadly, in the long-term, providing the extra material is essentially futile. It may, to some extent and temporarily, slow down the decline, but over time, the chain reaction will always prevail. Replenishing PUFAs does not eliminate the toxicity problem of their constantly forming oxidation products. Also, where would the replenishment come from? Logistically, there is about 1 billion km^3 of ocean on our planet. For a current population of 8+ billion, that is only 1 km^3 of ocean per 8 people, and there is every reason to believe that this will soon decrease to 1 km^3 per 10-15 people. Depleting the oceans of fish – for their DHA and protein, or as a discarded side-catch, or due to pollution and global warming, or all of the above – sounds increasingly like a foregone conclusion. Particularly as truly cutting-edge, progressive bottom-trawling technologies designed to scrape the ocean free of metal nodules covering the seabed (deposited at 1 cm per million years, and never mind the collateral damage of sacrificing trillions of bottom dwellers, and whole ecosystems) are already being approved by global-warming-fighting governments. (What's an antonym for "green tech"?) The nodules contain pretty much every element from the Mendeleyev table, so getting them should help satisfy the desperate need of the global population for a new smartphone every year, or every six months if at all possible. It certainly looks helpful (for fish) that the volume of water is increasing, owning to global greenhouse-effect-driven thermal expansion. Whether it can keep up with the demand for fish oil is doubtful. Replenishment of PUFAs on a global scale therefore seems tenuous, unless the existing biotech processes are scaled up to produce it on land.

As PUFAs always get damaged, replenishing damaged PUFAs by brand new ones is challenging and will not eliminate the problem of oxidized PUFA toxicity anyway. The only course of action remaining is antioxidants. Unfortunately, this intervention will not work either, and sometimes may be outright dangerous, as explained in *ch.2.8*. So, can the inevitable PUFA oxidation-driven decline be postponed or slowed down? Are there any options left? There might be. Read on.

1.2. Chain reaction: a portrait in oils

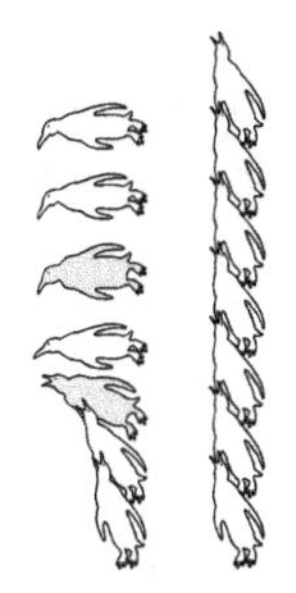

"In classical oil painting, there seemed to be a radical turn..."
— Henry Flynt

"The work of the individual still remains the spark that moves mankind ahead, even more than teamwork."
— Igor Sikorsky

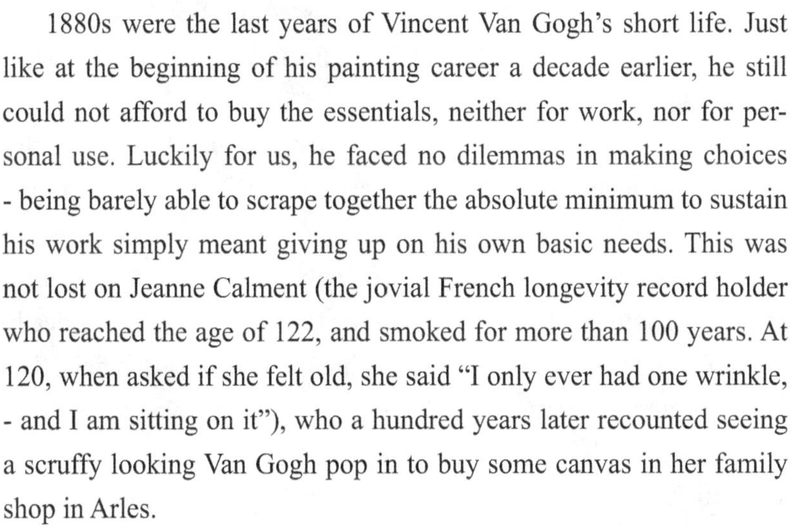

1880s were the last years of Vincent Van Gogh's short life. Just like at the beginning of his painting career a decade earlier, he still could not afford to buy the essentials, neither for work, nor for personal use. Luckily for us, he faced no dilemmas in making choices - being barely able to scrape together the absolute minimum to sustain his work simply meant giving up on his own basic needs. This was not lost on Jeanne Calment (the jovial French longevity record holder who reached the age of 122, and smoked for more than 100 years. At 120, when asked if she felt old, she said "I only ever had one wrinkle, - and I am sitting on it"), who a hundred years later recounted seeing a scruffy looking Van Gogh pop in to buy some canvas in her family shop in Arles.

When he stepped in, Calment's healthy and robustly wired psychological makeup, built to last for a dozen decades, was understandably revolted: "Oh my, do you call those rags "clothes"!? Blimey, look at his shoes, perhaps the very ones he used as models when painting his "The pair of shoes" a few years earlier? And what a reeking, rancid smell!" (Vincent had a bottle of past-the-shelf-life linseed oil in his pocket, bought at the shop next door - having the good fortune to turn up when they were just about to dump the fetid liquid as unfit for purpose, he got a good discount). The young lady's intemperate desire for the man to naff off without delay worked out well for Van Gogh, as he managed to get the canvas cheaply without much negotiation.

Those were Vincent's last pennies. But what about the pigments?

Well, they could be wangled, for the umpteenth time, from his housemate Paul Gaugin, couldn't they? But then again Paul was lately growing increasingly wary of this one-way relationship, so Van Gogh had to tread gently. Shipped from overseas, the blue lapis lazuli pigment was one of the most expensive dyes, and Gaugin was more than a little parsimonious about it. No problem, there would simply be less starry sky in the upcoming landscape painting, the image of which was urgently demanding to be spilled from Van Gogh's fantasy-inflamed brain (a PUFA-dependent quantum machine for generating such images, see *ch.2.5.3*), straight onto canvas. Routine and automatic prep work of grinding pigments, taking care to leave no lumps, and then adding the linseed oil. Boy is it rancid… Hurry, the **(t)rain** will not wait… Up to high doh and finally all set, Vincent dashed off to the countryside…

I first saw the results of this, if a little reconstructed, day of Van Gogh's activities in the mid-70s, in the Pushkin Museum in Moscow *(Fig.2)*. Without further extolling the much-recounted brilliance of the artist, I will just pick out one feature: the chaise in the middle. From up close, you only see a few seemingly arbitrary brushstrokes, remotely resembling a harness. But step back and magically there appears, in full detail, a dejected horse pulling the gig across the field. The true embodiment of impressionism! And that humble bottle of expired oil holds the secret to why we can still enjoy Van Gogh's landscape.

At around 150 years old, Van Gogh's are relatively young as oil paintings go. Hieronymus Bosch was born almost 600 years ago, and Pieter Bruegel the Elder in 1525. The pandemic of the rye-infecting fungus, ergot, was ravaging through the villages of Brabant and beyond for centuries, taking a high death toll. This blight helped fungus-immune potatoes, recently shipped in from the New World, to get accepted. As windmills were shared by peasants, whole villages would get their flour contaminated with the vibrant family* of LSD alkaloids. No one was spared from this "Saint Anthony's fire," and the vividly hallucinating folks fought the monsters, all these Lucies in the Sky with Diamonds, just like the artists themselves did. Look at Bosh's and Bruegel's work to see how powerful the lysergic acid effect can be. Sadly, consequences were even more insidious than the creepiest of monsters, as other members of the ergot family are long-term vaso-

---

* Ergothioneine was first discovered in ergot, see footnote on p.138

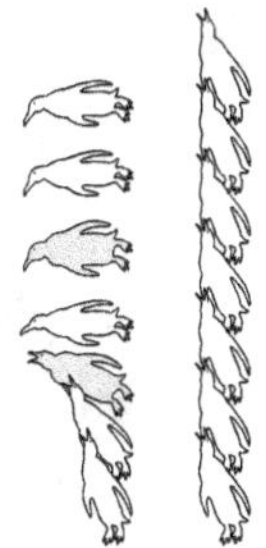

Figure 2. Vincent van Gogh, 1890. Landscape with Carriage and Train in the Background. Not much color blue here, right? Flaxseed oil's LNN underwent the LPO process, preserving the picture.

constrictors, causing people's limbs to dry out and fall off. The only remedy back then was the mandrake berry decoction made by friars. The monks would set up outlets along the roads, handing the drink over to the miserable cripples who were limping along… Mandrake berries were deemed so important that they were frequently painted oversized, often taking up the central space, as in Bosch's Garden of Earthly Delights. The red berries in the Garden look fresh and well preserved, considering the age of the painting – as well preserved as the bewitching lights which shine on, off the paintings of another colossus, Rembrandt van Rijn, who lit them up 400 years ago. I bet they are visible in his Night Watch, gleaming like LEDs right through the canvas in defiance of the laws of conservation of energy, even when the electricity is off at night in the Rijksmuseum in Amsterdam.

Oil paintings adorn art galleries the world over. Created over hundreds of years, they continue to inspire awe, conveying a sobering message: life and human bodies were remarkably similar back then and now. Generosity and avarice, abstinence and indulgence, strength and weakness, good and bad, life and death. But which unifying force was strong enough to carry the message across the centuries, preserving and keeping it intact for us, and hopefully for generations to come, to admire?

Oil paint is a suspension of fine pigment particles in a drying oil, which on exposure to air hardens into a solid film. While in Europe the earliest oil paintings date to the early 12th century AD, the oldest known example, which had been found in Buddhist paintings in Afghanistan's Bamiyan Valley, dates from 650 A.D., with walnut and poppy seed oils employed as drying oil[(1)]. The secret of this remarkable preservation lies in the strength of highly cross-linked polymeric material, formed when PUFAs making up the major fraction of drying oils on exposure to oxygen form the glass-like varnish which can trap and preserve the pigments for millennia. Some pigments may bleach, but the lacquer itself would be stable, if a little darkened. So, the key keeper of magic here is a dollop of the humble linseed (flax) oil.

Hans Baldung, as well as many of his Renaissance fellow artists, was enthralled by the process of (female) aging, particularly at the intersection with sexuality, strikingly juxtaposing the young against the old for a better contrast *(Fig.3)*, and the inevitability of death. There is a certain irony in his famous series – depicting aging bodies in a comparative way in paintings well preserved for more than 500 years – because the process that preserves the paintings is the same one that makes human bodies age and die. The rancidity of that stale flaxseed oil snapped up by Van Gogh; the off-putting smell of the past-its-expiry-date fish or a bag of walnuts; the ravages of aging, mercilessly distorting a female body; and the varnish preserving the artist's efforts for posterity – they all stem from oxygen driving the same chemical transformation: the chain reaction of lipid peroxidation, LPO. And the LPO process is basically the topic of this book.

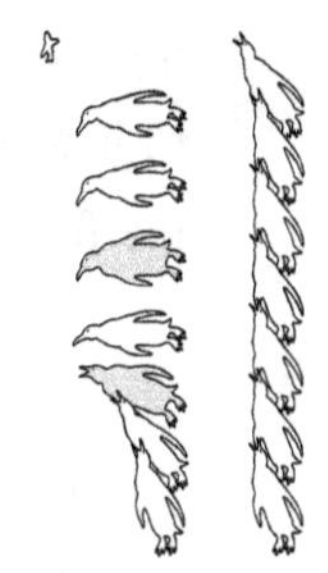

1.3. Linoleum brain

"Fortune favors the prepared mind."
— Louis Pasteur

"Hundreds of galleons lost in the sea..."
— Jim Morrison, the Doors

Novel technologies emerge at the interface of traditional disciplines. The process is unpredictable: we never know which unknown unknowns might be required to percolate through the neuronal networks and bump into each other, and over what period of time, for our brains to spot the opportunity. Who knows how it works? (*ch.2.5.3, 2.5.4*) Yet as always, everything is easy to explain in retrospect.

Growing up in the USSR, I can testify that most kitchen floors in dwellings across the 11 time zones were covered with a very practical, cheap and resilient material: linoleum. From Leningrad to Vladivostok, from Latvia to Belarus, from Ukraine to Georgia, from the Black Sea to the Pacific Ocean. I had no idea back then that, unusually for a construction material, our exhausted planet did not have to be tortured and pillaged to supply the fossilized-carbon starting components, as the stuff was made from renewables.

Frederick Walton (1834-1928), the guy who put LPO on solid footing (pun intended), was an accomplished painter. He once had an art dealer come to inspect his prized art collection. All the items were pooh-poohed bar one, and the following exchange took place:

Dealer: *"Now that is a real gem. Very fine indeed. Who is that by?"*

Walton (in a bashful tone of voice): *"I did that myself."*

It is tempting to speculate that his painting skills – honed over years and involving a lot of prep work, such as grinding the pigments and taking them up in linseed oil and then, most important of all, watching paint dry (productively!) – helped generate his key insight in 1855: the solid resin formed a couple of days later as the paint solidified could have applications other than in art. Excited by the possibility of having discovered a substitute for India rubber, Walton kept optimizing the process, among other things by accelerating the cross-linking using metal salts (*ch.2.4*) and heat. He developed various other products, including more resilient linoleums for floor tiling, and the dainty lincrusta, to be used as wallpaper, and obtained over 100 patents for his new material, establishing the Linoleum (From Latin, linum (flax), oleum (oil)) Manufacturing Company in 1864. So successful and widely used was the new material that it became the first ever generic term to be derived from a product name. More than 150 years after its discovery, the material is still in high demand. Its sustainability and environmental friendliness, resilience and affordability keep it competitive with vinyl, laminate and hardwood. Applications have diversified over the years, and today linoleum is even added to some types of concrete to improve their water tightness, freeze-thaw durability and abrasion resistance. The global linoleum market growth for 2024 is projected to be 5.8%. The US, which traditionally would much prefer materials and processes that require using and burning plenty of fossilized fuels while treating other options with suspicion, has about a 50% share of the global linoleum market (around $5Bn in 2025).

And all there is to this commercial success is LPO of PUFAs. LNN from oil is exposed to atmospheric oxygen, and the rest is the chain reaction. Oxidation, more oxidation, cross-linking, polymerization, done. But look, we have plenty of PUFAs in our membranes too, through which oxygen is constantly percolating *(Back-of-the-envelope-1)*, and many of us are concerned with having insufficient iron so we take it as supplements – well, what about us?

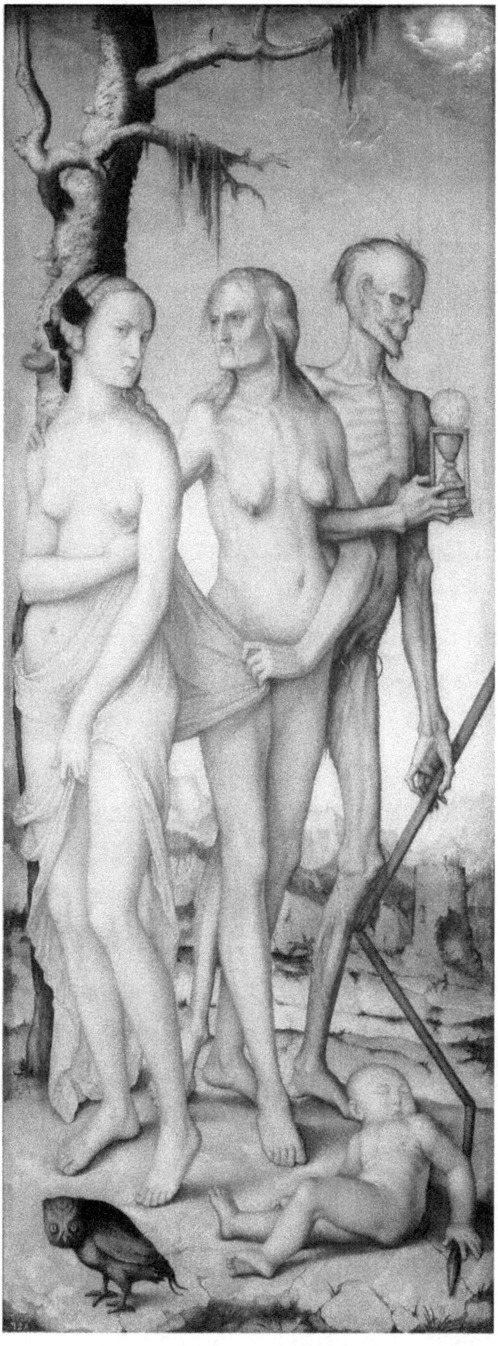

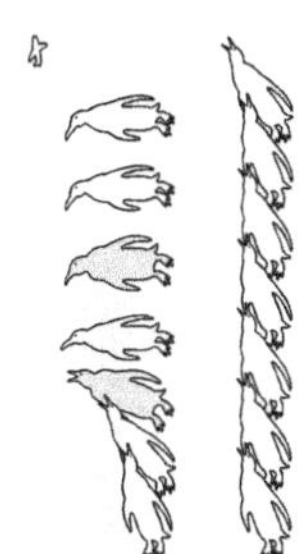

Figure 3. Hans Baldung, 1541-1544. The Ages and Death. The chemistry of human aging is similar to the way oil paint sets, as both depend on LPO.

We will talk about how lipid peroxidation affects the living brain, and what can be done to mitigate the problem, in later sections. However, the dead brain probably belongs in this non-biological introductory chapter, so here we go: Oil paintings and linoleum (just like aging and disease) both rely on LPO. But Alexandra Morton-Hayward of Oxford University is concerned with a more macabre outcome of the same process. Of all our organs, the brain tissue, a soft, squiggly-squirmy gel-like mass that fills up our decision-making cockpits, seems least suitable for long term preservation. From the medical perspective, this is one of the least resilient organs, vulnerable to mechanical trauma and all sorts of stresses and deficiencies. Cut the oxygen supply for one minute, and neurons start dying. However, human brains are preserved far more often than should be anticipated, when all other tissues are long gone, and for much longer than can be expected – sometimes for thousands of years. Special conditions that favor preservation, such as sunken shipwrecks, could be key[(1)], although in many instances the brain is the only soft tissue to survive, with some specimens preserved as long as 12,000 years! As we shall see later, our brain is the fattiest organ of the body save for adipose tissue, and most of that fat is highly oxidation-prone PUFAs. Morton-Hayward suspects that certain conditions, such as abundant iron, favor PUFA cross-linking, creating a stable covalent matrix, although the detailed mechanisms are still unknown. Essentially, our dead brain can self-immortalize by turning into a linoleum oil painting of itself.

A more gruesome question can be asked, which would beat the horrors of LSD-inspired monsters hands down: Is it possible that this "linoleum oil painting" process in the brain starts while we are still alive and well, and not onboard a sunken shipwreck? And if so, what could then be done, at least hypothetically, to slow down or undo this damage, which inexorably ages our bodies and eventually does us all in? We may not be able to "undo" an old, dead sheet of linoleum, but might it be different for cells and tissues which are still alive and kicking, even if a little unwell?

According to Aubrey de Grey, there are 7 pillars of aging which would be good to fix for us to get a chance to live longer[(2)]. Three of the pillars – extracellular aggregates, extracellular matrix stiffening, and intracellular aggregates – are likely directly impacted by PUFAs' tendency to turn into linoleum, as a combination of (a) the result of Denham Harman's age-promoting oxidative stress insight[3], and (b) the chain-reaction aspect of PUFA oxidation (*ch.2.6.1*). But even the remaining four pillars – senescent cells, mitochondrial mutations, cancer, and cell loss – likely all include an element of lipid peroxidation in their advance. Lipid peroxidation could have its tentacles out trying to smother just about every process that we would like to have running smoothly *(Fig.4)*.

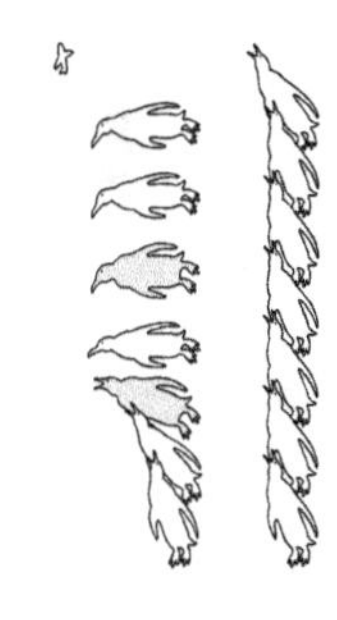

Can we tame this wayward and perfidious LPO, preserving our PUFAs? Perhaps there might be the way.

Figure 4. There is no hiding from, nor defending against the LPOctopus. Its underhanded tentacles easily outnumber the defence systems, and the chain reaction always wins, overwhelmingly. See Fig. 9 for more information on the beast's orbs.

2. Troupe of LPO actors performing in our body

"Find in yourself those human things which are universal."
— Sanford Meisner

"Acting together as a group can accomplish things which no individual acting alone could ever hope to bring about."
— Franklin D. Roosevelt

2.1. Oxygen

"In the presence of oxygen, everything burns."
— Natalie Angier

"Oxygen has been a trouble-maker since the very beginning."
— Doris Abele

When the first life-forms appeared on Earth about 3.5 billion years ago, nitrogen was likely the main ingredient of the atmosphere. 2.5-2 billion years ago in the late Archean (from Greek *arche*, beginning), the Great Oxygenation Event started, and the levels of atmospheric oxygen, produced by anaerobic photosynthesizing cyanobacteria, began to rise. The bugs used solar energy to split water, generating reducing hydrogen to be used in biosynthesis, and puffing out toxic molecular oxygen as waste to make sure it would not kill them. The images of the fossilized main actors are still carved, for posterity, in 2.7-billion-year-old Australian stromatolites. Considering how long ago it all happened, the following short reconstruction may be a little off the mark, but hopefully it still conveys the gist of events. Initially, this oxygen waste product oxidized huge quantities of soluble metal ions such as "ferrous" iron (II) into insoluble deposits of "ferric" iron (III), precipitating them out as red beds. Similar ore deposits were, and still are, also formed directly, by the maverick chemolithotrophs (from Greek, feeding on rock) "eating" the minerals, by oxidizing metals to higher oxidation state, while reducing suitable electron acceptors such as sulphur. For example, polymetallic (manganese, etc) nodules which cover the seabed are populated by a diverse prokaryotic community dominated by manganese (IV) – reducing, and manganese (II) – oxidizing bacteria. The nodules grow at a rate of 1 cm per million years (see *ch.1.1*), horrifying the environment protection activists at the prospect of a large-scale industrial dredging of these deposits, irreversibly destroying the seabed and numerous habitats*.

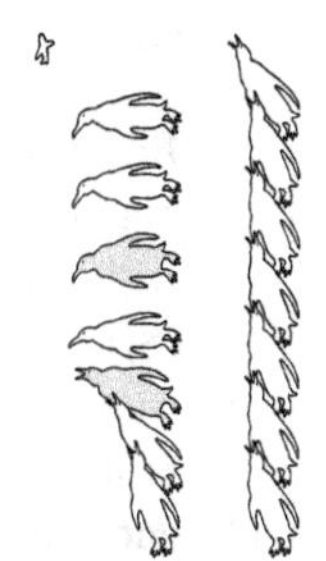

Meanwhile in the upper atmosphere (*Box 1*), gradually amassing oxygen was forming a protective ozone layer, which, by filtering out lethal UV irradiation, allowed life to migrate from ocean to land.

Eventually, oxygen got its hands on most of the atoms it could combine with, and then atmospheric O_2 tension reached a level that required organisms to make a choice: hide from this toxic new fact of life, embrace it, or perish. Countless languid, lazy or negligent types vanished, poisoned to death. Too bad we would never find out which wonderful biochemical gizmos they had evolved to help occupy the environmental niches available back then. The timid and conservative types picked the "hide" option. Today, they are still with us, and they did not change much along the way. They are still single-cell based and still occupy anaerobic, oxygen-free ecosystems like hydrothermal vents, or deep underground, or deep inside us, often using unorthodox sources of energy. Such as methanogens, feeding on hydrogen and carbon dioxide, belching out methane. And some of the timid ones are basically immortal(2)**.

---

* A curious recent paper(1) reports on high voltage potential associated with these polymetallic nodules, so that each one acts like an AA battery, with the source of energy still a mystery, electrolysing water at the abysmal depth, in pitch dark, to yield "dark oxygen". According to the authors this finding may force rewriting the oxygen history, but the opponents say photosynthetic oxygen is needed to produce manganese oxide in the first place. Perhaps some bacteria are involved? We shall see.

** Apart from the permafrost bacteria, microbes, in a state of suspended animation, were found in 4 – 100 million years old seabed sediments.

Box 1. Oxygen atmosphere is very rare, if not impossible, in the Universe, as oxygen is too reactive. And very abundant, at 54 atomic % in the Earth's crust. With time, it would react with almost anything. And it did, so that the main elements in the crust, such as silicon (Si), aluminium (Al), calcium (Ca), iron (Fe), magnesium (Mg), sodium (Na) and potassium (K), not to mention hydrogen, are all in forms that contain oxygen. Nitrogen is one of a few exceptions to still hold its ground (or volume?), but maybe there simply wasn't enough time, or lightning bolts. James Lovelock, the Gaia hypothesis visionary who died in 2022, aged 103, proposed to NASA that to discover life on planets, one had to look at persistent atmospheric disequilibrium, or the presence of chemically incompatible substances, to detect low entropy, characteristic of life. On our planet, this disequilibrium is maintained by an equilibrium of the two main systems used by most life forms. Oversimplifying, organisms either "eat sun" and carbon dioxide to produce biomass and oxygen, or eat "sun eaters" and oxygen to regenerate carbon dioxide.

The organisms which did not hide, but were the active, inventive and entrepreneurial types, - distant ancestors of the American people perhaps, - adjusted, and underwent a spectacular change. Predation was not really an option without oxygen, because weak, flaccid electron acceptors like sulphur would be very inefficient as oxidizers in converting the prey's biomass "fuel" into energy. Oxygen, on the other hand, yields much more bang for the buck or, in practical biochemical terms, more ATP. Using oxygen instead of sulphur would be like using a nitrous oxide canister instead of air in a "hot-rod" internal combustion engine. This opportunity was too attractive to pass, and ancient bacteria jumped on it - and on each other. Excess energy to be gained from eating a neighbour led to predation, and predation led to multicellular organisms, because in fighting, the size does matter. The process took eons: gradual accretion of the atmospheric oxygen levels required incremental evolutionary adjustments. It all peaked during the Cambrian (from Welsh *Cymru*, Wales) period (540-490 million years ago), when oxygen reached levels of 15-30%. The lower boundary of the period is associated with the first appearance of complex life, - arthropods including trilobites, although the exact beginning of the Cambrian explosion is currently being pushed into an earlier, Ediacaran (Ediacara Hills, South Australia) period. Evolving a total reliance on oxygen (try to stop breathing for a minute) produced marvellous innovation and diversity*,

* Consider people living at high altitudes on different continents. Similar to how

culminating, millennia later, in the arrival of Homo sapiens, who then consumed a lot of time, oxygen and fuel, first reconstructing the whole story, and now trying to figure out how to mitigate the oxygen-inflicted damage. One aspect of weaving this poisonous oxidizer into the biochemistry of eukaryotes while oblivious to its ulterior motives particularly shocked and horrified (some representatives of) the H. sapiens species: evolving to depend on oxygen cemented our mortality *(ch.5, Back-of-the-envelope-1)*.

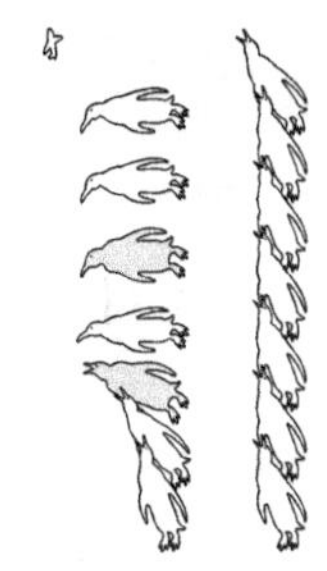

2.2. Reactive oxygen species

"The dose makes the poison."
— Paracelsus

Hydrogen peroxide, H_2O_2, is formed in cells through multiple pathways. It is a side product of various oxidase enzymes, which use two-electron reduction of molecular oxygen, an electron acceptor, to oxidize (attach oxygen) to various atoms, with xanthine oxidase and glucose oxidase being important examples. Insects employ glucose oxidase, the "Oxidase Ferrari" (a nod to its speed and precision) to generate hydrogen peroxide to kill bacteria, and bees use it as a natural preservative in honey, illustrating just how toxic hydrogen peroxide is. Still, H_2O_2 formation is often employed by cells to protect themselves from an even more toxic ROS species: superoxide (*Box 2*).

Electrons, released amply by fuel-burning mitos, are supposed to be passed on along the respiratory chain in a precise step-by-step fash-

---

fish from the South and North Poles never crossbred, these individuals had to develop unique adaptations to ensure sufficient oxygen delivery to their tissues. People from the Himalayas evolved to produce more nitric oxide (NO) to help widen their smallest capillaries, allowing red blood cells (RBC) to penetrate deeper into tissues, thereby compensating for the lower oxygen levels. The Andean dwellers, however, were less fortunate. Their adaptation was to have more RBCs in their blood. And thicker blood means an increased risk of cardiovascular diseases (CVD).

ion, gradually releasing bits of energy in a piece-meal way.

Box 2. It takes two electrons to make up a single chemical bond. An atom, molecule or ion that has just one unpaired valence electron is known as a radical. Radicals are highly reactive, striving to rip off another electron from other species to get a stable pair. Some radical and non-radical Reactive Oxygen Species (ROS) relevant to LPO, and their pecking order, are shown in the Table. Let's publicly name the who, the how, and the what of the brutal oxygen gang:

Name	*Structure*	*Comment*
Oxygen based radicals		
Hydroxyl radical	$HO^{\cdot}$	The most reactive radical, can initiate LPO
Lipid alkoxyl radical	$LO^{\cdot}$	Can initiate LPO; forms from PUFAs
Lipid radical	$L^{\cdot}$	Can initiate LPO; forms from PUFAs
Lipid peroxyl radical	$LOO^{\cdot}$	Can initiate and sustain (propagate) LPO; forms from PUFAs
Superoxide	$O_2^{\cdot -}$	Cannot initiate LPO, but gives rise to other radicals; water soluble
Protonated superoxide	$HO_2^{\cdot -}$	Can possibly initiate LPO; lipid soluble; exists below pH 4.7
Oxygen and nitrogen-based radicals		
Peroxynitrite	$ONOO^{\cdot}$	Forms from $O_2^{\cdot -}$ and NO, decays into chain-initiating $HO^{\cdot}$ and $^{\cdot}NO_2$
Non-radical ROS		
Singlet oxygen	$^1O_2$	Forms from molecular oxygen; cannot initiate LPO but reacts with double bonds
Hydrogen peroxide	H_2O_2	Cannot initiate LPO, but decomposes into other radicals that can

Because an instant release of all the energy, - an explosion, - is a very inefficient method of retrieving every last morsel. But enzymes are not perfect, and even a 99.9% efficient enzyme would still make 0.1% errors. That's only one error per 1000 cycles but imagine how many of those each enzyme performs per day, and how many copies of that enzyme are in a cell. Quite often the electrons go astray from their intended pathway "ruts", and when they bump into oxygen molecules *(Back-of-the-Envelope-2)*, one-electron reductions ensue, forming superoxide ($O_2^{\bullet-}$), a pernicious oxygen molecule with an added electron. As mentioned, according to some estimates this befalls 1-2 % of all the oxygen we inhale.

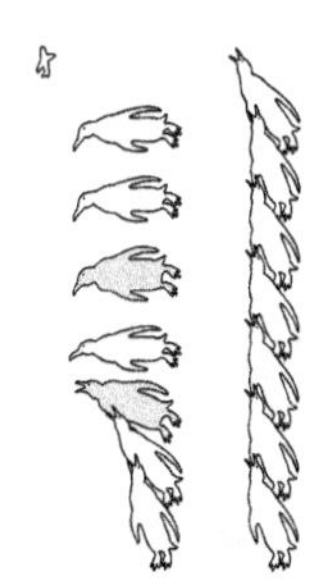

$$O_2 + \text{electron} = O_2^{\bullet-}$$

In response to this challenge, ubiquitous enzymes called superoxide dismutases (SOD) had evolved to detoxify these evil superoxide ROS into merely obnoxious hydrogen peroxide (H_2O_2) molecules. Catalases, another major type of antioxidant enzyme, would then step in to mop this H_2O_2 up, converting it into water and oxygen*.

It is at the intersection of these pathways that the transition metals reveal their dark side. Technically, transition metals form the d-block of the Mendeleyev table (groups 3 to 11), and their peculiar properties are down to the incomplete d - (or f) electron shells, hence the name "d-elements". Transition metals are essential for life. They are an ingredient in a diverse group of proteins comprising metalloproteins and metalloenzymes, and up to one third of all proteins require metals to perform their function. Cytochromes rely on them, as do enzymes that perform Lewis-acid catalysis, iron-sulphur proteins, oxygen transport and oxygen activation proteins, radical-forming and radical-processing systems and many more. All this machinery needs transition metals as they have two or more oxidation states, between which they can easily switch by oxidation or reduction.

---

* In biology, balance is everything. The gene encoding SOD1 lays on chromosome 21, that has three copies in Down's syndrome patients, leading to a 50% excess of SOD1 compared to catalase, encoded by a different chromosome. Hence the life-long elevated level of H_2O_2 in every cell. Amyloid precursor protein (predecessor of amyloid plaques) is also encoded by chromosome 21, and about a half of Down's syndrome patients would develop AD…

2.3. Ionising radiation

"There is no safe amount of radiation.
Even small amounts do harm."
— Linus Pauling

"Space isn't remote at all. It's only an hour's drive away
if your car could go straight upwards."
— Sir Fred Hoyle

If you think that this has nothing to do with you because you do not fly often, are not a member of a nuclear sub crew, live far away from the nearest nuclear power station, do not have a radon-filled cellar, and harbour no plans to relocate to Mars, here is the rub: ionizing radiation fills the universe, and daily ionizing particles and rays collide with molecules of about 1% of our cells, so whatever number of cells we are made of, - somewhere between 36 to 100 trillion *(back-of-envelope-3)*,- the number of hits is still astronomically large.

Radiation of any type – photons, neutrons or charged particles, - may interact directly with target molecules, triggering a biological effect[(1)]. However, as water makes up most of a cell (70-80%), it is more likely that it will be hit first, quickly (10^{-5} – 10^{-4} sec) falling apart to give ROS including $HO^{\cdot}$ which will then do the "indirect", secondary damage elsewhere in the cell. Just to give an idea of the processes that may occur upon irradiation, a water molecule would get hit and fall apart losing an electron and producing a radical cation, which as a very strong acid will immediately lose a proton to give our nemesis, the hydroxyl radical:

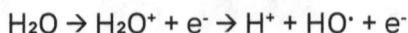

$$H_2O \rightarrow H_2O^{+} + e^{-} \rightarrow H^{+} + HO^{\cdot} + e^{-}$$

While we will shortly learn more about what to expect of hydroxyl radicals, the free electron, surrounded by stabilizing water molecules, will float around until it bumps into a suitable target, - a proton (more likely), or an oxygen molecule (less likely):

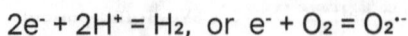

$$2e^{-} + 2H^{+} = H_2, \text{ or } e^{-} + O_2 = O_2^{\cdot -}$$

The stable guy here is molecular hydrogen, the others are ROS, and from this point onwards they would avidly jump right into the ROS–transition metal melee, eager to do mischief. And lipid peroxidation damage would be more substantial than the other types, because of the chain reaction format, leading to an increase in membrane permeability, disruption of ion gradients, and altered activity of membrane-associated proteins, followed by the emergence of a smorgasbord of peripatetic toxic carbonyls (see *ch.2.6*).

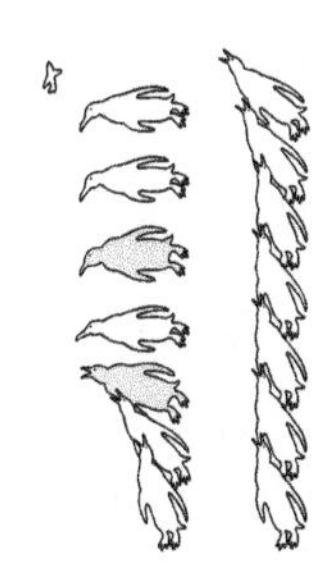

Tardigrades (water bears, moss piglets), the tiny eight-legged arthropods, are ultra-tough bordering on indestructible. In 2007, they went up for a ride on a spaceship and, having spent 10 days in the open space, were found to be just fine upon landing back on Earth. Their other feats – withstanding extreme heat and cold, zero gravity, etc, - pale in comparison. Considering this resilience, perhaps they should make the preferred animal stock* of the first Mars settlers, supplying the inhabitants with fresh meat, garbage disposal, and as pets, provided they are fed some growth hormone upon arrival to beef up their size and protein content[(2)]. The secret to their resilience is cryptobiosis (from Greek *crypto*, concealed, *biosis*, mode of life). They lose most of their water, entering into a suspended metabolic state. With the water removed, the ionisation mechanisms described above would slow down, providing protection against radiation. Perhaps such an extreme dehydration, followed by deep freezing, could be the way to send humans on long space missions without a lethal side-effect of ice crystals tearing through cells and tissues, - provided we could bring the water bears' knowledge (of how-to re-hydrate) to bear, so that we could again start bearing some internal water, a few parsecs away from Earth. Huge savings on central heating for these long missions, too.

There is a poorly understood phenomenon known as bystander effect: irradiating just one cell in a highly targeted way mysteriously

---

* According to Steven Austad, breeding warm-blooded animals for meat is an utter waste of fodder, as most of the energy would be squandered maintaining high body temperature.

causes its neighbour cells, separated from the target cell by some distance, to show the signs of damage. As ROS would not be able to transition that far, LPO products are likely to be playing a role of the signalling species, travelling across to damage their neighbours[(3)]. Tantalizingly, this signalling may turn the neighbour, non-irradiated cells into senescent phenotype, suggesting the effects of radiation can extend way beyond directly hit cells, contributing to various pathologies and aging[(4)].

High levels of oxygen (hyperoxia) can aggravate the ionizing radiation-induced damage to cells, likely through peroxyl radical formation[(5)]. Plasma membranes play a key role in ionising radiation-induced apoptosis[(6)]. Vit E has shown a modest irradiation-mitigating effect on cells, implicating LPO in the process[(7)].

Histological alterations from hyperoxia as well as from elevated temperature look very similar to the radiation-induced damage[(8)*].

2.4. Ferro-cious metals in oxygen environment

"Heavy metal is immortal, but we're not."
— Rob Halford

"Mad, bad, dangerous to know."
— Lady Caroline Lamb (about Lord **B-iron**)

Iron (II) can easily oxidize into iron (III), and this can really challenge the enzymatic antioxidant machinery, because this oxidation, called the Fenton reaction, can be carried out by hydrogen peroxide, which in the process gets reduced into

---

* Breathing pure oxygen at atmospheric pressure for a few hours leads to chest and throat soreness while longer exposure results in irreversible lung damage. The toxicity of pure oxygen was tragically learned the hard way when decades ago, premature babies kept in incubators would develop lung and retinal damage. The higher is the oxygen level, the less well do the animals survive. To connect the dots, high temperature and high oxygen look similar to radiation damage, - and radiation damage, as noted by George Sacher, looks (superficially) like aging[(9)].

the ultimate, apex ROS predator: HO•, the hydroxyl radical:

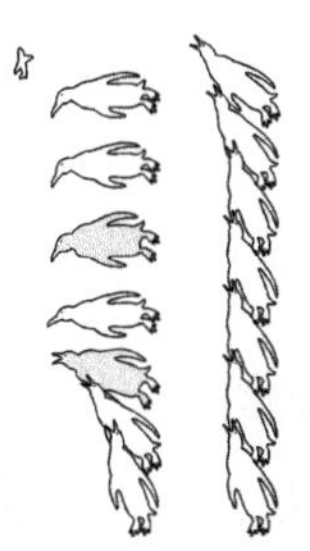

$$Fe^{2+} + H_2O_2 = Fe^{3+} + HO^{\bullet} + OH^{-}$$

HO• is so reactive that any molecule is fair game for it. Oxidizing anything it touches, it is looking hungrily to provide a partner for its unpaired electron. A related complex multi-step process called the Haber-Weiss reaction then commissions superoxide to recycle the oxidized metal back to its more reduced form, simultaneously detoxifying a superoxide molecule. While we are still in the dark on the exact pathways, here is the established key step:

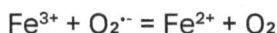

$$Fe^{3+} + O_2^{\bullet -} = Fe^{2+} + O_2$$

Even though in living cells there may not be enough superoxide around to provide the major reducing power for oxidized transition metals, other reducing agents may be available (at concentrations higher than the metals) to do the job – vitamin C, quinones (like CoQ), cysteine, GSH and NADH to name a few. But even at low levels, superoxide can still do mischief by helping release iron from Fe-S clusters or ferritin. On inspecting their structures, you may want to know if LOOH and hydrogen peroxide are related in any way, for don't they look similar: LOOH and HOOH. I am glad you asked! Indeed, they are, and transition metals elongate the LPO chain even more by branching it. They would react with non-radical, stable lipid peroxides just like they react with hydrogen peroxide, converting them into new lipoxy radicals (LO•), which would then branch out into new chains:

$$LOOH + Fe^{2+} = Fe^{3+} + LO^{\bullet} + OH^{-}$$

Pernicious effects of transition metals on LPO come from different oxidation states. For example, various copper ions can decompose LDL peroxides into highly toxic, chain-propagating radicals:

$$LOOH + Cu^{+} = Cu^{2+} + LO^{\cdot} + OH^{-}$$
$$LOOH + Cu^{2+} = Cu^{+} + LOO^{\cdot} + H^{+}$$
$$LOOH + Cu^{2+} = Cu^{3+} + LO^{\cdot} + OH^{-}$$

Iron is the major transition metal *(Back-of-the-Envelope-3)*, but there are many other d-elements in circulation*.

There is only about 0.05-0.12 g of copper in a human body versus 3.5-4 g of iron, - a thirtyfold difference – but copper is actually more active in the Fenton reaction. Worse, when present together in a binary system, iron can help reactivate copper to further enhance its potential in Fenton. While it is challenging to extrapolate from the test-tube data to living cells, the activity of ROS-generating transition metals** is:

$$Cu > Cr > Co > Fe > Mn > Ni$$

Even cobalt can be more active than iron in the Fenton reaction, so do not overdose on your Vitamin B12 supplements. Injecting cobalt in animals led to damage consistent with hydroxyl radical formation – and, this was not known in the early 1960s, when dishwashers became a kitchen staple. A thin film of detergent left on washed dishes and glassware is a pretty nasty thing to ingest, as surfactants interfere with lipid droplet processing in the intestine (rinse your dishes after the dishwasher!), but unexpectedly the residual detergent on the glass surface compromised beer foam formation. A techno fix followed, - sadly, not to remove the chemicals, getting rid of the nasty film, but to add more chemicals, as beer foam-stabilizing cobalt salts. The heart is low on enzymatic antioxidants, which are not of much help against LPO anyhow (see *ch.2.8*), so the ensuing car-

---

* Andy Wahrhol, a modern art genius, used this fact to create unique **masterpisses**. Taking advantage of transition metals in his urine, which would undergo colour-changing redox processing on exposure to air and various substrates on canvas, at the **peeak** of his career he generated an entirely new class of paintings www.warholstars.org/warhol_piss_oxidation.html. I am grateful to Prof. V.E. Kagan for this reference.

** There are exotic forms of transition metals with rare oxidation states which would not fit these series. For example, ferrate (VI) ions are stronger oxidizers than MnO_4. They can oxidize ammonia (NH_3) to nitrogen! Cyt 450 uses Fe^{4+} to activate oxygen.

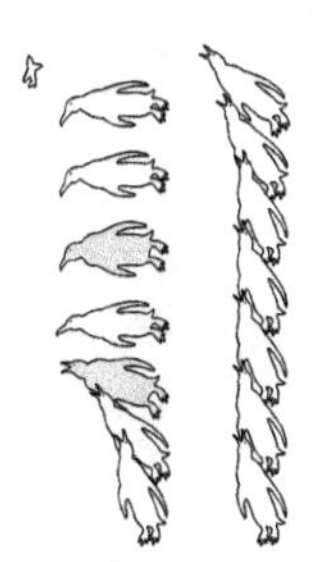

diomyopathy among beer drinkers led to a high mortality rate. While we are on the beer topic, storing acidic beers, or other drinks in copper, iron or steel containers could lead to transition metal overload in regular drinkers ("African iron overload"). To be on a safe side, consider storing and cooking in containers made of gold, if at all possible. This is similar to side-effects associated with piercing (see *ch.4.7*). Gold is not active in Fenton. It is also protected against rust, breakage and inflation. And it looks good!

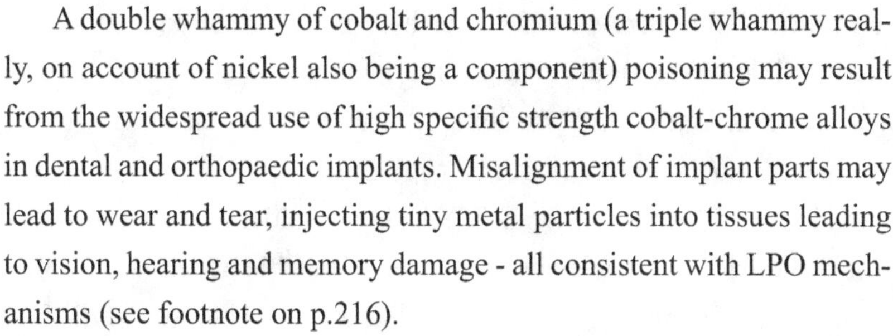

A double whammy of cobalt and chromium (a triple whammy really, on account of nickel also being a component) poisoning may result from the widespread use of high specific strength cobalt-chrome alloys in dental and orthopaedic implants. Misalignment of implant parts may lead to wear and tear, injecting tiny metal particles into tissues leading to vision, hearing and memory damage - all consistent with LPO mechanisms (see footnote on p.216).

Chromium is (almost certainly) not a part of any biochemistry in humans, yet according to a poorly understood mechanism, chromium (III) may enhance the action of insulin (similar claims were made about vanadium). Other oxidation states are quite toxic and carcinogenic. An accepted explanation cites the ability of chromium (VI) to generate hydroxyl radicals. Again, the Fenton-Haber-Weiss chemistry and antioxidants like vitamin C see to it that some of these different oxidation states are fairly interchangeable. On a personal note, exposure to chromium of various oxidation states was, admittedly, one of the highlights of my work in a chemical lab: washing in chromic acid would almost instantly produce the crystal-cleanest glassware imaginable, however messy it was prior. If one is not careful, it would also produce beautiful assortment of holes large and small, in, and sometimes under, one's lab coat. This washing method has now been almost completely phased out on account of "chromium toxicity" (citing Health and Safety).

Nickel (from German *Kupfernickel*, "copper demon", as the mineral was mistakenly thought to contain copper) is less reactive in the Fenton reaction, but as an essential element for the gut microflora, some intake of this probiotic metal may be warranted. There is another side to this nickel coin however: bacterial pathogen levels may be boosted by nickel intake anywhere in the body, creating various problems. And this may be difficult to control, for nickel from piercing, cutlery or steel cookware leaches out helping exceed the 100 microgram/day norm.

Chelating (from Greek, *chele*, claw) agents form complexes with metal ions which can then be moved, or removed, as desired. Most of them are metal-acid salts, but other electron-rich functional groups can interact with metals too, such as amines, ethers (including Crown ethers), or hydroxyls. A well-known household example is adipic acid (not all of it goes to make nylon stockings) used to remove limescale from kettles and plumbing. A tricarboxylic citric acid forms very strong bonds (dissociation constant is a negligible 1 x 10^{-12}) with iron (III), while still being highly soluble in water, hence the household method for rust-spot removal. Protein and small molecule chelating agents, capable of "arresting" metal ions to put them out of commission, may seem like a reasonable way of dialling down on transition metal toxicity, and are widely used by nature. Sadly, the issue is complicated and the opposite is sometimes the case, as some chelate-metal-ion complexes are more, not less, active in Fenton! For instance, nickel (II) ions react slowly, if at all, with H_2O_2, nor with LOOH. However, chelation with proteins or citric acid alters the reduction potential, facilitating hydroxyl radical generation and organic peroxide decomposition:

$$Ni^{2+} - \text{chelate} + H_2O_2 = Ni^{3+} - \text{chelate} + HO^{\cdot} + OH^-$$
$$Ni^{2+} - \text{chelate} + LOOH = Ni^{3+} - \text{chelate} + LO^{\cdot} + OH^-$$

Such enhancement of Fenton chemistry by chelating agents is known for various transition metals and various chelating agents. There are still many unanswered questions at this major intersection of Fenton, ROS generation and transition metals*. Indeed, haem, the iron-complexing cofactor group of haemoglobin,

---

* Some investigators go as far as suggesting that contrary to the mantra of the field, species other than $HO^{\cdot}$ are formed when iron reacts with hydrogen peroxide in the absence of ligands, and that it must be in a form of a complex with a chelating agent for hydroxyl radical production to occur[(4)].

actually promotes ROS generation and oxidative damage.

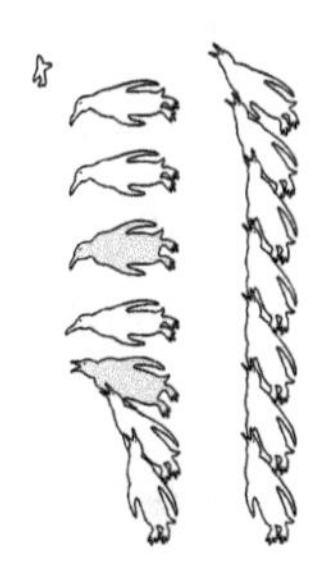

To emphasise, ROS generation by transition metals has the direct, and massive, impact on the fate and well-being of our PUFAs, all through their ability to convert less reactive species into more reactive ones*. And the list of molecules that can be turned into highly reactive and toxic actors by a misfortune of having bumped into a transition metal contains thiols, ascorbic acid, hydrochlorous acid (HOCl), various super important phenethylamines, including adrenaline and dopamine, the omnipresent NAD(P)H, and many more, - but not to get too far into the weeds here, the reader should look up the reference[(5)] for more detail.

Consider a popular if macabre idiom, "to smell blood". Predators are attracted to the odour of blood, while the scent is often aversive to the prey species. But blood itself has no smell, so what is going on? Well, you are actually smelling LPO, and the Fenton reaction. When exposed to air, iron in haemoglobin generates ROS, which oxidise PUFAs in red blood cells into lipid peroxides, which decompose into volatile secondary PUFA oxidation products *(Fig.9)*, giving off the characteristic smell, even when present at disappearingly low levels. Alternatively, LPO can be initiated by other mechanisms, including enzymatic, exposure to oxygen, bad environment, light, or all of the above, and the peroxides formed would then be decomposed by iron. A related "metallic" smell emanates from metal objects - but tellingly, not before we touch them. A strong whiff of the "musty" odour would waft around if a finger were soaked in a solution of iron (II). As with blood, the good old Fenton reaction generates species that get to work oxidizing, or further decomposing, our skin PUFAs (see *ch.4.7*), with several highly volatile molecules, including fungal-metallic-smelling 1-octen-3-one. And the smell is the same if, instead of touching metals, some blood is smeared over skin. And it works the same way with copper.

Chelating agents that trap metals may play mischief with Fenton,

---

* On manganese, see footnote (**) on p. 165.

but at least they could "theoretically" help excrete the excess metals, for which there does not seem to be a dedicated biological mechanism (see *ch.4.2*). And the transition metal overload is a dangerous state to be in, in view of the above. Indeed, numerous pathologies are linked to genetic, or environmental excess of metals. Wilson's disease involves copper overload, while well-known pathologies associated with excessive iron include haemochromatosis (a body-wide build-up of iron levels with some tissues, such as insulin-making pancreatic beta-cells, being particularly sensitive), thalassaemia (faulty haemoglobin synthesis), optic neuropathy, Friedreich's ataxia and numerous orphan diseases. Both iron deficiency and iron overload have profound consequences for neurons. But we ought not to worry too much about the former here, for, from middle age onwards iron levels only increase. Transition metal-overloaded patients show elevated levels of LPO, urinary and plasma isoprostanes, oxidized LDL and HDL, and decreased levels of vit E, CoQ and vit C, - all the hallmarks of the LPO chain going out of control. While the disease implications of this will be addressed in other chapters, one example seems to be a particularly relevant case in point.

A hundred years ago, bloodletting, by physicians and leeches alike, was a staple medical procedure, employed to "rebalance the humours". Now debunked, the procedure, nevertheless provided a process through which, outside of battle injuries, men would be losing iron, - as otherwise there is no reliable mechanism for iron to leave the body, let alone the brain[*]. Over a lifetime, the levels of iron, and copper, in the brain keep growing. And this, according to the Australian neurobiologist Ashley Bush[(7)], coincides with the increasing probability of developing Alzheimer's. The greater incidence of AD in females was explained by a greater activity of synaptic zinc transporters. But additional hypotheses may be put forward. In pre-menopausal women, the natural bleeding mechanism keeps removing some transition metals. Perhaps there is less pressure then, up to that point, to neutralize the adverse effects of transition metals. And with no natural bleeding and hence elevated metal levels compared to women, may be engaging some compensatory mechanisms. But when the period abruptly ends (at an average age of 51 in the USA), women may simply not have enough time to adjust... We cannot live without metals, but they are toxic, mostly through the lipid peroxi-

---

* See also footnote (*) on p.165.

dation path, yet antioxidants do not work (see *ch.2.8*), and the chelating agents are controversial at best. Go figure…*

The Oxidative theory of aging, proposed by Denham Harman in 1950s, has received religious following. Intuitively, it makes sense: the more intense the process of fuel burning in oxygen, the more oxidation products. So, from the 1990s, attempts were made to merge the rate-of-living and oxidative stress theories. This seemed to work in some cases (broadly comparing mammals of different sizes) but not in others (birds versus mammals). Other discrepancies kept cropping up, such as no shortening (and perhaps even extension) of lifespan in rats nor humans following a mild exercise-associated increase in metabolic rates. Likewise, there was no lifespan difference within the same species, when some individuals had a very different metabolic rate compared to others (see *ch.5*). Low levels of ROS are generated in a controlled way and play a vital role for various signalling pathways. And when the delicate balance is thrown off kilter by transition metals, faulty defences,

---

* While zinc is a d-block metal, it is, just like gold, inactive in Fenton because its d-shell is full. However, it can exchange with other transition metals, thus affecting the Fenton process in a non-direct way. The statement "d-shell is full" here and earlier in the chapter sounds as if it were the explanation. Far from it! Gold, with one electron in the outer shell, belongs in the first column of the Mendeleyev table, to the super reactive alkali metals. One would expect it to behave accordingly, yet it is one of the least reactive metals. What if we add an electron to it? That might be like adding an electron to hydrogen, - the resulting helium is not reactive at all. Yet adding an electron to gold "gives" mercury – one of the most reactive metals. Rough models of electron shells are used because we simply cannot calculate wave functions for complex atoms. The chemical bond energy range is 10^0-10^1 eV. Nuclear energies are in the range of 10^7 eV. Why is it, then, that most of the funding is "tunnelled" into the LHC tunnel in Geneva? At 10^9-10^{12} eV, what is a practicality of that research? What urgent, or future, chemical or biological problems can be solved using the LHC? Perhaps a new bomb could be created, 10^6-10^9 more powerful than the Tsar Bomba, but do we need it? And another niggling point. There are two pieces of kit, one costs $100, and another $100Bn to build. Which one, do you think, is more likely to produce a "positive result"? In 2008, this concept, admittedly in drug discovery, received an Ig Nobel prize (more expensive placebos give bigger effects). And still, we have no idea what makes gold – gold[8]…

or myriad other factors, that's the time for LPO to attack PUFAs.

But what exactly are these indispensable yet vulnerable PUFAs?

2.5. Polyunsaturated fatty acids (PUFA)

2.5.1. Were PUFAs chosen by the Blind Watchmaker?

"The imagination of nature is far,
far greater than the imagination of man."
— Richard P. Feynman

"Nature does nothing uselessly."
— Aristotle

We cannot make PUFAs, yet our cells cannot survive without them. But why? No one knows. Even the nature itself may not, as some of us suspect it assembles its clockwork mechanisms blindly. PUFAs are not always abundant in food, so nature goes an extra mile to preserve them in species from cephalopods to vertebrates. If you still want to take a deeper look at what's known or surmised, the rest of the chapter is an attempt to hold together a bunch of disparate facts using various speculations.

At first, FAs were not considered to be essential nutrients because it was thought that they could be synthesized from dietary sugars. PUFAs were initially discovered in 1923 as vitamin F, required to prevent a deficiency disease in rats fed a fat-free diet. By 1929 the husband-and-wife team of George and Mildred Burr realised it was actually made of fats, a paradigm-changing insight. They renamed it as "essential fatty acids". Strictly speaking, only LIN (n-6) and LNN (n-3) belong to this category of truly essential. There are enzymatic mechanisms in place to convert LIN and LNN into long chain (LC) n-6 and n-3 series, such as ARA and EPA/DHA, by desaturation (adding double bonds) and elongation (adding carbon atoms), as shown on *Fig.5*. However, in real life some of these transformations (particularly from LNN to DHA) are inefficient. In young men, only about 8% of LNN is converted into EPA and 0%-4% to DHA[1]. It is higher in young women, at 21% conversion to EPA and 9% to DHA, presumably as reserves for pregnancy and breast feeding[2]. But this conversion level is still too low, in addition to LNN intake being inadequate in the Western diet, so an extra LC-PUFA supply is required. It is rather unclear how this eclectic mixture of PUFAs, coming from sea (n-3) and land (n-6), ended up being essential for us. There are no PUFAs with more than 3 double bonds in plants, perhaps in response to high oxygen levels, unlike in the marine environment[1]. PUFAs are vital for cells yet their intake varies depending on geography and food, so animals treasure their PUFA stocks in tissues that rely on LC-PUFA. According to Tom Brenna at UT Austin, PLs are constantly renewed through the Lands cycle and retina-specific renewal (to be discussed later) to keep membranes in good shape. Antioxidants are pumped in, and various cleaning, detoxifying and ROS-scavenging enzymes patrol the area. But with no direct control over the chain reaction itself, all these secondary measures are just not good enough to keep cells free of LPO products, which accumulate over time leading to disease and aging. However, the pool of LC-PUFAs in the body is significant, and with recycling, due care and preservation may last for quite a long time without replenishment. If rats (excluding pregnant animals) are fed on diets totally devoid of PUFAs, in adult animals, it would take about 15 weeks before the symptoms of essential PUFA deficiency became apparent[4], and two generations before the deficiency becomes lethal. Applying allometric (from Greek *allos* other, metron measurement) scaling, a conversion parameter for metabolic rate difference between different species, suggests it would take about two years of essential PUFA deficiency for the symptoms to become apparent in humans[1]. But why are PUFAs so important?

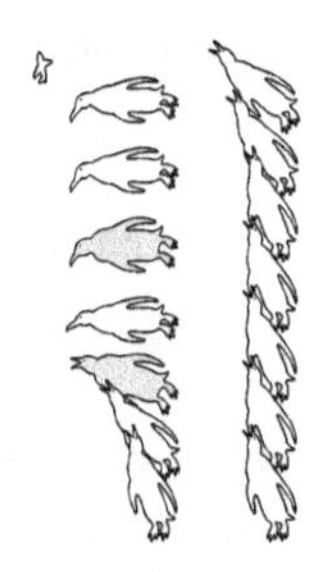

Water is a great medium, yet some biochemical reactions run better in organic solvents. Lipids readily provide such an environment,

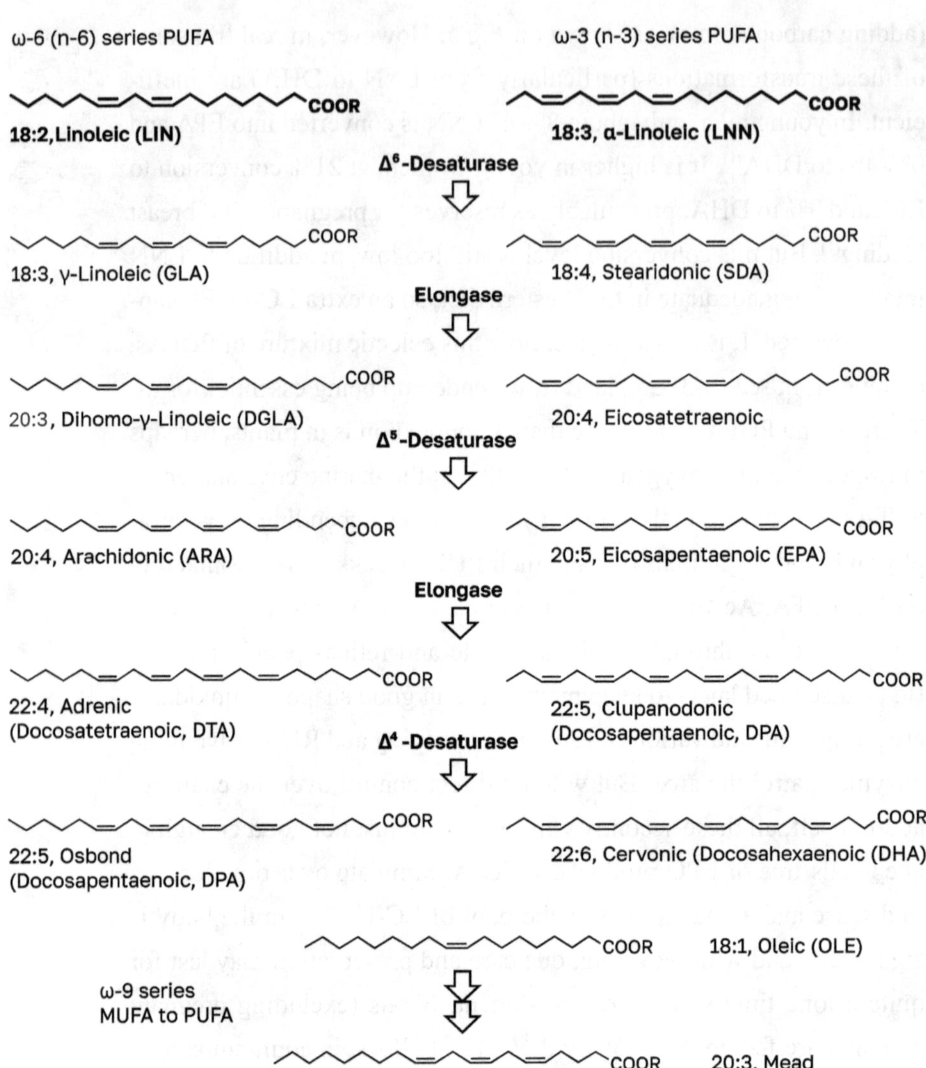

Figure 5. Last carbon in a chain is ω-carbon, counting from the carboxyl group. The first double bond is placed at the 3rd (ω minus 3, or n-3) or at the 6th (ω minus 6, or n-6) carbon. LIN and LNN are essential (cannot be made by mammals). Other PUFAs can be made from LIN and LNN by enzymatic desaturation (adding double bonds) and elongation (adding carbons), with various efficiency. The ω-6 and ω-3 series compete for elongases and desaturases, so too much dietary ω-3 PUFA would result in less LIN converted to ARA, downregulating inflammation, and vice versa. Likewise, GLA-rich primrose oil would reduce the conversion of LIN to GLA. Dietary EPA inhibits its synthesis from LNN. PUFA deficiency forces the organism to convert OLE into ω-9 Mead acid. GLA, DGLA, DPAn-3, DPAn-6, Mead, etc give rise to a number of enzymatic and non-enzymatic signalling species (PG E1 from DGLA or ω-9 eicosanoids from Mead) LC-PUFA have 20 or more carbons.

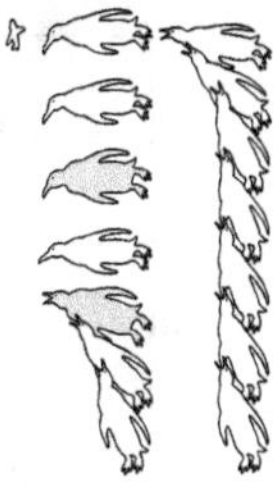

with a caveat: solutions should be liquid and fluid (low melting temperature), allowing for easy diffusion and movement of reactants. As not much movement would be taking place in a piece of solid dairy butter made of saturated fats, PUFA's double bonds may be key here (*Table.1*). Roughly speaking, the more double bonds, the "fluidier" the membrane, made of those fatty molecules. PUFAs make membranes more "liquid", elastic and flexible, while decreasing their order parameter and this is all good*. However, as you can see from the melting temperatures, fluidity of PUFAs only increases up to a point, such that DHA is less fluid than ARA! And here is another but. Having more than one double bond makes PUFAs vulnerable to the LPO (*see 2.6.1*). Somehow nature puts up with this nuisance, yet the way out could have been so easy! If the double bonds are needed for fluidity, but their spacing by one methylene (CH_2) group makes them vulnerable to LPO, then why not move the double bonds further apart? Forgetting about the cis-trans configuration for the moment, a two double bond fragment like:

$$...-CH_2-CH=CH-CH_2-CH_2-CH=CH-CH_2-...$$

would have a melting temperature similar to that of LIN:

$$-CH_2-CH=CH-CH_2-CH=CH-CH_2-,$$

yet would be as stable to LPO as OLE, because it would have no oxidation-susceptible bis-allylic CH_2 units. So why not space the double bonds by two methylene groups instead of one? There are precedents for this in biology. Fatty acid desaturase 3, which puts double bonds into fatty acids, makes sphingolipids with double bonds many carbons apart. Cats are obligate carnivores partially because they are

---

* This relates to cis-double bonds, not to trans-fats (*Table 1*). The latter disturb the bilayers and are difficult to remove from membranes once incorporated. They also negatively affect the LDL-HDL ratios. Curiously, trans-PUFAs are more resistant to LPO than their cis-relatives (Prof. Chryssostomos Chatgilialoglu, personal communication).

Table 1

Fatty acid	Melting point, °C, free acids	Structure
Stearic, 18:0	+70	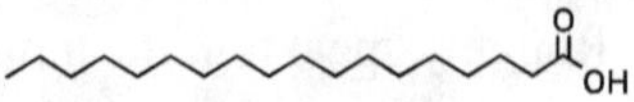
Tuberculostearic, 18:0, 10-methyl-stearic	+13 (pure l or d)	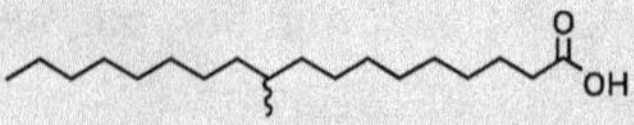
Oleic (OLE), n-9 (cis) 18:1	+13	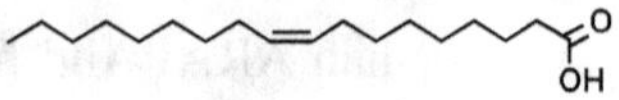
Elaidic acid n-9 (trans) 18:1	+45	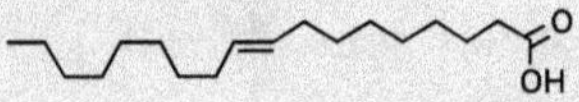
Linoleic (LIN), n-6 18:2	-5	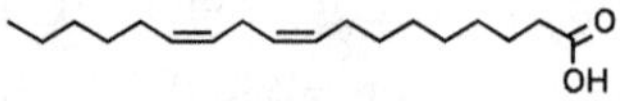
Linolelaidic, n-6, (trans, trans) 18:2	+29	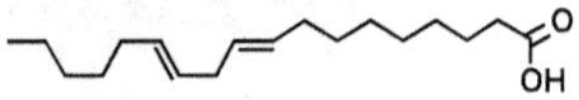
Linolenic (LNN), n-3 18:3	-11	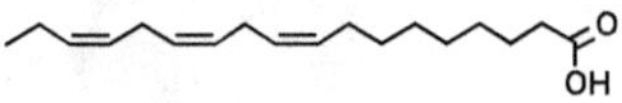
Arachidonic (ARA), n-6 20:4	-49	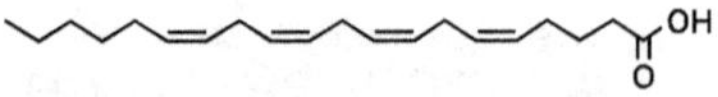
Eicosapentaenoic (EPA), n-3 20:5 timnodonic	-54	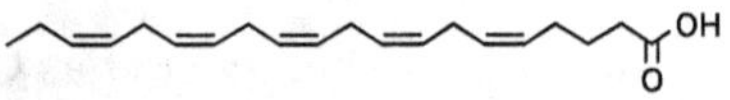
Docosahexaenoic (DHA), n-3 22:6, cervonic	-44	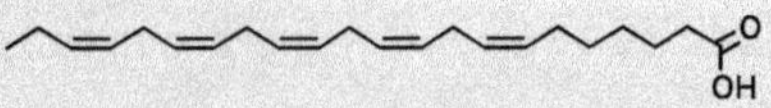

deficient in 6-desaturase, so cannot make ARA from LIN, nor EPA from LNN. However, when their diet is deficient in LC-PUFAs, they will make sciadonic acid with four CH_2 groups between two adjacent double bonds (*Fig.6)*.

Some types of breast cancer lose 6-desaturase and so end up producing sciadonic acid, - perhaps to reduce the toxic LPO? And finally, pine nuts may taste delicious because up to 20% of their fat is pinolenic acid, with two CH_2 groups be-

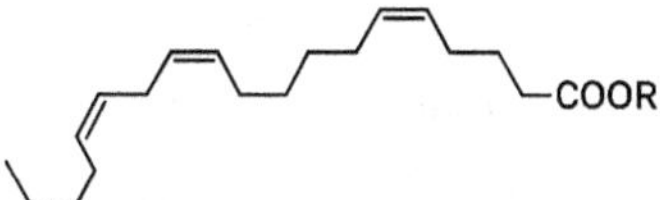

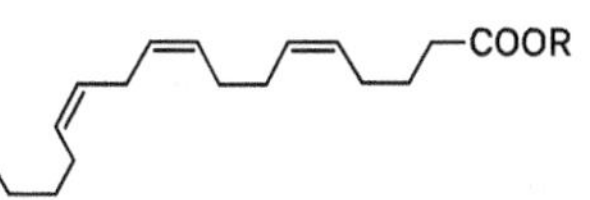

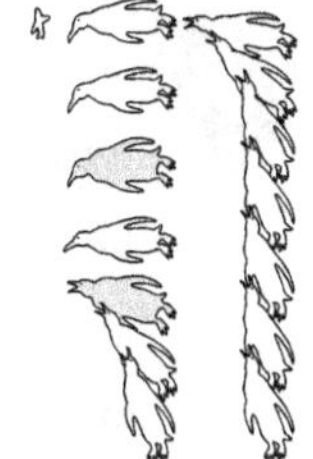

Sciadonic acid, all-cis-5,11,14-20:3n-5 **Pinolenic acid,** all-cis-5,9,12-18:3n-6

Figure 6. Examples of PUFAs where some double bonds are so far apart that there is no bis-allylic positions between them, and hence no LPO at those sites.

tween double bonds (*Fig.6*). So why still use oxidation-prone PUFAs with bis-allylic methylenes?

Karl Hampe's quip "History does not tolerate the subjunctive mood" overlaps somewhat with a menagerie of winged concepts including the butterfly effect (Ray Bradbury) and the Black swan theory (Nassim Taleb), basically meaning that randomly made choices have consequences but cannot be undone or explained away with hindsight. Questions like "what for" and "why" should only be asked with great care if at all. Why do we have an eye structure where the light has to cross a dozen irrelevant layers including blood vessels before hitting the photoreceptors (see *ch.4.1*), in the most illogical and wasteful arrangement possible, – just to confirm that the designer was indeed blind? Or, why do we have a navel*? Was it a "pre-programmed", designer feature? Perhaps eons ago, people had been mostly relying on radishes for food, eating them at night when the predators were asleep. And of course one would find no better place, eating radishes in bed in the middle of the night, for some table salt to dip them in, than one's own navel, wouldn't one. Giving the advantage of more radishes eaten to people with prodigiously concave navels, hence more vigour, and more affection from the opposite gender, = more progeny, hence discrimination against the poor "small navel" folks who died out as the result, - or could the emergence of navel have just been an accident? Too bad we can neither prove nor disprove this experimentally by looking

* Inspired by John Steinbeck's interview in Moscow, USSR, in 1963.

at ancient skeletons.

We will never know for sure why PUFAs were picked. The choices were made on the spot, based on the materials available at that moment, and an unimaginable number of variants were tried and discarded over time. Every evolutionary innovation had been a random accident, rather than a logical choice. This continued for eons, with the only driver being a (slight) reproductive advantage over the comrades who failed to inherit that innovation, thus helping the early adopters of the new gimmick to generate (slightly) more progeny. Take a well-known example of a monkey typing randomly hoping to generate Shakespeare's Hamlet (130,000 letters excluding punctuation). If every proton in the cosmos (10^{80}) were a monkey with a typewriter, typing from the Big Bang to the end of the universe, the chance would essentially still be zero. For a one in a trillion (it is a mere one in 45 million for Lotto) chance of success, you would need $10^{361,000}$ universes, each holding 10^{80} monkeys typing away non-stop. A rather low chance. But what if each time a correct word was typed, it would be selected and put to some use? What if implementing just that one word would already give reproductive advantage? And what if a single correct letter typed would give a tiny benefit? Then just for one monkey hard at work the chance would be one in 26. But once taken on board, it would be difficult to undo the change, as other letters will keep piling up, for – remember, - monkeys can type but not erase… All this is just speculation of course, and we shall never be able to fully reconstruct why PUFAs were taken on board. Jumping the gun, it is unlikely that PUFAs were picked as precursors for eicosanoid type signalling molecules. That must have come much later, when PUFAs were already fully incorporated into the fabric of life, for too many enzymes, like COX and LOX, had to evolve for this job, which would have no use without PUFAs. Double bonds in PUFAs decrease the melting temperature, or increase fluidity, which is important when the bilayers are used by biochemical reactions as a medium, or essentially as a solvent. Curiously, more carbons in FAs increase the chain length while more double bonds shrink it, maintaining the thickness of membranes at about the same value (*Back-of-the-envelope-6*). Many enzymes perform their duty whilst submerged in such solvents, particularly in retinal, neuronal and mitochondrial membranes. LC-PUFA would also help keep membranes liquid at low temperatures, a property rather irrelevant for us warm-blooded animals. Yet, unsaturated FAs and saturated branched FAs both make membranes fluid[6], thus ruling out any unique role for double bond-forming π-electrons in generating membrane fluidity. Tuberculostearic acid and OLE both have T_m of +13°C, compared to stearic at +70°C, so one methyl group is an equivalent of one double bond in reducing the T_m. The first double bond (Stearic to OLE) makes the largest impact on membrane T_m). Bacteria living on ice pump more branched FAs into their membranes when it gets colder[7]. Had logic been used to design the fluid membranes rather than chance, perhaps it would have made more sense to use oxidation-stable branched fatty acids, just like the microorganisms do. Unless it were a cost-saving exercise on the Nature's part, getting fluidity at a cheaper price by desaturating rather than making costly branched chains, as less carbons would be needed? Or, if the main purpose was not the membrane fluidity? The number of double bonds in a PUFA affects membrane fluidity (through intermolecular bonding, as each double bond introduces a kink on a FA backbone. These prevent the FA residues from tight packing loosening up the membrane, hence lower T_m), but is it "the more double bonds the fluidier"? Melting temperature, a parameter increasingly neglected in these days of NMR and MS techniques, is quite revealing here. Melting temperatures (*Table 1*) of stearic (no double bonds; +70°C), OLE (one double bond; +13°C), LIN (two double bonds; -5°C), LNN (three double bonds; -11°C), ARA (four double bonds; -49°C), and EPA (five double bonds; -54°C) form a nice series. However, DHA (six double bonds), this darling of highly functional membranes and nutrition gurus alike, bucks the trend. It is a massive outlier: at -44°C, it is even less "fluid" than ARA. This anomaly suggests that DHA is not used because of its fluidity.

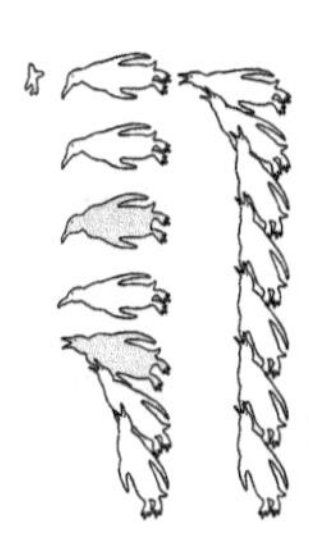

Fish living in freezing waters of Polar Regions do not have more DHA than the natives of temperate habitats, further questioning the antifreeze fluidity purpose of DHA. Indeed, the membrane fluidity in fish from cold and hot waters is controlled by the levels of OLE, while DHA stays constant! For many key tissues, having DHA is non-negotiable and highly conserved[8]. What is it then that makes DHA so indispensable?

Perhaps this has to do with a peculiar geometry of PUFAs. To count as essential, they must have their double bonds in cis-configuration. Having just one double bond in a trans form removes them from this category. They are still taken up by the body all right, but then just quickly get burned down for energy, unlike their treasured cis-brethren[9]. But some trans may get into membranes. Feeding trans-LNN to rats led to detectable levels of trans-n-3 PUFAs in the retina. In humans, a link was proposed between trans-LNN in dietary sources (pre-2002) and higher incidence of intermediate AMD, a trend which disappeared when the level of trans-LNN in food was substantially reduced after 2002[10]. The eye goes an extra mile removing substandard or damaged PUFAs (*ch.4.1*) as do other organs, mostly by relying on the Lands cycle, so the reason for this trans-LNN toxicity is unclear. The selection criteria for essential PUFAs seem to have been based on spatial uniqueness of cis-double bonds as the number one priority, followed by some fluidity considerations.

Upon ingestion, gut lipases release free FA, including PUFAs, from dietary TGs and PLs, for processing by the small intestine and liver. The newly formed PUFA-triacylglycerides (PUFA-TG), PUFA-PCs and free PUFA are bound to chylomicrons, LDL, albumin, and other carriers. There are tissues and cells very conservative in their choice of PUFAs. They go out of the way, sometimes cannibalising the rest of the body to maintain the homeostasis of their PUFA composition. Examples already mentioned include brain (neuronal plasma membranes have DHA to ARA ratio fixed at 55 to 45), eye, sperm tails, skin, etc. Other tissues and structures are more flexible, taking in whatever PUFAs are available in the diet. Red blood cells and non-cellular structures like LDL and HDL* would be good examples[11]. This implies that at least in some compartments, there are delivery systems in place to make sure the right PUFAs are distributed, as passive diffusion of whatever is sloshing about in a bloodstream would not provide such selectivity. Cell culture experiments suggest that cells will take any PUFAs from the medium, and a cell membrane PUFA profile will be very similar to the one of the medium. Where, then, are the control centres, selecting the right PUFAs for their subordinate tissues? Even within cells there is often a difference in PUFA composition.

---

* Which makes sense, considering their lipid transport role.

Optical discs in rods and cones have more DHA than their mitochondria and so do sperm tails, but how is this achieved? For the brain, selection may be taking place at the level of blood brain barrier (BBB). PUFAs dissociate at the BBB through various mechanisms involving endothelial lipases, fatty acid binding proteins (FABP) and apolipoprotein E (ApoE), and free FAs freely pass the BBB. The plasma pool of free FAs is a major contributor to brain uptake of ARA and DHA. Experiments with n-3 and n-6 PUFA-deficient diets confirmed that different mechanisms exist to maintain brain concentrations of DHA and ARA, independent from one another(12). Within CNS, DHA is transported by FABPs and astrocyte-produced ApoE. Membrane-bound DHA cycles out of PLs in membrane to intracellular free FA through Lands cycle, maintaining the membrane homeostasis(13). In 2014, a new DHA transporter protein was reported, with an easy name, Mfsd2a(14). According to the authors the receptor, expressed exclusively in the endothelium of BBB, specifically recognizes a DHA-containing PL molecule with another acyl chain missing, a so called lyso-lipid. This is stated in the title of the paper, "Mfsd2a is a transporter for the essential omega-3 fatty acid docosahexaenoic acid"(15). Yet the authors also report that the same protein carries lyso-lipids (PLs missing one acyl chain) with OLE and even with palmitic residues attached, in fact any FA longer than 14 carbons, so it is not clear what is so DHA-specific about this transporter. A similar uncertainty relates to the DHA-rich eye. While it was claimed that Mfsd2a receptors might be actively transporting DHA from blood into RPE, it fails to explain how the outer segments end up with several fold more DHA in their membranes than their own plasma membranes (see *ch.4.1*). All in all, the selective PUFA delivery question is very much a work in progress, and many questions remain unanswered.

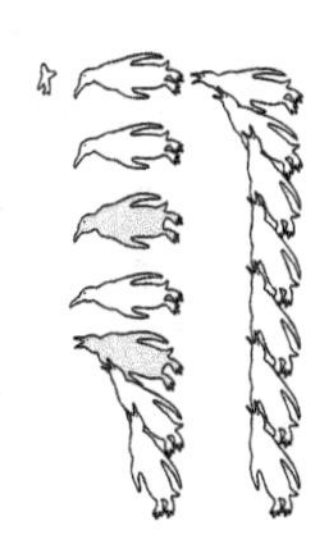

As in the earlier examples of an EPA-dependent subtype of glia and 3000 different types of neurons, we may very well be missing the trees for the forest, because our best analytical techniques only give us average PUFA composition data. What if in the optical discs there is a gradient of DHA from centres to edges? What if this is also the case for axonal membranes or mitochondrial cristae? What if there are special key PUFA points, or bottlenecks, in neuronal membranes? On a single cell level, spatial resolution of our instruments is limited, and nothing still works on a single molecule level, so we are forced to average the PUFA (and other biomolecule) compositions across the membranes*. Let's meet some PUFAs personally.

2.5.2. Portrait of a lipid membrane, in eight PUFAs

"To me the interesting main character is never the one without flaws."

— J. J. Abrams

Linoleic acid (LIN, 18:2n-6)

LIN is my least favourite PUFA, but some major distortions in LIN supply and abundance, causing a lot of global problems, need to be covered. LIN was isolated from linseed oil (hence the name) in Justus von Liebig's lab in 1844. Its structure was worked out 1889 by Alexander Reformatsky, although he did not separate the isomers. The emergence of LIN corresponded to the arrival of seeds, through flowering plants (angiosperms: most trees, grasses, vines and shrubs). They first appeared 360 MYA, went mainstream about 135 MYA and gradually outcompeted the previously predominant gingkoes and ferns. The key feature of them as a group was an abundance of LIN in seeds. Nobody understands how the seed oil content and fatty acid composition determine plant fitness and help adaptation[16]. This coincided with the emergence of mammals, and without getting into the ancient angiosperm weeds, LIN-derived ARA seems to have been the key for evolution of better blood vessels, and made a placenta possible. To this day,

* Just to illustrate the danger of averaging, 8 billion people on this planet, on average, have one breast and one testicle. How helpful is that? Or, how useful would it be to know that the average body temperature across all hospital wards, inclusive of pulmonology, morgue, and infectious diseases units, is 36.6°C?

ARA is the main placental PUFA, while LIN is the main PUFA in skin, responsible for the barrier function and wound healing (see *ch.4.7*).

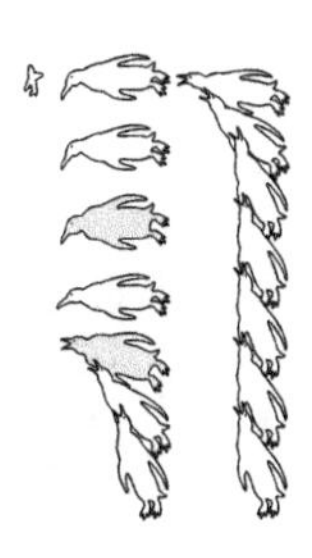

LIN is also the major component of most industrial "vegetable" oils, though the name is surprising as oils are pressed from seeds not vegetables. Why not call them seed oils? A trick to lure people in, considering our mums told us vegetables were good? Canola (65% LIN), corn (50% LIN), soybean (60% LIN) and sunflower (70% LIN) oils totally dominate the industrial world food chain (as the darlings of the processed food and restaurant sectors), although there is currently a very welcome (fingers crossed) trend to genetically engineer those oils to have more LPO-stable OLE. Is the agricultural lobby finally waking up to the dangers of excessive LIN? These new "high oleic" varieties may have in excess of 80% OLE, besting OLE content in olive oils (olive oils are still better as there are other components like flavonoids too, - but more expensive). Due to the huge commercial value of "vegetable" oils sector plus agricultural subsidies, various misleading claims were pushed by the industry to expand the customer base (remember the high fructose corn syrup and tobacco slogans? Just compare: "Discover the flavour" (Marlboro) – and "Taste the flavour miss the fat" (Argola). Sound familiar?). Statements on taste are intriguing, considering gasoline is often used for "vegetable" oil extraction. Cunningly, while making the health claims the lobby indiscriminately bundles n-6 and n-3 (n-3 are really good, see below) together as PUFAs, and then continues extolling the virtues of PUFAs - while totally lacking the n-3 part. One of the most notorious is the claim that "vegetable" oils improve cardiovascular health by reducing cholesterol. It is perpetually perpetrated by the industry, although the best and most rigorous clinical study ever conducted, the Ancel Keys Minnesota Coronary Experiment(17) which ended in 1973, revealed that "vegetable" oils did not provide the expected benefits, and no positive effects were reported. It also seemed to increase risk for cancer. Sadly, Ancel Keys, who lived to 100, for some reason did not publish the results as one report, until 1989. The data was only revealed in a piecemeal fashion over years, giving the "vegetable" oil lobby an opportunity to misrepresent it. Multiple other clinical studies looking at replacing saturated fats with "vegetable" oils were carried out. Invariably, meta-analysis (study of studies) showed no effect on the risk of major CVD events. These developments were nicely analysed recently[(18)]. While we will focus on cardiovascular conditions later, increased levels of LIN may indeed decrease the levels of cholesterol. Fancy ways of explaining this include (1) upregulation of LDL receptor and redistribution of LDL-cholesterol from plasma to tissue, (2) increase of cholesterol catabolism, (3) decreased VLDL to LDL conversion, etc. Put simply, all there is to it is more LIN leaving less room for cholesterol in LDL and HDL particles. But the benefits of that are indeed dubious, as increased LIN will cause LPO in LDL and HDL particles (see *ch.4.6*). In addition, high LIN intake would generate more pro-inflammatory ARA, exacerbating the problems caused by LPO and inflammation (see *ch.4.9*). But most of the excess of LIN is stashed away in the adipose tissue.

If the LIN-rich "vegetable" oils ceased to exist tomorrow, that would not fix the distorted n-3 to n-6 ratio, as livestock would still be fed on grain, the "vegetable" from which the oil is pressed[(19)]. LIN-rich soybeans are grown to feed pigs and chicken, and the soy bean oil is a by-product, which is then fed to humans. Even farmed fish get a hearty chunk of their diet in a form of grain. But at least that would be a step in the right direction. Still, LIN is an essential nutrient and a precursor to ARA, so we will continue to rely on the dietary supply. Just practice moderation.

Linolenic acid (LNN, 18:3n-3)

LNN keeps a relatively low profile compared to the other PUFAs. Discovered in 1887 by an Austrian, Karl Hazura, it was isolated in a pure form in 1909. While in most seeds the predominant PUFA is LIN, there are exceptions. Chia seeds (64% LNN), kiwi seeds (62% LNN; and don't laugh at kiwis: a single fruit can have up to 2500 seeds), perilla (58% LNN), flaxseed (55% LNN), lingonberry (49% LNN); cranberry (35% LNN); sea buckthorn, blackberry and raspberry (30% LNN); walnuts and canola[*] oil level at 10% LNN. Even cucumbers provide a decent helping of LNN[(20)]. Photosynthesising, green parts of plants hold

---

* Canola ("Canadia Oil, Low Acid") is erucic acid-depleted rapeseed oil.

more LNN than seeds because chloroplasts (particularly their chlorophyll-holding thylakoid compartments), on account of their lipid membrane activity, require higher degree of fluidity, although the LNN levels may still depend on geography (the further north the more LNN). LNN is much less suitable for frying than LIN (even though LIN itself is pretty bad (see *ch.3.1*), as it readily oxidizes even at room temperature (see *ch.1.2*).

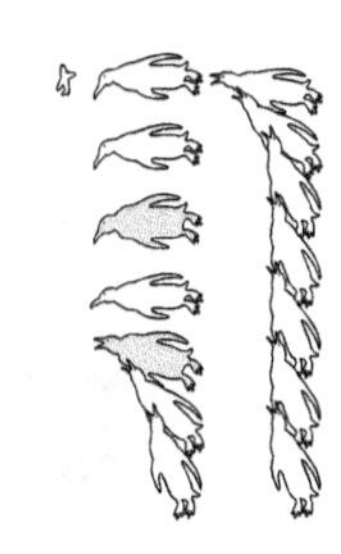

LNN can be enzymatically converted into n-3-LC-PUFA EPA and DHA to some extent, but we do not have to do all this work ourselves. Simply feeding livestock on grass would change the omega balance of beef, lamb, pork etc. in favour of n-3 PUFA (LNN an LC-n-3-PUFA) [4], although not linearly as microflora in ruminants play a role in PUFA processing.

LNN can be enzymatically converted into n-3-LC-PUFA EPA and DHA to some extent, but we do not have to do all this work ourselves. Simply feeding livestock on grass would change the omega balance of beef, lamb, pork etc. in favour of n-3 PUFA (LNN an LC-n-3-PUFA)(4), although not linearly as microflora in ruminants play a role in PUFA processing.

Multiple studies found that higher circulating levels of LNN and LC-n-3-PUFA (EPA, DPA, DHA) correlated with a reduced risk of fatal coronary heart disease. LNN consumption lowers the risk of CVD and improves lipid profiles by reducing triglycerides, total cholesterol, HDL and LDL, decreases blood pressure, and reduces the risk of mortality from all causes[21]. This may have to do with both “washing out” the cholesterol from various lipoproteins (see LIN) and anti-inflammatory properties of LNN, EPA and DHA.

In schizophrenia, mutant desaturases, the enzymes that extend LIN and LNN into the corresponding LC-PUFA, favour the n-6 series. The conversion of LNN into EPA and DHA is therefore reduced, while the conversion of LIN into ARA is increased, depleting the levels of ARA precursors, LIN and DGLA. All these outcomes are undesirable, as LC-n-3-PUFA are anti-inflammatory, and DGLA gives rise to series one anti-inflammatory prostaglandins which also promote vasodilation (blood vessel widening). This pattern can be modelled by depriving rodents of essential n-3[22].

Arachidonic acid (ARA, 20:4n-6)

Named after its saturated relative found in peanuts (from Latin, *arachis* peanuts; not to be confused with *arakhne*, spider), - arachidic acid (20:0), ARA wears many hats. It is a structural component of lipid membranes including neuronal plasma membranes, and is abundant in liver, skin (second most abundant after LIN) and muscle, including heart, although curiously not in the photoreceptor outer segments, - perhaps to help keep the eye inflammation low? In skeletal muscle, ARA makes up 15% - 20% of the PUFA content*.

Feeding rats on ARA as the only dietary source of n-6 would result in some ARA being retro-converted into LIN(23). It is unknown if this mechanism operates in humans. Too much ARA may not be a good thing, as it is a precursor for pro-inflammatory eicosanoids (see *ch.4.9*). One of the key enzymes for converting LIN to ARA, - fatty acid desaturase 1 (FADS1) - is elevated in various cancers[24]. Different ethnicities have different levels of FADS1, but this area is poorly investigated. Rather than being incorporated into PLs during the de novo synthesis, ARA is attached to lyso-PLs through Lands cycle[25,26]. ARA is enriched in some PL classes, while the mechanism behind this selectivity is unknown.

ARA is the starting material for the Arachidonic acid cascade, which refers to the release and enzymatic conversion of ARA by several enzymes, into an immensely important group of active lipid mediators called eicosanoids, which regulates numerous biological functions and is one of the first major steps of turning on the inflammation (see *ch.4.9*). As a free acid, even an intact, non-oxidized ARA is cytotoxic**[27], and pro-apoptic, although it may still get oxidized first through LPO into derivatives that elicit cell death (see *ch.4.8*). Pro[28] -and anti-cancer[29] properties of ARA may both rely on its penchant to get oxidized, through LPO or

* Curiously, kangaroo tails seem to have the highest load of ARA of all animal species. Massive and powerful, the tails essentially convert kangaroos into three-legged tripods, - yet such high levels of ARA are still not understood. Nor is it clear why farmed, but not wild, tilapia have so much ARA (Tom Brenna, personal communication).

** Bodybuilders taking ARA as a muscle growth supplement, beware!

enzymatically.

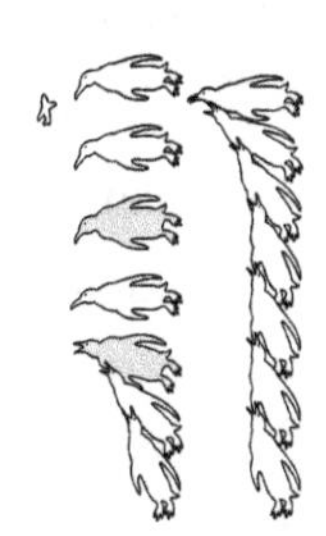

ARA is particularly important when brain, muscle and blood vessel networks are being formed, through its conversion into signalling molecules, by affecting numerous membrane ion channels, but also in necrosis, apoptosis and cell death, - the pivotal events during embryogenesis and early development. ARA is also the dominant PUFA in the inner cell membrane of endothelium that lines the arteries. No surprise then that it features prominently in the mother's milk, both as PL and TG. As a global demand for baby formula (artificial milk) grows, so does the need to find good sources of ARA, which by law has to be an ingredient of the baby formula, particularly for preterm babies[30]. Soil fungus Mortierella alpina is the species of choice for the industrial scale production of ARA[31], for baby formula, even though for adults, taking up to 1.5 g per day showed no benefits[32]. ARASCO, the commercial form of ARA-enriched TGs, is added to the formula to a level of 0.5 % of ARA, or 70-85 mg ARASCO/kg/day. The adult brain consumes about 18 mg of ARA daily[33].

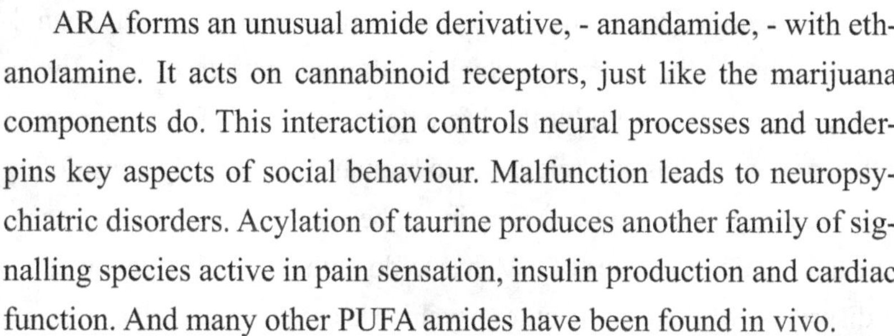

ARA forms an unusual amide derivative, - anandamide, - with ethanolamine. It acts on cannabinoid receptors, just like the marijuana components do. This interaction controls neural processes and underpins key aspects of social behaviour. Malfunction leads to neuropsychiatric disorders. Acylation of taurine produces another family of signalling species active in pain sensation, insulin production and cardiac function. And many other PUFA amides have been found in vivo.

Eicosapentaenoic acid (EPA, 20:5n-3)

As noted earlier, the whereabouts of EPA, also known as timnodonic (from Greek, *timnos*, cod) in the body are shrouded in mystery. We have to consume it, or its precursor, LNN. Yet we do not know which membranes it resides in*. There is some in lipoproteins in blood (EPA getting damaged by LPO is therefore quite possible, see *ch.4.6*),

---

* Except for a possible subgroup of glia, were allegedly EPA takes up the place of DHA[34].

and a well-publicised property of EPA is its ability to thin the blood, preventing platelet aggregation and hence the blood clotting. Hugh Sinclair, an Oxford don, had a seal delivered to him from Canada to engage in a 100-day-long self-experiment, into which his wife was also coerced. Sinclair's peers were much annoyed by stench as the college cook prepared daily seal steaks. He confirmed the anecdotal observations that Inuit and Eskimo people hardly ever had cardiovascular diseases feeding on an n-3-rich seal meat as a staple. A witness account described how Hugh's shoes filled up with blood when he pruned roses in his garden, on account of his clotting factors being disturbed(35). Platelet aggregation prevention is due to the action of prostaglandin-3, an enzymatic oxidation product of EPA*, whom we shall meet again (see *ch.4.9*). This property of EPA was put to a good use by commercializing** a highly pure ethyl ester of EPA, which was approved by the FDA to treat patients with statin-resistant hypertriglyceridemia(36). It seems the only established role for EPA is to sire the series three eicosanoids and related oxidized derivatives. Still, the combined enzymatic and non-enzymatic EPA oxidation products do not account for the quantities taken up(37). Beta-oxidation may balance the books, but would be a waste considering the essential nature of EPA…

Similar to anandamide (see above), EPA also forms an amide with ethanolamine, eicosapentaenoyl ethanolamide, which, in the **workhorseworm** at least, when supplied in diet, was shown to inhibit the life extension effect of dietary restriction(39). Whether or not this has any relevance to humans (particularly considering that caloric restriction probably hasn't) remains to be seen.

One of the enigmatic properties of EPA has to do with its good effect on autistic spectrum disorders, while in some other psychiatric conditions, such as depression, it fared even better(40), outperforming DHA(41). Higher levels of plasma EPA (but not DHA or DPA) were associated not just with the lower risk of depression, but also with decreased dementia and cognitive decline in the elderly(42).

Produced by algae in the marine environment and then working its way up

---

* Seal meat is very high on EPA and DPA, but low on DHA.

** As mentioned, pure ethyl ester of EPA has been approved as a CVD drug. The mechanism may be anti-inflammatory, involving enzymatic conversion to prostaglandin-3 (platelet aggregation inhibitor), thromboxane-3 and leukotriene-5 eicosanoids, although side-effects including musculoskeletal and joint pain, peripheral edema, gout and rash would rather point to its pro-inflammatory properties(38).

the marine food chain, EPA is also made commercially using various fungi such as Yarrowia lipolytica, and various microalgae. Daily intake of EPA for adults should be around 220 mg/day, or 2-3 servings of fish per week, unless your staple food is walnuts with fresh greens under chia seeds, sprinkled with flax oil. In this case, you can reduce your fish intake to once a week.

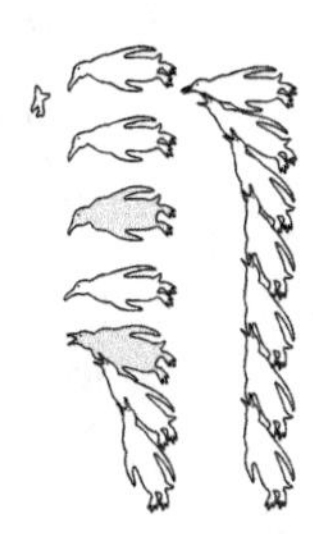

Docosahexaenoic acid (DHA, 22:6,n-3)

When talking about EPA being shrouded in mystery, what descriptor should we then reserve for DHA? Hold this thought for the next chapter. DHA (cervonic acid, from Latin, either *cervus*, deer, stag, or *cerebrum*, brain) makes up over 90% of the n-3 PUFAs in the brain and 10%–20% of its total lipids. At 60% fat, brain is the fattiest organ which consumes 4-5 mg (this may be an underestimate) of DHA per day[43,44], with an estimated half-life of brain DHA of 2.5 years, much longer than in other tissues except the eye *(Back-of-the-envelope-4)*. A lot of DHA "changes hands" during gestation. Mother's milk can be as high as 1.4% DHA depending on her diet, etc*. Brain accretion of DHA reaches a plateau in early adulthood, and the infant brain acquires five times the level of lipids daily as the adult brain. While fish oil is mostly squeezed from anchovies (as TGs) or krill (as PLs), DHA is also made commercially using microalgae such as Schizochytrium[45]. Daily intake of DHA for adults should be around 200-300 mg/day (2-3 servings of fish per week), or about 0.5 g/day of EPA and DHA combined (fish oil). Just like ARA, DHA is a component of baby formula, particularly for premature babies. The commercial form of DHA, a mixture of TGs called DHASCO, is added to the formula to a level of 0.5 % of DHA,

* And none of that DHA is burned for energy production. To enter a mitochondrial fat-burning beta-oxidation process, fatty acids have to be converted first into carnitine derivatives. As of today, DHA-carnitine has not been detected! Damaged DHA can no doubt be chopped down, likely in peroxisomes, but the intact one is treasured, and only used for membrane building. It could be interesting to see if this may change in animals fed on a huge excess of DHA.

or 70-85 mg DHASCO/kg/day.

An average DHA concentration in RBCs in Americans (~4% of total FA) requires 4-6 months of oral dosing to reach a steady state, and the concentration is dose-dependent (9% for 1g/day; 5% for 0.2g/day). Blood levels of DHA can reach up to 5% of the orally ingested quantity, and 10% of that is delivered to CNS. Organs with the highest level of DHA are brain, eye and, somewhat surprisingly, mito-free sperm tails[(46)]. The mechanisms responsible for enrichment are unknown.

Similar to ARA and EPA, DHA also forms an amide with ethanolamine, called synaptamide, and other N-acyl amides have been found in vivo. The observed effects include neuropathic pain attenuation, reduced inflammation in the CNS (in rats), mitigation of TBI symptoms and associated neuroinflammation[(47)] and in general, synaptogenic and neurogenic effects. However, these are early days, and it is not yet clear how the background level of DHA may be affecting the results.

PLs containing EPA and DHA have opposing effects on membrane structure. By X-ray analysis, the structures of membranes made entirely of PL-EPA or PL-DHA were similar: both EPA and DHA residues were bent, looking like hybrids between a hook and a bow. When PLs with OLE and cholesterol were added, the membranes started looking more natural. EPA in such mixed bilayers was found to straighten up into taut extended shafts, and the electron density was mostly located about half way through the lipid leaflet, 10 Å away from the centre of the membrane. DHA, however, stayed bent, pushing electron density closer to the PL head group space, and disordering the inner membrane structure. ARA behaved similar to EPA, also assuming an extended configuration and concentrating the electron density around 12 Å away from the centre of the membrane[(43)]. Strikingly, DHA is again an anomaly here compared to ARA and EPA, just as we say comparing their T_m. The two features are no doubt connected. The "lensing" of electron density closer to the membrane surface seems unique for DHA, and we shall come back to this shortly.

Opposing trends were reported for EPA and DHA on various developmental aspects of neurons, with DHA, unlike EPA, consistently looking indispensable for neurogenesis[(48)]. Whether these are direct effects of EPA and DHA, or their metabolites, is unclear. In human clinical trials of LIN, EPA and DHA in the elderly people with mild cognitive impairment (MCI, an early step of AD), EPA and DHA groups improved their depression scores, but only DHA improved memory and verbal fluency[(49)].

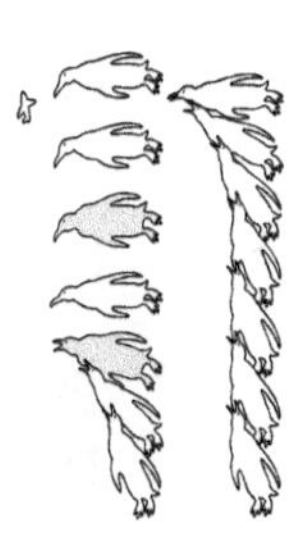

The main claimed benefit of DHA is cognition. The theory of DHA driving the human brain evolution is being trailblazed by Michael Crawford[(17)] and Stephen Cunnane[(50)]. But how did it come to it? Stephen Cunnane may have been the first to propound the idea that our ancestors gaining access to sea food, for the first time in the history of primates, when the sea levels dropped due to the ice age making swathes of shallow intertidal areas available for foraging. Specific archaeological dates include a 1.95-million-year-old proof of seafood exposure in Turkana Basin, Kenya, and a 0.78 MYO finding, in Gesher Yagov, Israel, of cooked seafood.

Meta-analysis of meta-analyses of numerous PUFA supplements trials (LIN, LNN, ARA, EPA and DHA) may be trying to tell us a very simple thing. There may not be any curing effect of any particular PUFA, except if it were deficient before the supplementation started. In other words, maintaining a good omega balance would be the way, rather than distorting it with modern diets and then trying to fix the discrepancy with supplements.

Other PUFAs

Inspecting our PUFA inventory, we occasionally bump into some pretty rare species. There are many intermediate PUFAs shown on the (*Fig.5*). They all form PL, TG or rarer groups such as sphingosine and plasmalogen derivatives. Rare PUFAs often come from the outside*, such as gamma-linolenic acid (GLA, 18:3,n-6) which makes up 10% of primrose oil. ARA can be made from it, suppressing the LIN to ARA

---

* Mammals cannot make n-3 and n-6 PUFAs and so rely on the dietary supply. In rare cases when neither is available for prolonged periods, mammals will start desaturating oleic acid to produce a unique PUFA known as Mead acid (20:3,n-9), to make membranes more fluid. It is used in animal husbandry to assess deficiency of n-3 and n-6. When available, it mostly accumulates in neutrophils and eosinophils, giving rise to leukotrienes C3 and D3[(51)].

conversion.

Docosatetraenoic (Adrenic, 22:4,n-6) acid is a two-carbon extension of ARA, sometimes found in neuronal membranes, its as is its cousin with one more double bond, - docosapentaenoic (DPA; Osbond; 22:5,n-6) acid. An n-3 isomer of Osbond acid is more common as it is an intermediate product of EPA conversion to DHA (clupanodonic acid (from Greek, *klupos*, herring) or DPA, 22:5,n-3). Not much is known about this PUFA except that it is metabolized to resolvins of D and T series which we shall meet later (see *ch.4.9*). Adrenic acid and DPA may play a role of storage reservoirs, to be converted back to ARA or EPA and DHA, as needed. Very long chain PUFA (VLC-PUFA) will be discussed in the (*ch.4.1*).

2.5.3. Or by the Quantum **Mechaniac**?

"With quantum physics, who needs drugs?"
— Richard P. Feynman

"Nobody understands quantum mechanics."
— Richard P. Feynman

Magnetoreception was long suspected in birds. In 1963, Salvatore Bellini at Pavia University discovered magnetotactic bacteria that align themselves with magnetic field of Earth. Wolfgang Wiltschko at Frankfurt reported in 1968 on the role of static magnetic fields on the flight orientation of robins[1], later proving that birds respond to the direction and inclination of the field. In 1977, iron-containing magnetite receptors were found in the snouts of trout, and multiple species such as turtles were since recognized to have evolved the magneto-sensitivity. With the advent of Big Data, the large-scale observations, often unintentional, continued. A recent example: domestic cattle preferentially orient themselves along Earth's magnetic field, as detected using Google maps[2].

50 years ago, Anatoli Buchachenko at the Institute of Chemical Physics shed some light on the possible mechanisms by discovering the magnetic (mass-independent) isotope effect[3], of magnetic fields discriminating atoms by nuclear spins

and nuclear magnetic moments. Peter Hore at Oxford worked out the basics of magnetoreception, which centres on radical pairs, the magnetically sensitive intermediates formed by photoexcitation of the retinal chryptochrome proteins[(4)]. The prime function of cryptochromes, however, is to act as central circadian oscillators in animal brains and as receptors generating the "go to bed" signal in response to blue light, as shown by Steven Reppert at Nebraska. Cryptochrome genes are expressed in the ganglion cells and inner nuclear layer, so blindness resulting from retinal diseases would not affect the sleep, whereas a trauma or surgical eye removal would. All this machinery is immersed in membranes highly enriched with DHA. A coincidence?

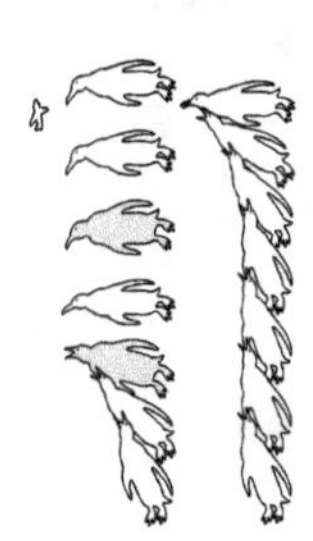

The idea of quantum effects in consciousness* is speculative. We are focusing here on biological phenomena which so far failed to be explained in any other way. And leading the way is the hard problem of consciousness, which overlaps with concepts of conscious free will and is the subject of vast literature. Reducing this arcane math-replete field to a Twitter style outline, the first proposal of QM algorithms being involved in cognition seems (as nothing is certain in the quantum world, including timing) to have come from Efim Liberman in the USSR. Roger Penrose at Oxford and Stuart Hameroff at Arizona suggested that consciousness is driven by QM inside neurons and involves quantum vibrations in microtubules, naming the conjecture thus: "Orchestrated objective reduction", or Orch OR. They proposed that consciousness arises with the collapse of the wave function, associated with making the choice[(5)].

Anaesthetics (*Box 3*) interact with DHA bilayers, reversibly depolymerizing microtubules[(10)]. The fact that unconsciousness would coincide with microtubular disturbance (the latter, however, occurring at much higher concentrations of aenesthetics than is needed to cause unconsciousness) and vice versa was taken as generally supportive of the Orch OR hypothesis. Still, Orch OR continues to draw criticism.

---

* How do physical and chemical processes in the brain give rise to subjective experiences.

Box 3. The importance of anaesthesia in modern medicine cannot be overstated, yet the mechanism of action is still a mystery. What has been clear for the last 100 years is that the key anaesthetics, such as isofluorane, xenon, sevoflurane, halothane and propofol somehow disrupt ordered lipid membranes and lipid rafts(6). Known as membrane-mediated anaesthesia, it has a well-recognized side-effect. Three hours of general anaesthesia would reduce the IQ of school children for up to three months, while the IQ of elderly patients might never recover(7). Considering the PUFA composition of neuronal membranes, might it make sense to zoom in on the interaction of anaesthetics with DHA? A recent observation validates this suggestion: supplemental DHA decreased memory and learning impairments in rats exposed to repeated treatment with anaesthetics(8).
The sedative action of xenon has been shown to involve electron transfer. More intriguingly, different isotopes (we shall meet isotopes again soon, see *ch.6*) of xenon – same electron shells, but different numbers of neutrons in nuclei, - have different anaesthetic strength(8)! But which electron interaction would that be, considering xenon is a non-reactive, inert gas? The nucleus must be involved. Magnetic isotope effect involves interactions between nuclear spins (different between isotopes) and valence electrons, when an external magnetic field is applied. But in the brain, what magnetic field could that be? Whatever the source, it is known that microtubules, the seat of "quantum consciousness" according to Roger Penrose, are very sensitive to magnetic fields(9). Could the field be in any way connected to the numerous, neatly arranged double bonds of DHA? I am getting deeper and deeper into explaining the unexplained with very poorly understood, but this riddle wrapped in a mystery inside an enigma is just begging to start getting addressed experimentally. How about testing, for starters, if DHA-depleted brains would react to anaesthetics similarly, or differently, to well-DHA-stocked animals? See *ch. 2.5.4.*

Questioning the ability of microtubules to be in a fragile state of superposition, not at a habitual close-to-absolute zero temperature in vacuo but at messy physiological conditions. However, a proof of QM not being relevant is also missing. Perhaps Orch OR and magnetoreception have more in common than meets the eye, with electromagnetic fields playing a role, acting through electron and nuclear spins to keep multiple neurons entangled in simultaneous contact, a spooky action at a distance? Could our brain be a quantum computer? Would the microtubules be the right, or the only, compartment to look at though? Anaesthetics do not just affect microtubules but also reversibly disturb lipid bilayers. And one feature that magnetoreception, Orch OR and anaesthesia have in common is the vicinity of highly DHA-enriched lipid bilayers of retinal or neuronal tissues.

Enter Michael A. Crawford, a nonagenarian who since 1970s trailblazed the concept of DHA being vital for brain development(11) from ground zero to the mainstream knowledge, to many accolades. In a true spirit of Renaissance naturalist explorer, he even moved into a house next to the London Zoo, to be closer to his precious lipid samples, donated by all kinds of Zoo animals. Access to brain samples helped him establish beyond doubt: brain size is defined by availability of DHA[12]. (*Back-of-the-envelope-8*). A contribution from Benolken and Anderson in 1973 cemented the apex status of DHA for neuronal function by revealing its vital and unique, yet mysterious, role in visual processing[13].

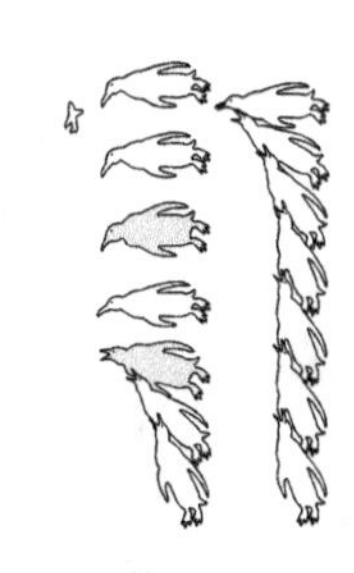

Entanglement of some neuronal elements would nicely explain the total unpredictability, as well as speed (see *ch.2.5.4*) of our creative thinking. Rather than travelling across 100,000 miles of synaptic wires from neuron to neuron, perhaps our thoughts may be shaped by a simultaneous connection between all neurons, then collapsing into a new idea. I cannot wait for someone to disprove this. Or not…

2.5.4. Thought bearings: grid, magnetic, or all at once?

"I insist upon the view that 'all is waves."
— Erwin Schrodinger

"If the brain were so simple we could understand it, we would be so simple we couldn't."
— Lyall Watson

The long-standing mystery of the visual processes in the eye, as mentioned in George Wald's Nobel Prize lecture in 1967, was that the events following the photoactivation of rhodopsin, namely the transduction of visual excitation, were too slow to explain visual reception. It should take substantially longer for a visual signal to get from the retina to visual cortex based on the speed of the spike travelling along neuronal membranes and synaptic contacts, than it does in reality. Michael Crawford suggested that the energy released from a vitamin A molecule after it underwent 11-cis to 11-trans isomerisation would be trapped by DHA (see *ch.4.1*), with ensuing hyperpolarization, through some electron processing, leading to membrane depolarization carrying the photon energy to the brain[1]. The signal would be carried across from the outer segments, along the DHA-packed plasma membranes of rods and cones, through multiple other sections including amacrine, bipolar and horizontal cells and ganglia to the visual cortex of the brain, - with DHA-based lipid bilayers along the entire chain, interrupted by synaptic junctions*.

Retina is outrageously complex (see *ch.4.1*). Between the outer segments and ganglion cells, which function as the output neurons, the messaging is analogous, changing continuously. Ganglion cells then convert this into digital spikes, at the cost of losing a lot of detail[2]. This is further complicated by the inverted nature of photo detection: rods and cones signal in the dark, and an event of photon capturing actually stops the signalling. So, another function of ganglion cells is to "flip" this, such that the spike reaching visual cortex conveys a positive message on a photon captured, not on darkness.

DHA may have been the staple key element of the photoreceptor development, and conversion of photons into electricity may have stimulated the evolution of the nervous system and brain. DHA is highly conserved as the principal membrane component in photoreceptors, synapses and neuronal membranes in the cephalopods, fish, amphibian, reptiles, birds and mammals, species whose common ancestor lived 400 MYA. According to Michael Crawford, this extreme conservation despite great genomic change suggests it was DHA dictating to DNA rather than the other way round[3]. And so attractive were the unique properties of DHA that nature neglected the LPO-associated danger…

Nature had to choose from what was available, and having picked something from the environment, it would then incorporate it looking for an immediate reproductive advantage, improving things here and now, not caring about the long-term consequences. The fancy phrase for this way of adopting innovation no

---

* In neurons, electric current is combined with chemistry, as the signal crosses the junctions by vesicular transport, and then continues as current along the next neuronal membrane. But also see footnote (*), p.150.

matter what, like there is no tomorrow, is "antagonistic pleiotropy". Proposed by Peter Medawar in 1952, it suggests that some genes may be beneficial early in life, enhancing reproductive success, while also causing detrimental effects later in life, contributing to aging. There is ever only one ultimate criterion for biological innovation to persist: help leave more offspring. It would be just fine for the parents to die a horrible death as long as they had sired more offspring first. And given enough time, the species producing more progeny but suffering in later life would outcompete and push out the less fertile others destined to live long and healthy post-reproductive lives. Huntington's disease is an example[(4)], as the carriers tend to have more kids, on average. Tantalizingly, Huntington's has elevated LPO involved in its aetiology[(5)]. We are but mere vehicles for passing the software on to the next generation. Life is wandering across an endless, marshy hunting ground, rifle in hand, Wellies boots squelching through the impassable mud. Partridge! Aim! Bang! Out charges the buckshot, down plops the empty cartridge. The hunter squints, tracing the trajectory, while mindlessly stomping on the empty cartridge, trampling it into the mud to rot. The empty casing here is but a used up, battered post-reproductive organism, having performed the function of releasing its charge. Rusting in the mud is the end stage of a brief life, preceding the big atom recycling process. There is life out there, waiting for your carbon, nitrogen and hydrogen atoms... Remember factories for human phosphorus recycling in Huxley's "Brave New World"? What an abominable destiny for our bodies and brains. But I digress.

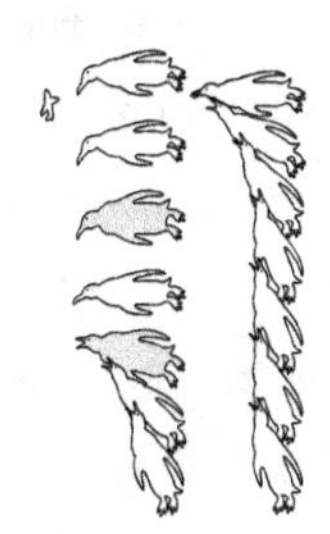

Attempts continue to refine Roger Penrose's idea of linking consciousness with the collapse of the wave function. Turning it upside down, a recent suggestion postulates that a new experience is created when a system goes into a quantum superposition state, - rather than when it collapses[(6)]. Unusually for the field, the authors propose experimental testing, using xenon isotopes with different nuclear spins in the anaesthesia experiments on DHA containing membranes (*Box 3*). So, in case you want to ask what exactly it means to equate a state of superposition to consciousness, - would that mean perceiving and not perceiving the colour red simultaneously or something, - just wait till these guys publish their follow-up study.

Quantum effects are associated with a couple of tiny, itsy-bitsy whits, like photons, getting into a superposition state. Perhaps as a step in the right direction, objects with more than a billion atoms (a 100 nm glass bead) can get into a state of a single quantum wave(7). We seem to be edging nearer to the brain size objects.

The striking thing when looking at a brain, human or cetacean (the two I saw), is the unbelievable quantity of the folded pleats. Technically called "gyri" (singular, "gyrus"), such convolutions make up an area of 2000 cm^2. And many of those gyri are positioned facing (almost touching) each other. What is the importance of this arrangement for the grey matter? Why might neurons want to be sitting on a large folded surface? Is it just the way to increase surface area – or could that be beneficial for connecting with multiple other neurons in a contactless way? Perhaps consciousness itself is a simultaneous connection – superposition, - between many parts of the brain. QM offers a way: entanglement. Many molecules can be kept in synchrony, whatever the distance. But how to make things entangled? A recent paper speculates that myelin sheaths, the fat-rich insulating sleeves which coat brain cables, may be the source of entangled photon pairs. Would be interesting to treat those with xenon isotopes! Electromagnetic radiation, perhaps coming in a form of infrared photons from neuronal mitochondria, would excite DHA, compelling it to produce pairs of entangled photons(8). DHA antennas in every brain cell would be tuned to receive the signal, and then…

Perhaps wavelengths other than infrared, - radio waves? - may also create the coherence or entanglement, - the parameters that may be operationally equivalent. Related processes may be behind the too fast to be explainable visual signal transfer*.

The quantum consciousness area seems to be heating up. Worth keeping all our yet-to-be-discovered DHA antennae in a superposition state with all the new developments. If all that is true, might the brains of two or more people theoretically be able to enter into a superposition state, entangling their shared thought for

---

* According to Michael Crawford, DHA might be capable of fulfilling the role of a dipole, required for radio wave processing, on account of its unique structure.

an enhanced creativity and giving a new meaning to the "brain the size of a planet", or would that be prohibited by quantum encryption? And what if you do not want it? Tinfoil hat, anyone?

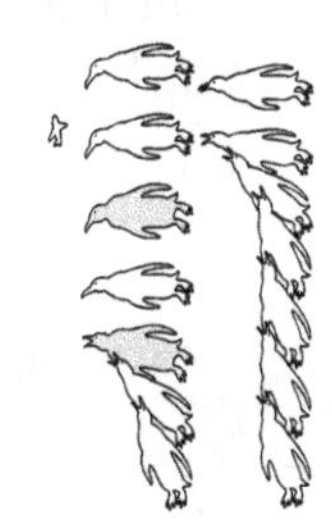

Applying the Occam razor, maybe there is a much simpler explanation. Perhaps brain DHA just makes our heads itch. And occiput scratching is a well-known way to boost our creativity, thank you very much DHA.

Let's wait and see what the xenon experiments reveal.

2.6. The unstoppable chain reaction of PUFA damage

"Freedom is living without chains."
— Indra Devi

"The only thing necessary for the triumph of evil is for good men to do nothing."
— Edmund Burke

2.6.1. The basics of lipid peroxidation (LPO)

"You go back to fundamentals when things start to go awry"
— Bill Cowher

"Explosions are not comfortable"
— Yevgeny Zamyatin

Various metaphors can be used to describe the concept of chemical or nuclear chain reactions (CR). A domino effect, matchsticks set aflame in a matchbox, an avalanche, a snowball effect, or a car pile-up on a busy motorway. My favourite analogy, if technically less strict, formed based on the alleged reports by British pilots during the Falklands war. There are million-strong colonies of penguins in the Southern Atlantic. "Penguin" sounds deceptively close to "pinguis" (from Latin for fat, oil). This false connection stubbornly refused to erase itself from my brain even when I learned the correct meaning of the bird's name (from Welsh *pen*, head, *gwyn* white), - as penguins do look rather pudgy. In 1982, several Royal Air Force pilots reported penguins staring, with curiosity, at fighter jets taking off at some distance, headed in the direction of the tightly packed, million-strong huddle of a colony. By the time a jet were right overhead, their beaks would be facing straight up, agape. As the plane would be passing over, the birds would continue gawping at it, all as one tilting their heads further backwards. The first rows would then start falling on their backs like dominoes, pushing on the neighbours and so on, until the whole colony would end up in a supine position, save for an odd extra-chubby specimen who would manage to keep his balance for a bit longer, even helping prop up some of his weaker brethren, - but ultimately succumbing and losing his ground, too. A chain downfall of millions of puffy fatty units, making up colonies so massive they are visible on satellite images. Talking of satellites. They can also be hit by the chain reaction, on the planetary scale. Known as Kessler syndrome, the threat stems from a large, and proliferating, quantity of debris in low to middle Earth orbit. One more unfortunate collision may cross the threshold, cascading exponentially to so high a density of objects that certain orbital regions might become unusable for a very long time…

Let's look under the hood of the chain "reactor". The concept of chemical CR was conceived by Max Bodenstein in 1913, and in 1956, a Nobel Prize was awarded to Nikolai Semenov and Cyril Hinshelwood for working out a quantitative chain reaction theory*. CR is the key underlying process in various fields

---

* In the late 1970s, I chatted with Prof Semenov, us lads having bumped into him a couple of times as he was enjoying his postprandial walks in Chernogolovka, postprandial cigarette in hand. Our conversations were limited to his recollections of nice little explosions he was entertaining himself with, as a chemistry graduate. When I started taking chemistry more seriously in the early 1980s, one of the first things I tried at home (do not try this at home) was drying some nitrogen triiodide, which he mentioned. Kicking myself

of chemistry, including fire, explosions, polymers, and various aspects of photochemistry. A chemical CR involves initiation, - often the rate limiting step (other steps are much faster), which kick-starts the chain, for example by forming the initiating radical. This is followed by a multistep propagation. CR comes to an end with the termination step, which happens when all the material is used up, - or when quenching of the propagating radical occurs, putting out the fire. Another type of termination is when two ravenous chain-propagating radicals bump into each other (a fairly rare event). The radicals

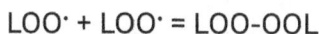

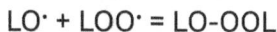

$$LOO^{\cdot} + LOO^{\cdot} = LOO\text{-}OOL$$

$$LO^{\cdot} + LOO^{\cdot} = LO\text{-}OOL$$

$$LO^{\cdot} + LO^{\cdot} = LO\text{-}OL$$

shake hands and form a bond, removing themselves from circulation. The scientific name for this is homologous recombination. The compounds formed can be cleaved back into ROS by metals, as explained in *ch.2.4*.

The only example of a non-enzymatic CR in biology is the chain reaction of lipid peroxidation (LPO), essentially an explosive, radical process of burning PUFAs in oxygen. As it is not controlled by enzymes, nature could not have evolved, through DNA mutagenesis, any type of a brake pad to keep it in check directly. And roundabout, non-direct means, - throwing in antioxidants, mopping up the damage, - are not efficient (see *ch.2.8*). And it gets worse. Falling dominoes imply a linear sequence of events, one following the other. Sadly, LPO can branch sideways, just like a nuclear reaction left without control, so the damage will grow exponentially, spurring more and more PUFAs into the avalanche of damage. This is as bad as it sounds: the chain reaction, with no direct control over it. When this combination happens in different settings, the results are often catastrophic (Three Mile, Chernobyl, Fukushima).

---

now for not asking him about chain reactions in biology…

PUFAs are sitting, densely packed, in a lipid bilayer like trees in a forest, in a tight, direct contact with one another. Various metabolic processes run inside membranes, and oxygen is never far away. Several mechanisms, involving light (see *ch.4.7*) and metals (see *ch.2.4*), lead to generation of ROS (see *ch.2.2*) in membranes. The process is random, and the newly formed ROS species would most likely find themselves surrounded by PUFAs, with the nearest antioxidant, like vit E, likely being some distance away. What unfolds then is shown on *Fig.7*. The weakest CH bond of a PUFA molecule is at the so-called bis-allylic position, squeezed between two double bonds. A radical rips away a hydrogen atom from the bis-allylic CH_2 group. This is called a homolytic bond cleavage, with each atom retaining its electron. Using a hydroxyl radical as an example,

$$-CH_2-CH=CH-CH_2-CH=CH-CH_2- + HO^{\cdot} \rightarrow$$
$$-CH_2-CH=CH-HC^{\cdot}-CH=CH-CH_2- + H_2O$$

Having satisfied its intemperate cravings for an unpaired electron, the hydroxyl radical forms a bond with the abstracted hydrogen radical, creating a stable, non-radical water molecule.

$$HO^{\cdot} + H^{\cdot} \text{ (abstracted from a PUFA molecule) } = H_2O$$

Radicals, like charges, cannot disappear, so we now have a new radical, - sitting on the carbon backbone of a PUFA. Surrounded by double bonds, the radical can shift left or right performing castling with the double bonds, giving more stable resonant structures:

$$-CH_2-HC^{\cdot}-CH=CH-CH=CH-CH_2- \leftrightarrow -CH_2-CH=CH-HC^{\cdot}-CH=CH-CH_2- \leftrightarrow -CH_2-CH=CH-CH=CH-HC^{\cdot}-CH_2-$$

Molecular oxygen is much better soluble in lipids than in water, by about 10-fold. The concept of "bags in bags" for cells and organelles may have something to do with that. Compartmentalisation, achieved through separating smaller and smaller volumes within each other using lipid membranes, something Bruegel

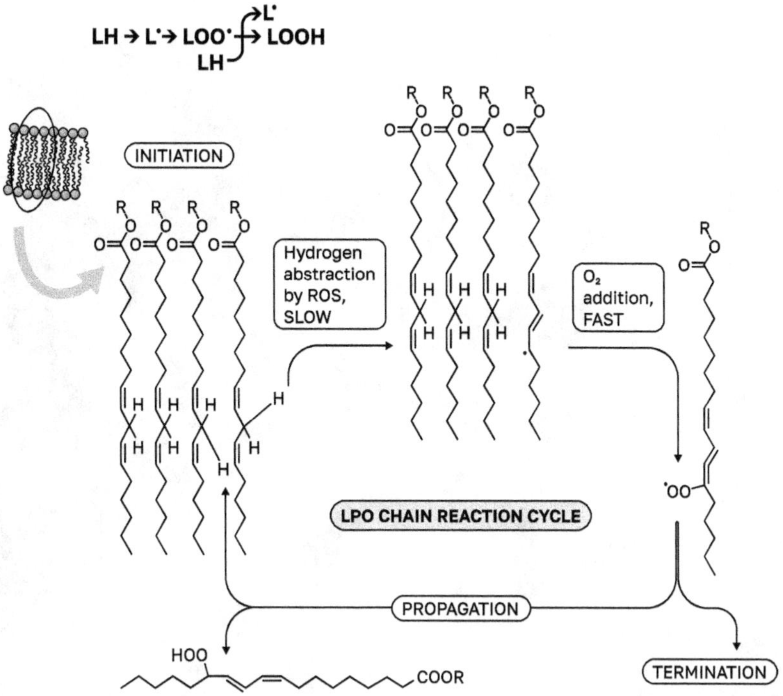

Figure 7. Chain reaction of LPO· In the first, slow step, a bis-allylic (between double bonds) hydrogen atom of a PUFA (LH) inside a membrane is removed (abstracted) by an ROS, forming a PUFA radical, L·. L· quickly reacts with oxygen O_2, forming a peroxyl radical, LOO·, a new ROS, which abstracts a bis-allylic hydrogen from the next PUFA, propagating the LPO cycle. Newly formed PUFA peroxides (LOOH) can decompose through Fenton reaction, giving more ROS, branching the chain. Termination (quenching of LOO·) can happen through various mechanisms.

---

illustrated well 500 years ago (*Fig.8*), would have depleted the oxygen content in the inner compartments, were it not for the ability of lipid membranes to concentrate oxygen, serving as relay stations for forwarding it into smaller compartments (*Fig.1*). The elevated level of oxygen in membranes sees to it that there is always plenty available to quickly convert the fatty acid radical into a lipid peroxy radical (LOO·) by reacting it with oxygen, O_2:

$$-CH_2- HC(OO^\cdot)-CH=CH-CH=CH-CH_2- \text{ or}$$
$$-CH_2-CH=CH-CH=CH- HC(OO^\cdot)-CH_2-$$

Figure 8. Pieter Bruegel the Elder, 1556. Big Fish Eat Little Fish. Another illustration of the "bags in bags" membrane compartments concept.

This newly formed reactive oxygen species can do the exact same thing as the hydroxyl radical did, to the undamaged PUFA molecule next to itself in the membrane*. It abstracts a hydrogen from the neighbour molecule, turning itself into a "stable" peroxide (LOOH):

$$-CH_2-HC(OOH)-CH=CH-CH=CH-CH_2- \text{ or}$$
$$-CH_2-CH=CH-CH=CH-HC(OOH)-CH_2-$$

The radical now migrates to the next molecule, - completing one step of the chain. Note the quotation marks on "stable" - because it is not, really. Even though it is not a radical and so is unable to damage its healthy PUFA mates by abstracting their bis-allylic hydrogens, it can, however, give rise to free radical species that can. Different bad actors can convert it into a fully-fledged ROS. We considered one such group of bad actors in the previous chapter (another type of a bad guy, - light - is discussed in *ch.4.7*). Our good old friends transition metals can do it, through Fenton or Haber-Weiss processes, branching the chain:

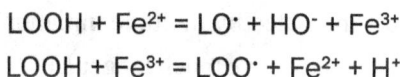

$$LOOH + Fe^{2+} = LO^{\cdot} + HO^{-} + Fe^{3+}$$
$$LOOH + Fe^{3+} = LOO^{\cdot} + Fe^{2+} + H^{+}$$

Both lipid oxy ($LO^{\cdot}$) and lipid peroxy ($LOO^{\cdot}$) radicals duly initiate new chains, going sideways (branching), in the process turning into LOH (lipid alcohol, stable and inert), and LOOH (lipid peroxide, which can form more radicals). There is also an enzymatic way of initiating the non-enzymatic LPO. LOX and COX enzymes can also convert PUFAs into $LOO^{\cdot}$[(1)], and the propagation would then proceed as described above.

If this is the case, then how do we even manage to stay alive, having inhaled our first lungful of this oxygen poison at birth? How is this nightmare kept in check?

Well, the LPO process can just run out of its PUFA fuel. This out-

---

* Or sometimes even on the same PUFA molecule. See chapter 2.7.

come is least desirable, like a war ending with the last soldier killed. Antioxidants, while unable to prevent the LPO chain, can still somewhat reduce the chain length (number of steps), and lacking antioxidants altogether is bad, as described below. A definition of antioxidant is pretty loose and overlaps with other concepts like hormesis (see *ch.2.8*). There are plenty of proteins jiggling around in membranes, and when an LOO• radical bumps into one, it may rip off a hydrogen atom. If the newly formed protein radical is unable to continue propagating the chain, or unable to react with molecular oxygen, the chain reaction will stop. This protein would act as an antioxidant. If several LPO chains are ongoing at the same time, then two LOO• radicals can collide, terminating two chains at once resulting in a molecule with a four-oxygen atom bridge, L-OO-OO-L. Metals can turn it back into ROS though. The currently accepted, canonical LPO mechanisms* have been polished to perfection by Ned Porter at Vanderbilt(2). How many steps would a chain take before getting terminated? It depends. In many densely packed PUFA membranes, such as in mitochondria, according to various estimates every tenth, if not every fifth PUFA molecule is in contact with some transmembrane protein or other non-lipid body. This, of course, would lead to a quick chain termination. On the other hand, mitochondrial membranes generate so much ROS that this short chain length may be compensated by multiple initiation events. Other lipid bilayers or other arrangements might have much longer LPO chain length, such as retina, sperm tails, LDL/HDL particles, skin, etc.

Answering these questions experimentally is far from straightforward. Most LPO studies look at bulk oils, which may be relevant to LDL/HDL oxidation but rather unrelated to LPO in bilayers. Liposomes are very poor models too, for they cannot possibly reproduce the complexity of cell membrane composition. Working with cells at atmospheric oxygen exposes them to orders of magnitude more O_2 than they would have been exposed to in their natural environment, deep inside of tissues (*Back-of-the-envelope-1*). It is also very difficult to reproduce natural PUFA composition in cell cultures (*Box 8*). Moreover, LPO studies in test tubes or cells are often carried out at room temperature. This also makes the observations rather irrelevant looking from a homeothermic (warm-blooded) animal's perspec-

---

* Minor additional nuances, such as Criegee intermediates or the Schaich pathway(3), do not change the principle.

tive. As was already mentioned, LPO is a non-enzymatic process which proceeds "out of its own free will", without much control from the host cell. Enzymatic reactions are under much stricter control. A well-known rule of thumb, the van't Hoff principle, postulates that each 10°C increase in temperature generally accelerates chemical reactions by a factor of two or more, applicable to both enzymatic and non-enzymatic reactions. While non-enzymatic processes just unmurmuringly obey the van't Hoff law, cells can circumvent the temperature-driven acceleration by using various approaches, like reducing the concentrations of enzymes, or up/down-regulating the levels of substrate, etc, essentially maintaining the equilibrium. Studying the non-enzymatic process of LPO, intricately surrounded by enzymatic mop-up activities, would then arguably only make sense in models kept at physiological temperatures, ideally in animals. Hydrophobicity and van der Waals forces are muscular enough to corral intact PUFA chains together, but when oxidized, the hydrophilic oxygen groups attached break the arrangement and stick out, compromising the membrane barrier function*. Membranes grow (lipid) whiskers and become leaky, but bigger problems are yet to come...

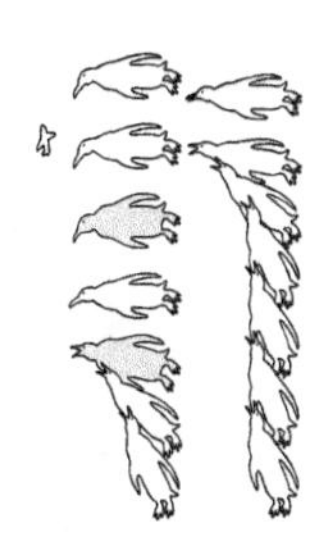

2.6.2. Cell-smothering smithereens

"Stopping bad things is a significant public service."
— Ted Cruz

"Bombs do very, very bad things to human bodies.
It's incredibly shocking to see."
— Phil Klay

Having spread through the PUFA membranes like wildfire, frying

---

* Various analytical techniques (see *ch.2.9*) can be used to observe the detrimental physical changes to the membrane structure during LPO[4].

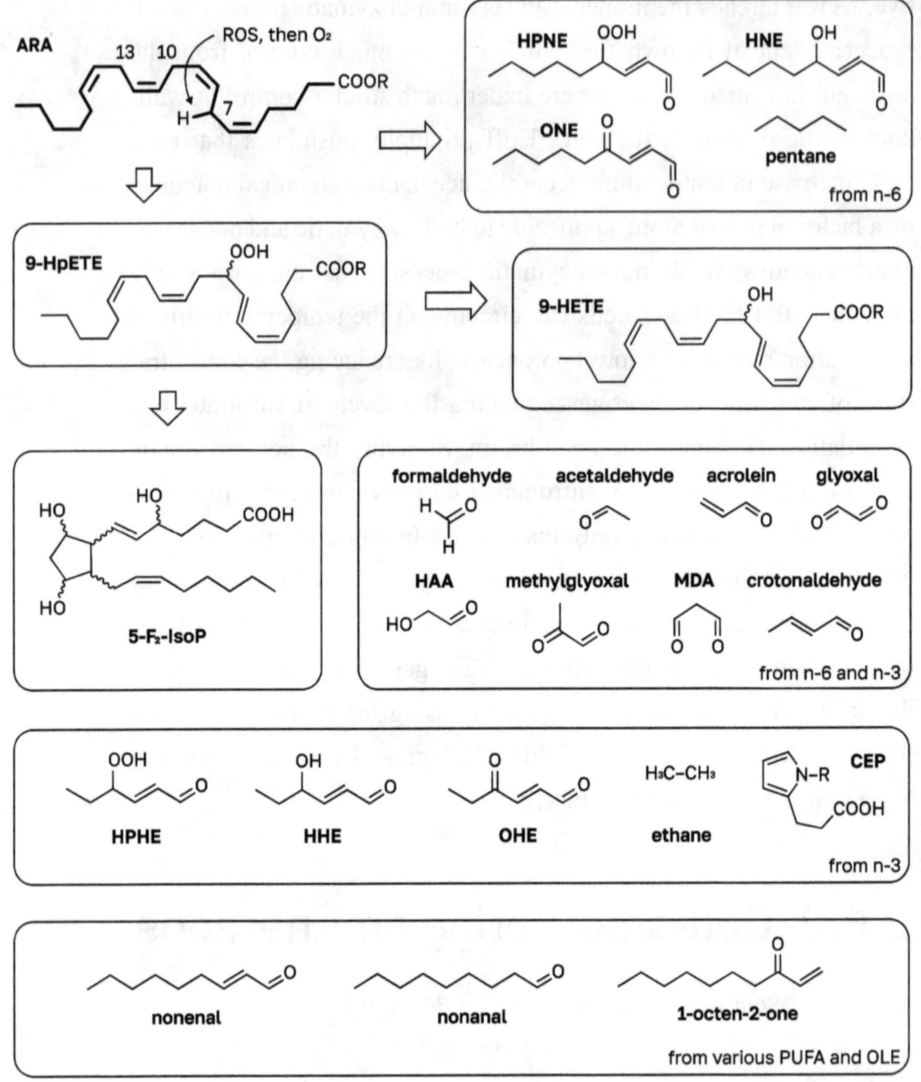

Figure 9. Non-enzymatic oxidation of ARA at pos. 7, 10 or 13 gives 9 different chiral (wiggly bond) peroxides, hence 18 LOOH derivatives. 9-HpETE (shown) rearranges into the 5-F2-IsoP series. Oxidation at pos. 10 and 13 generates 8-, 12- and 15-F2-IsoP series. 64 ARA-derived F2-isoPs are known. Delightfully complex LOOH decomposition chemistry gives many end-products. Alpha, beta-unsaturated RCS are highly reactive and do many bad things (note that n-3 and n-6 PUFAs yield different RCS). LOOH rearrangement pathway also yields smaller RCS. Their reactivities and quantities vary, complicating the detection. For example, the amount of HNE formed is up to 80 fold lower than that of MDA, but MDA is less reactive than HNE or ONE. Inert LPO end-products include ethane (from n-3) and pentane (from n-6 PUFA). Aged-skin-smell LPO products include 1-octen-2-one. Compare this to Fig. 26 to appreciate the complexity of the analytical detection challenge..

them from within and destroying the barrier function, eventually the LPO chains cool off. The trouble, however, is just beginning. Oxidized PUFAs are unstable and rapidly break down through multiple pathways, many not well understood, reducing them to smithereens. Dozens of fragments are generated, some inert (like isoPs or alkanes), but many more highly reactive, hence toxic (*Fig.9*). And this, arguably, is the main danger of LPO. Too numerous to even mention, never mind describe in detail, this gang is much safer when viewed from a distance, - so here is a bird eye's view.

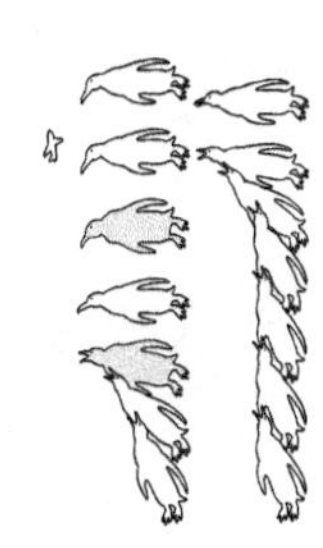

Carbonyl group (C=O), the core element of aldehydes and ketones, looks like an oxygen atom "leech" biting into a helpless carbon atom, grabbing it tight with both hands (a double bond), while sucking out its electron density. In terms of oxidation state, carbonyls are between alcohols and acids. This intermittent, neither-here-nor-there position, and the partial positive charge on the carbon atom, make the carbonyl croup highly reactive. And in gets worse. As PUFAs have multiple double bonds, the newly formed fragments retain some of them. And if a double bond is positioned next to a carbonyl, like shown for this alpha, beta-unsaturated aldehyde:

$$-CH_2-HC=CH-HC=O$$

this additionally activates the compound ("activated carbonyl"), because the two double bonds are "conjugated", - they "talk", increasing each other's reactivity, making the molecule even more toxic. As discussed in *ch.4.7, 5.6*, molecules containing such fragments can react at either of the two ends (*Fig.10*). Depending on the exact structure, the rough actors can be more (HNE from n-6 PUFA) or less (HHE from n-3 PUFA) hydrophobic. Accordingly, they may prefer hydrophilic (cytosol) or hydrophobic (membrane) environment to carry their mischief. They act like super glue*, and no space within or outside of

---

* This is not an irrelevant comparison. Superglue, based on alkyl cyanoacrylate, is structurally very similar to acrolein (acryl aldehyde) and other

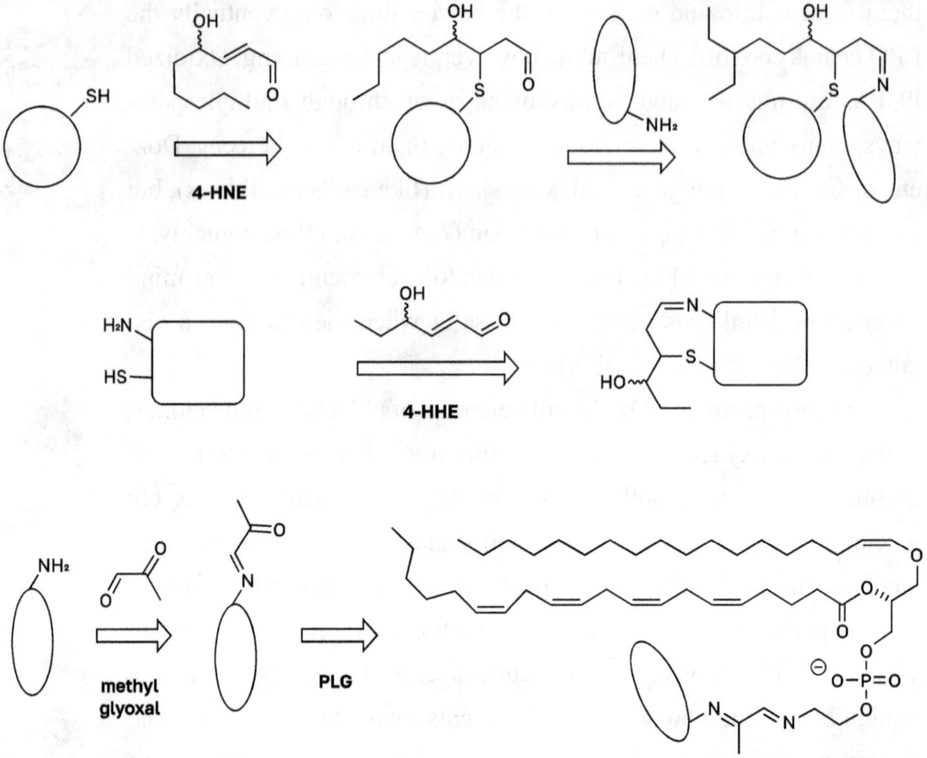

Figure 10. Cross-linking by LPO-derived RCS (Fig. 9). Different shapes represent proteins with various functional groups. An n-6-derived 4-HNE cross-links proteins by reacting with amine and thiol groups. The same protein molecule can get cross-linked internally, as shown for an n-3-derived 4-HHE. P-Tau, implicated in AD, PD, TBI and PSP, undergoes this modification, preventing it from dephosphorylation and leading to misfolded fibril aggregates. Methylglyoxal reacts with two amines, for example linking together a protein and a plasmalogen (PLG). RCS cross-link DNA or RNA (Fig. 19). Various combinations are possible: DNA-protein, lipid-lipid, etc. Such links require special conditions in lysosomes and peroxisomes to be cleaved. The process often falters, leading to the accumulation of the material, such as in lipofuscin, filling up space with this toxic sticky goo which attracts metals, catalysing more LPO. ONE and OHE (Fig. 9) can link three different molecules together.

---

cells is safe from their attack.

Cells routinely digest biomolecules. They disassemble proteins, by cleaving peptide bonds between amino acids, or disconnect nucleotides of DNA, by hydro-

---

LPO products, and glues things together using the chemistry very similar to that utilized by the LPO-derived species. We get glued, solidifying into stiff linoleum-like polymers (see *ch.1.3*). What an atrocious thing the LPO is…

lysing phosphodiester bonds. However, covalent bonds formed when unsaturated aldehydes cross-link proteins and other biomolecules are much more insidious as they cannot be easily undone by the mammalian enzymatic machinery. And the places where the waste recycling is carried out, like lysosomes and peroxisomes, struggle to cope and can easily get overwhelmed. This proceeds in tandem with Maillard and Amadori processes (*Box 5*), generating even more unrecyclable waste, both inside and outside of cells. The goo piles up, chelating the transition metals and catalysing the formation of more goo, - cellular spaces get clogged up, metabolism slows, - yet more waste accumulates, - cells choke, - aging, - death. Here, look at the lipid peroxidation gang again, on *Fig.9*, and the book cover.

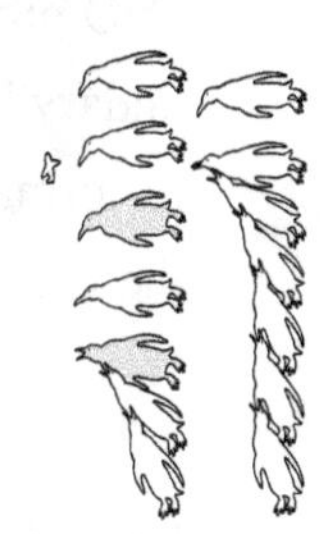

2.6.3. LPO is enhanced by unfavourable genetics

"Your genetics is not your destiny."
— George Church

"Biology always beats will power."
— Mehmet Oz

Cells must have PUFAs in their membranes. The PUFA composition of many cells is tightly controlled, irrespective of the dietary intake, except in cases of severe deficiency. And antioxidants, as we shall see (*ch.2.8*), cannot stop the LPO, and sometimes they make the matters worse. So LPO always festers in the background. Many LPO products cross-link with other biomolecules forming difficult-to-remove conjugates, which gradually accumulate. These are often the hallmarks of aging and age-related diseases. And this is inexorable, yet "normal".

But are there disorders which result from an even more imbalanced

LPO process? LPO is not controlled by enzymes in a direct way, but many "secondary" systems transport and supply antioxidants, mop up the LPO products and re-format the membranes, trying to control the chain reaction non-directly. Are there pathologies that stem from any of those systems being faulty? Indeed, there are, as illustrated by a few examples below.

Vitamin E is a chain-breaking antioxidant. It cannot fully stop the chain reaction, but it slows it down. We cannot make vit E, but it is plentiful in both animal and vegetable foods, so there is never a dietary deficiency. However, the transportation of vit E around the body is tightly controlled, and when a faulty alpha-tocopherol transfer protein (TTPA) gene is inherited, individuals would be unable to retain and use dietary vit E. This results in Ataxia with Vitamin E Deficiency, an ultrarare neurological disease with a prevalence of 0.5-3 people per million in North Africa, Italy and France. It is similar to Friedreich's ataxia and affects central nervous system (brain and spinal cord), heart and eyes. The carriers have trouble coordinating movements (ataxia) and speech, loss of reflexes in the legs, and loss of sensation in the limbs. They suffer from retinitis pigmentosa and abnormalities affecting the heart (cardiomyopathy). This list suggests that LPO is the common denominator of these different neurological, retinal and cardiovascular symptoms. And the treatment? TTPA is not totally absent, it just has an impairment, so the patients are flooded with lifelong, daily, high-dose vit E oral supplementation. And early treatment can stop the progression of the disease, or even reverse symptoms.

Glutathione peroxidases (GPX) family are an enzyme family that neutralizes H_2O_2 and potentially toxic phospholipid hydroperoxides (LOOH, see *Fig. 7*). These are PUFAs with one or more peroxide groups, which are stable but on contact with transition metals can form chain reaction-initiating radicals. A different way for the unstable LOOH to do mischief is by decomposing into reactive carbonyls, a really nasty bunch. Those LOOH are still attached to a PL head group, and so still sit inside of lipid membranes. An important example is GPX4, that defuses these ready to fire, cocked rifles into peaceful stable lipid hydroxides. For its action, it requires glutathione, which is recycled afterwards by glutathione reductase. Disabling GPX4 gene is embryonic lethal, but even malfunction has tremendously

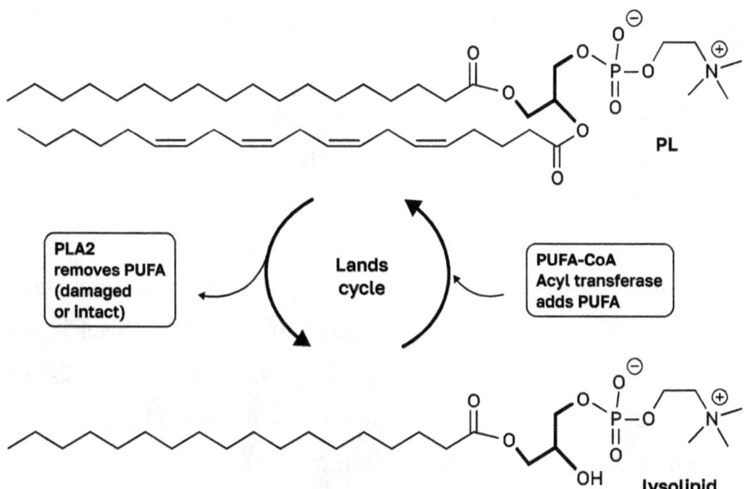

Figure 11. Lands cycle (Lands pathway in plants) is the remodelling process, which takes place in lipid membranes, yielding new PLs by deacylation and reacylation of PUFA-occupied position sn2, obviating the need to make the molecules from scratch. The cycle removes both oxidized and intact PUFAs but only reattaches the good ones, maintaining the PL integrity and removing the LPO products from membranes. The energy cost is huge. A quarter of all brain energy is spent remodelling neuronal PLs: brain spares no resource trying to keep LPO in check. Pulmonary surfactant PLs are made when saturated fatty acids are attached to the sn2 position instead of PUFAs. TGs in white adipose tissue mostly have LIN as PUFA, requiring less recycling. In brown adipose tissue, TGs have longer PUFAs. TGs do not form membranes, so to make them more accessible to lipases, the size of the droplets is small to increase the surface area. Neither sphingosines nor plasmalogens undergo the Lands cycle, while cardiolipin undergoes it indirectly.

---

bad consequences. Mutations in this gene result in various pathologies, such as Sedaghatian-type Spondylometaphyseal Dysplasia. Individuals carrying GPX4 mutations often die young, suffering from neurodegeneration and inflammation among other things. Again, this points to LPO being a common denominator of neurological conditions. There are currently no drugs to treat the GPX4 deficiency*.

---

* Peroxiredoxins are another important peroxide-detoxifying family of enzymes.

PLA2G6, calcium-independent phospholipase A2, is an enzyme that releases ARA and other PUFAs from PL, playing the major role in Lands cycle (see *ch.5.8*, and *Fig.11*) - based lipid turnover. Faults in this enzyme would lead to pathological accumulation of oxidized PUFA residues in membranes. Many diseases are associated with such faults. PLA2G6-associated neurodegeneration group of disorders affect basal ganglia and involve various movement impairments. The most severe form, Infantile Neuroaxonal Dystrophy (INAD), is characterised by axonal swelling and lipofuscin-like spheroid body deposits in the CNS. Onset is within the first 2 years of life, with death typically by the age of 10 years. Close siblings with the same mutation may have different outcomes, with one developing INAD, while another being disease free till early adulthood and then developing a subtype of Parkinsonism termed Parkinson disease 14 (PARK 14), again pointing to the LPO as a common denominator. There are currently no disease-modifying, or mitigating drugs for PLA2G6 deficiency.

ALDH2, a mitochondrial aldehyde dehydrogenase 2, belongs to the family of 19 human ALDH enzymes. It has a mutation with a dubious distinction of being shared by more people than any other mutation known. 8% of the world's population (600 million people, mostly East Asians, with record holders Japan and Taiwan at 50% and 45%) carry the ALDH2*2 genetic variant, which has reduced enzymatic activity. Imbibed ethyl alcohol (which gives us a high) is first catabolised to acetaldehyde (which gives us headache the morning after the night before) by ethanol dehydrogenase. This aldehyde is then picked up by ALDH2 for further oxidation into acetic acid, an inert compound which can be burned in the Krebs cycle furnace for energy. When ALDH2 works normally, some overindulgent types, known as alcoholics, become over-reliant on its services. A drug called disulfiram (Antabuse)* helps them hold their horses, by irreversibly inhibiting ALDH2. Upon alcohol consumption, disulfiram causes an immediate reaction of hangover, with flushes, headache, copious vomiting, shortness of breath, lock, stock and barrel. Sadly, this is exactly what the ALDH2*2 carriers experience,

---

* Metabolites of quercetin (p. 97), a flavonoid found in high concentrations in red wines, can also inhibit ALDH2 activity, explaining the "red wine headaches". Keep this in mind when considering quercetin as a senolytic supplement.

without any disulfiram, when they consume alcohol. With a faulty gene inherited from just one parent, the net total aldehyde dehydrogenase activity is 10-45%, but if a defective gene comes from both parents (homozygous), the activity drops down to 1-5% of the normal*.Up to 20% of the faulty ALDH2 carriers have both of their gene copies compromised. This mutation is relatively recent. The Bering land bridge between Eurasia and Americas was crossed over by East Asians migrating to Alaska by 16,500 years ago, and since there is no ALDH2*2 phenotype in Araucanians in Southern Chile, the mutation must have occurred after the last Ice age. The current theory posits that the mutation somehow conferred some resistance against widespread hepatitis, like the sickle cell anaemia against malaria. But inability to enjoy a glass of wine could be the smallest of the problems the carriers are facing. There is no ethanol in animals' natural diets (except for an occasional fermented fruit allegedly enjoyed by elephants and monkeys, or an entrepreneurial grizzly bear helping himself to beers in a tourist camp), and accordingly, ALDH2 did not evolve to detoxify acetaldehyde. Instead, it has likely evolved to detoxify LPO-derived aldehydes, as did the other members of the ALDH family.

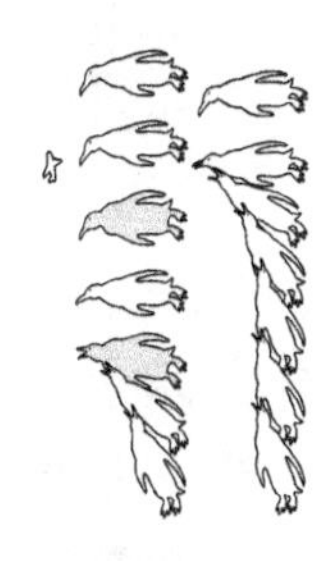

Over years, a very large number of publications reported on a possible connection between ALDH2*2 phenotype and various diseases. Databases are brimming with claims for major pathologies to have links to the mutation. To itemize, CNS diseases (particularly late onset AD), oncological diseases (elevated lung cancer in smokers, oesophageal cancer), cardiovascular diseases, gout, diabetes, obesity, osteoporosis, liver and kidney diseases(5). Yet statistically, the connections linking various diseases to the faulty ALDH2 phenotype are rather weak. Epidemiological studies, looking at how health and disease patterns are distributed, do not show statistically strong risks for the mutation carriers. Taking Parkinson's as the case in point, the incidence of PD,

---

* All four subunits of an operational ALDH2 tetramer have to be mutation-free for a full efficiency, and the units can come from either gene, hence reduced efficiency even in heterozygous carriers.

well known to have LPO linked to its aetiology[6], is not higher in Japan than in the USA. Why might that be, considering that the animal studies fully support the role of ALDH in PD[7]? Several factors may play a role. (a) There are major differences in the dietary habits between the USA and Japan. It is often reported that migration from places with longer average lifespan, including alleged Blue Zones like Sardinia and Okinawa, to other parts of the world leads to a quick lifespan reduction, implicating lifestyle, environmental and dietary factors as very important in longevity. (b) High prevalence of smoking in Japan could be important, as smoking is known to protect against PD[8]. (c) The ALDH2*2 mutation still allows for some enzymatic activity, particularly in heterozygous individuals, and with 18 other ALDH enzymes available, perhaps the body learns to compensate, in some ways, for the decreased activity (*Box 4*). (d) Perhaps the Asian PD patients drink less, compared to their Western counterparts, further skewing the statistics? (e) Western populations may be more prone to PD owning to the aSyn mutation[9]? (f) Homozygous carriers are not always analysed separately from the less affected heterozygous ones, which could be another reason. But when they are, the results often confirm that homozygous carriers are the worst affected, for example in a smoking-related cancer study[10]. So, the negative effect of ALDH2*2 could still be there, concealed by (a)-(f), which would be responsible for a rather weak statistics across multiple published studies. ALDH2*2 deficiency is not dissimilar to having one's own internal wok burning 24/7, cooking stir-fry (*ch.3.1*). And despite of the mentioned reservations, it is of course desirable to either have a fully functional ALDH2, or to compensate for the deficiency. But short of gene editing, how could that be achieved? For a while there have been attempts to activate the malfunctioning enzyme. It was claimed in 2008 that Alda-1 can enhance the efficiency of ALDH2*2. It may have a long way to go, as the jury is still out on how efficient the approach is[11].

Relationship between ALDH and cancer may be even more multifaceted. It could be argued that some cancers may benefit from ALDH2*2 and be retarded by the normal ALDH2 activity, while the reverse could be true for some other cancers[12]. The stages of cancer could matter: once cancer gets a foothold in the body, some anti-cancer drugs like cisplatin, whose mode of action generates bursts of cancer cell-toxic ROS, may benefit from decreasing the activity of ALDH2. But prior to that, ALDH2*2 may be procancerous. However, PUFA profile of cancer cells is different to that of the healthy surroundings, making it even more difficult to draw conclusions. Still, the preponderance of evidence seems to be in favour of having a fully functional ALDH machinery in the battle with LPO, and in the battle for longer life. Take the case of the North American beaver. It shows striking resistance to cancer and is a particularly long-lived species. One of the possible reasons for its evolutionary success could be a massively expanded Aldh1a1 gene. Multiple copies of the gene result in an enhanced level of the enzyme, and hence an enhanced capacity to mop up aldehydes[(13)].

ALDH family is vital to all aspects of metabolism, from clearing highly toxic formaldehyde[(14)], to neutralizing an aldehyde form of vit A into an inert acid (ALDH1a1-3[(15)]) to mopping up sleep deprivation-related accumulation of carbonyls[(16)] (see *ch.4.5*), and many, many more[*]. Assisting the extended ALDH family in its difficult task of mopping up the nasty carbonyls, by reducing the amount of work they have to do, cannot help helping to make things better for an organism. Read on, there might be the way (D-PUFA).

Multiple other pathologies may be linked to the unfavourable genetics. Some will be discussed in further chapters, but one group deserves a general mention here: a metal processing disturbance. As considered in *ch.2.4*, metals are one of the major ingredients of LPO because, when not tightly controlled, they provide a steady source of LPO-initiating ROS. Neurodegeneration with Brain Iron Accumulation (NBIA) is a large group of orphan neurological diseases all involving elevated levels of iron, with the ensuing LPO. Wilson disease is similar, but stems from excessive copper, a stronger Fenton catalyst than iron. Ashley Bush from Australia even argues that AD results

---

* ALDH2 malfunction reduces lifespan in animal models, yet a recent study surprisingly reported a (weak) association of ALDH polymorphism with longevity in Koreans. The paradox may be explained by the mutation carriers avoiding alcohol, consumed by 68% of Korean males, reducing the latter group's lifespan. In agreement with this, the longevity trend was not observed in Korean female ALDH2*2 carriers, - who drink much less.

from gradual, unavoidable accumulation of iron in the aging brain, - a mechanism that will be discussed in more detail in (*ch.4.2, 4.8*).

2.6.4. And by unfavourable environment

"Your genetics load the gun. Your lifestyle pulls the trigger."
— Mehmet Oz

"A bad system will beat a good person every time."
— W. Edwards Deming

Elevated LPO and the associated toxicity does not have to be genetically predetermined. Environmental factors can also initiate the chain reaction. While light and oxygen exposure are discussed in *ch.4.7*, multiple chemicals are known to initiate the chain reaction. There are chemicals in the environment deliberately designed to easily form radicals, in lipid or aqueous environment. They can initiate the LPO and are, therefore, highly toxic. As a personal recollection, a research group once planned to follow in the wake of the great Linus Pauling, aiming to detect traces of lipid peroxidation products ethane and pentane in the exhaled breath, just like he proposed in 1971[17]. This makes sense as ethane and pentane are stable, while other decomposition products are highly reactive (think aldehydes) and so unlikely to be detectable in the free form, but rather as conjugates with multiple other molecules, complicating the readout. However, there are several decomposition pathways which all happen in parallel, with only a small fraction of oxidized PUFAs getting converted into ethane and pentane, making their detection challenging (see *ch,2.9*). Two chemicals are very efficient in starting the LPO: bromotrichloromethane CCl_3Br*, and iodotrichloromethane CCl3I. They are highly soluble in lipids so once injected, go right into membranes, where they decompose into radicals, which initiate the LPO:

---

* This application is not unique. The compounds are used in the industry for exactly the same purpose: to initiate radical processes, for example in making polymers.

$$CCl_3\text{-}Br \rightarrow CCl_3^{\cdot} + Br^{\cdot}, \quad CCl_3\text{-}I \rightarrow CCl_3^{\cdot} + I^{\cdot}$$

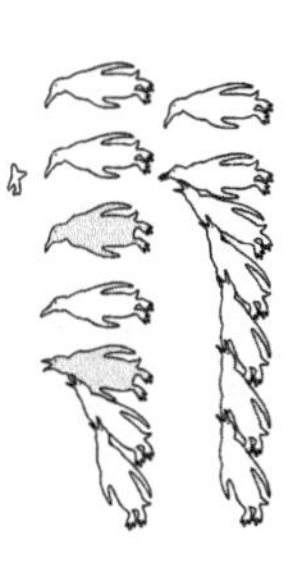

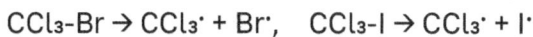

The C-I bond is "slacker" than the C-Br bond, so iodo- derivative decomposes faster, is more potent as an initiator of LPO, and thus more toxic, particularly to liver, where it arrives first. The plan was to inject the animals, collect exhaled breath a couple days later to measure the telltale signs of LPO, and then sacrifice the animals to correlate the level of LPO in tissues with the values in breath. An application was duly made to the University's Ethical Approvals Committee, for a permission to carry out the study. The University, to remain as anonymous Justin Case, wasted no time in analysing the proposal, and moved fast. It took the Committee about 18 months of deliberations to arrive at the thoughtful and logical decision: perfectly ethical to use the bromo derivative, two days prior to sacrificing the mice; categorically unethical with regards to the iodo derivative. Now you see how toxic some of these methane halogenides are.

Many other chemicals can also cause LPO. The ill-famed MPTP, a by-product in illegal manufacturing of synthetic opioids, was discovered when young drug addicts mysteriously started developing rapid onset Parkinsonism. MPTP is a rare case of a bioactive compound which in research went from humans to mice, and not the other way round. Easily crossing the blood brain barrier (BBB), MPTP is metabolized into the toxic cation MPP^+, which selectively kills the dopaminergic neurons in substantia nigra. It blocks complex I of the mitochondrial electron transport chain, causing cell death by a burst of ROS and LPO and ATP depletion. As a positively charged molecule, it lures electrons away from Complex I, generating plentiful ROS, which duly kick-start the LPO process in mitochondrial inner membranes. Comparing the toxic MPP^+ cation with paraquat, a member of a group of weedkillers (herbicides) called viologens, due to the structural similarity (see *Fig.12*) it is kind of obvious what to expect. Indeed, paraquat toxicity is also associated with Parkinsonism and leads to elevated LPO, as does flavonoid rotenone. They all affect Complex I, causing

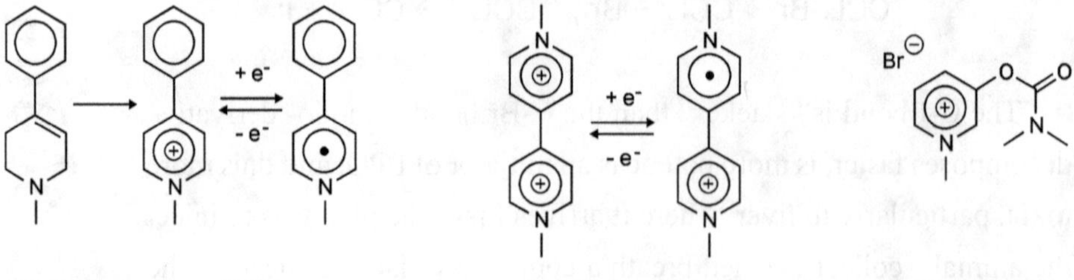

MPTP MPP+ Paraquat (viologen) Pyridostigmine

Figure 12. A similarity between MPTP, a neurotoxin, and paraquat, a pesticide. MPTP is oxidised, by monoamine oxidase, into toxic MPP+, a Complex I inhibitor in mitochondrial ETC. The mechanism of action relies on the redox properties of the alkylpyridinium fragment. Similar to MPP+ and paraquat, pyridostigmine bromide, a suspected neurotoxic agent behind the Gulf War syndrome, incorporates a redox-active methyl pyridinium moiety.

---

toxicity through LPO, although the exact mechanisms of action may be different. It should not therefore be surprising that there is a link between exposure to pesticides (likely from Russian, *писец*, the brutal end to something) and herbicides and the development of PD[18]. Remember about it when reaching out for an apple, or some strawberries (see *ch.3.3*). Heaven forfend you decide that washing them may not be necessary, for they seem to look clean enough*...

Gulf War Syndrome, a mysterious condition affecting veterans of the 1990-1991 Gulf War, involves a range of symptoms such as chronic fatigue, cognitive dysfunction, various types of pain, etc. Deployed troops were routinely injected with an acetylcholinesterase inhibitor Pyridostigmine, an antidot for the nerve agent Soman. Pyridostigmine is also used to treat some neuromuscular disorders. Inspecting its structure (*Fig.12*) reveals the same element, - the methylpyridinium moiety, - that makes MPP$^+$ and paraquat toxic and helps them induce LPO. LPO is indeed elevated, as judged by increased levels of RCS, when Pyridostigmine is used[19]...

Another common LPO-involving environmental condition which is not induced by chemicals is traumatic brain injury, to be considered in *ch.4.2*.

---

* A very poorly regulated international floral industry, in countries like Ecuador and Ethiopia, uses such high levels of fungicides and insecticides that the neurological issues affect the flower handlers not just in these countries but even thousands of miles away. One wonders if this also affects the herbal teas industry (see also *ch.5*)

To do justice to LPO-increasing environmental factors may require a dedicated book, so to finish up on these here, just one more example. Using tools or operating machinery, traffic accidents, explosions etc, even nowadays, at the age of Health & Safety, may lead to a surprisingly common trauma. Woe betide those who neglect the safety goggles! Ocular siderosis (from Greek *sideros*, iron) is caused by a foreign iron object penetrating the eye, and leads to a degenerative process called siderosis bulbi[20]. Pathological changes can be traced to iron kick-starting the Fenton process, with the resulting burst of ROS production damaging the PUFA-rich retinal environment through LPO. The iron rule to follow here is: don't be safety blinded, be safety minded.

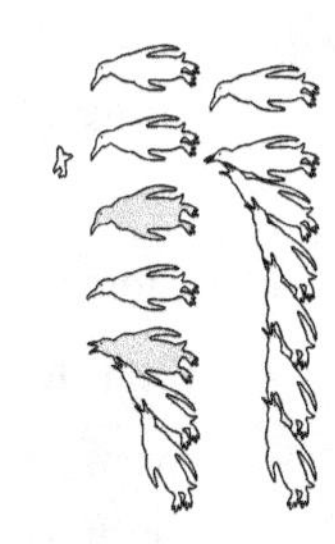

2.6.5. LPO is the driver of multiple pathologies

"The more specific we are, the more universal something can become."
— Jacqueline Woodson

"We are more prone to generalize the bad than the good. We assume that the bad is more potent and contagious."
— Eric Hoffer

A provoking conclusion follows from inspecting the examples above. In many pathologies – neurological, retinal, metabolic, inflammatory, and age-related, - LPO is likely one of the drivers. Almost all neurological disorders, main indications (AD, PD, TBI, etc), and orphan (rare) diseases (FA, INAD, ALS, etc) have an LPO component, detectable at very early stages of the disease. And the reverse is also true: implant a source (genetic or environmental) of LPO into a tissue, and a specific pathology will follow, depending on what cell/tissue, and what exact source of LPO. But if this works both ways, then surely a claim of LPO being the driver is justified? There are multiple other hints of this being the case. Concerning INAD (described above), when close siblings have the same phospholipase A2 mutation, INAD will be developed, typically before the age of two. Occasionally, the other sibling stays healthy till early adulthood- and then develops… Parkinson's! LPO is then what these two diseases have in common, suggesting LPO is the driver, - not only in PD and INAD, but in the majority of CNS pathologies.

2.7. Dr Miyashita's paradox

"Most things disappoint till you look deeper."
— Graham Greene

"The most exciting phrase to hear in science, the one that heralds new discoveries, is not 'Eureka!' but 'That's funny."
— Isaac Asimov

Just when we thought we understood the LPO process fairly well… Kazuo Miyashita at the Hokkaido University has been baffled by his observations for years[1]. The ease of PUFA peroxidation is linked to the number of bis-allylic positions between the double bonds, from which hydrogen atoms are abstracted by ROS: the more bis-allylic hydrogens on a PUFA, the easier to oxidize. The kinetics of the process had been worked out in fine detail by Keith Ingold and Ned Porter among others, typically in solutions of PUFAs in organic solvents, and is very predictive of the bulk oil stability, in relation to food industry oils, fish oils, etc (*ch.2.6.1*). However, Dr Miyashita kept observing that in micelles and liposomes, in other words in PUFA arrangements similar to biological membranes, the situation was not just different. It was the opposite! The more double bonds in a PUFA, the more stubbornly it would resist oxidation, while PUFAs with less double bonds would be just going up in smoke.

Lipid bilayer structure and its key building blocks, phospholipids (PL), are

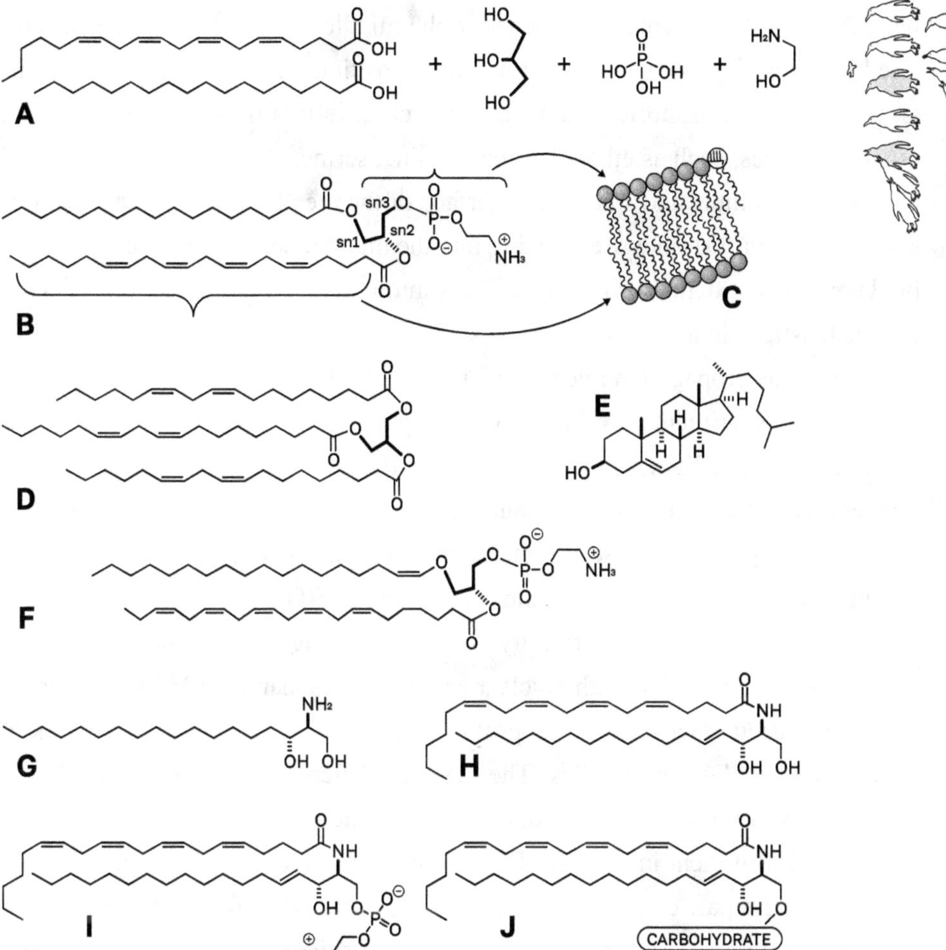

Figure 13. A, the PL-making kit: FAs, glycerol, phosphoric acid and ethanolamine (choline, serine, threonine and inositol instead of ethanolamine give different PL classes). **B**, a PL. PUFA is typically at pos. sn2, and phosphoric acid at sn3. **C**, a lipid bilayer. PL's hydrophilic parts face outside, as FAs snuggle inside, expelling water. **D**, triglycerides (TG) are not in membranes. They store fatty acids in the adipose tissue and blood. **E**, cholesterol is a vital component of membranes, adipose tissue and blood lipids. **F**, plasmalogen (PLG; heart and brain) contains a PUFA (typically, DHA) at sn1, and a more stable vinyl ether. **G-J**, sphingolipids have sphingosine (**G**) core instead of glycerol. Attachment of a PUFA (typically, ARA) to the amine gives a ceramide (**H**), while phosphorylation converts it into sphingomyelins (**I**). Carbohydrates, instead of phosphorylation, yield a diverse group of cerebrosides and gangliosides (**J**). **H-J** are key lipid blocks in myelin and axons. Cardiolipin (CL) is shown on Fig. 23.

shown on *Fig.13*. Two fatty acids* are attached to a triol, glycerol, through ester bonds, with a PUFA typically attached to the middle hydroxyl group, and a saturated FA taking up one of the two remaining positions. The third hydroxyl forms an ester with a phosphoric acid residue, which in turn is linked to a selection of other molecules, such as ethanolamine, choline, serine, etc. This phosphoric acid part is water soluble and sits at the interface of the membrane facing the aqueous side, while the FA strands are pointing into the membrane, forming the hydrophobic layer. This structure brings the PUFAs into a highly structured close contact, like matchsticks in a box, making it easy for the chain reaction to propagate. But how could this propagation rate be lower for the PUFAs with more double bonds?

Various experiments, by Miyashita group and others, have been carried out attempting to shed light on the mystery. That the ROS entering DHA-containing bilayers from the aqueous neighbourhood outside would preferentially oxidize the bis-allylic positions closest to the carboxy group was to be expected, as these would be the closest to the ROS entry site at the interface. More intriguing were the observations of relative "rigidity" of PUFA arrangements. Conformation of molecules can be studied with Nuclear Magnetic Resonance (NMR) spectroscopy, using various metrics such as proton relaxation time, comparing PUFAs in solution and PUFAs in micelles. The detected difference indicated that PUFAs in micelles were rigidly associated, compared to their solution state, again as fully expected. But then, in micelles DHA showed considerably higher mobility of its hydrophobic part compared to linoleic acid, with ARA falling somewhere in between. And this would translate, among other things, into more water molecules slipping along in between DHA strands, likely assisted by six double bonds, which are more hydrophilic compared to single carbon-carbon bonds. DHA may find it easier to interact with water molecules whilst making up the lipid bilayers, due to both its backbone properties ("angle iron-shaped"), and the number of water-interacting double bonds.

Paul Else at Wollongong University has inserted the final piece into this jigsaw puzzle[(2)]. His insight is related to a well-known chemical concept of creating cyclic molecules. Formation of five- or six-membered ring structures is favoured

---

* In triglycerides (TG), used for for storage purposes, three FAs are attached to a glycerol core.

over larger cycles, for the same reason for which, in normal circumstances, you might find it easier to stomp on your own foot (intramolecular interaction), rather than on someone else's (intermolecular), even in a crowded room. Because on average, your own feet are closer to each other. Similarly, for a jiggling and squiggling PUFA in a bilayer, it might be easier, on average, to bump into itself, rather than onto its neighbour PUFA. And with five bis-allylic positions on a DHA molecule, it might be easier for a peroxyl radical just formed at one position (*Fig.14*) to sting not a bis-allylic position of a neighbour PUFA molecule, which might have just happened to roll away this instant, but one of its own bis-allylic positions, always a fixed distance away. And our dear friend, poor "one-legged" LIN, with only one bis-allylic position, just has no such option, so the chain will have to spread laterally. And while it does, that clumsy DHA dancer would just keep stinging itself. Now stop the clock: while DHA was busy oxidizing itself multiple times, the LPO in LIN spread away laterally. Even though there might have been more abstraction events for DHA, the damage per square area would be more extensive for linoleic acid containing membrane, counting damaged PUFA residues. An interesting side-effect (no pun intended) of this mechanism is the predominant location of the LPO damage in just one out of two bilayer leaflets (halves). This is again fully consistent with the mechanism pointed out by Paul Else.

Could this, rather than some fuzzy quantum stuff, be the actual reason for the predominance of DHA in key membranes, in neurons and retina?

A simple fact that DHA membranes are less oxidizable? Could be, - but there are also studies reporting on the direct, not reverse, correlation between the number of bis-allylic sites in a PUFA and its oxidizability so, as with many other aspects of PUFA biochemistry, the jury is still out.

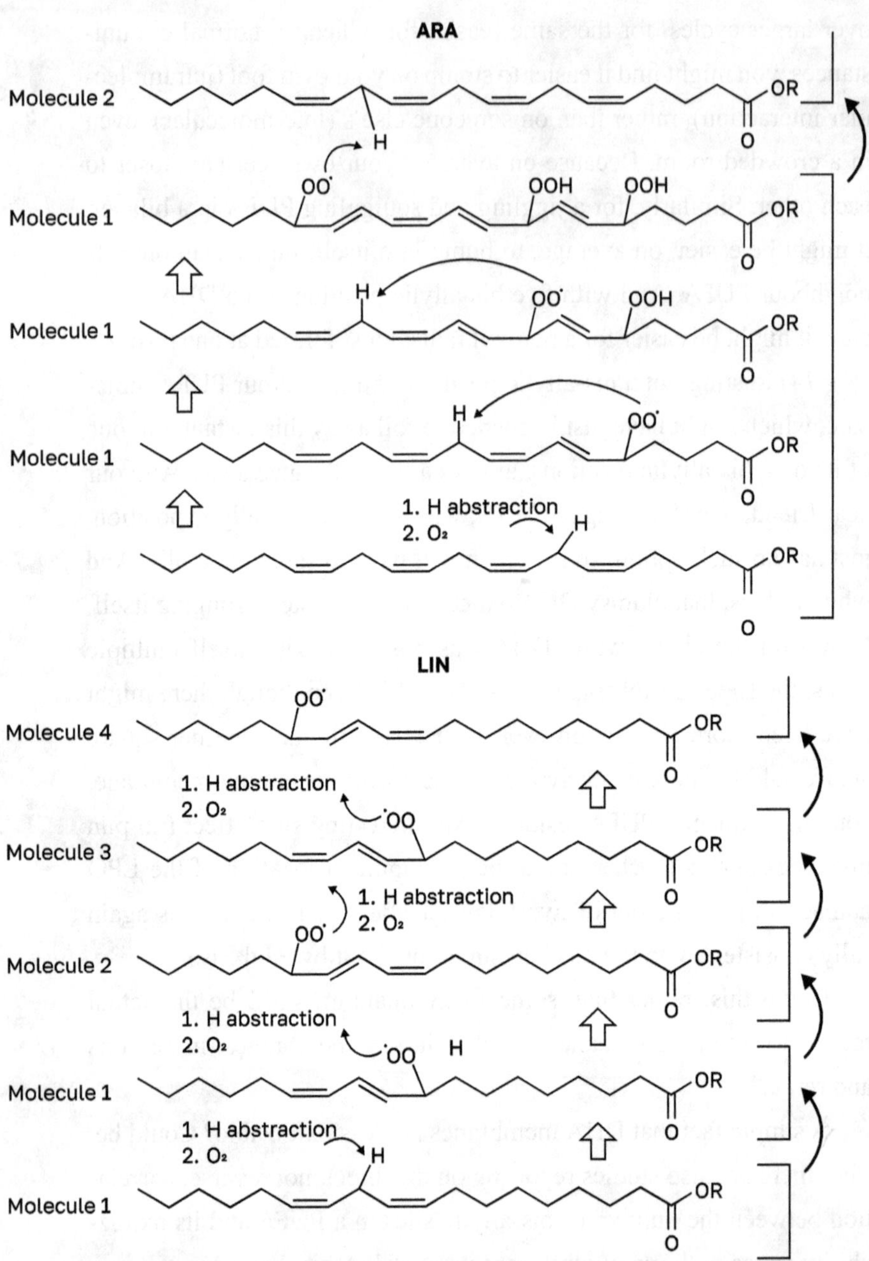

Figure 14. Inter – versus intramolecular explanation of the Miyashita paradox. The more bis-allylic positions a PUFA has, the more internal, self-oxidation steps will the molecule undergo, slowing down the lateral propagation of the LPO process. While the same ARA molecule (left) undergoes three consecutive oxidation steps, LIN (right) can only undergo one oxidation, so three molecules will be damaged over the same period of time. The process will be even slower for DHA, with its 5 bis-allylic positions.

2.8. Antioxidants cannot stop the chain reaction

"Friends don't let friends have Oxidative Stress."
— A concerned friend

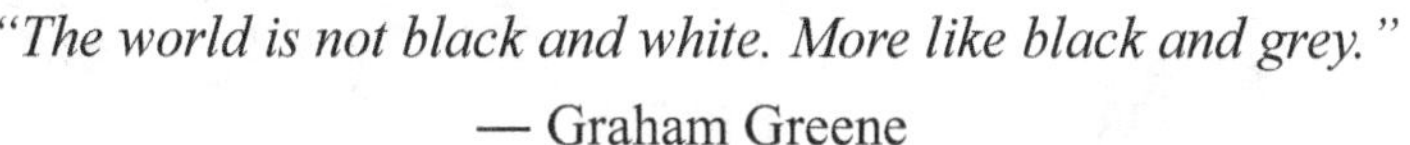

"The world is not black and white. More like black and grey."
— Graham Greene

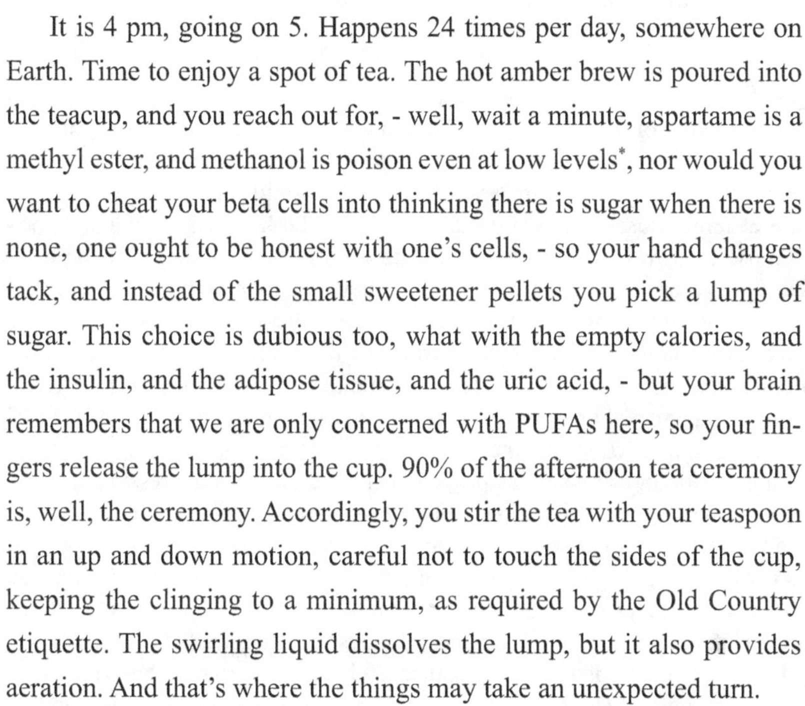

It is 4 pm, going on 5. Happens 24 times per day, somewhere on Earth. Time to enjoy a spot of tea. The hot amber brew is poured into the teacup, and you reach out for, - well, wait a minute, aspartame is a methyl ester, and methanol is poison even at low levels*, nor would you want to cheat your beta cells into thinking there is sugar when there is none, one ought to be honest with one's cells, - so your hand changes tack, and instead of the small sweetener pellets you pick a lump of sugar. This choice is dubious too, what with the empty calories, and the insulin, and the adipose tissue, and the uric acid, - but your brain remembers that we are only concerned with PUFAs here, so your fingers release the lump into the cup. 90% of the afternoon tea ceremony is, well, the ceremony. Accordingly, you stir the tea with your teaspoon in an up and down motion, careful not to touch the sides of the cup, keeping the clinging to a minimum, as required by the Old Country etiquette. The swirling liquid dissolves the lump, but it also provides aeration. And that's where the things may take an unexpected turn.

Teas are famous for their flavonoids, flavones and all the rest of a large family of plant antioxidants. As antioxidants, what do the likes of quercetin and catechin do? They reduce oxidants (ROS), by readily

---

* Methyl esters and high methanol levels is the reason why one of the most fragrant red grapes, - Isabella, and wines made from it, are banned in many countries. Steer clear of trace methanol in food and drink! Warning: this is not an endorsement for overindulgence in methanol-free ethanol-containing tipple.

donating electrons. The electrons which oxygen, an oxidiser, is avidly craving. Consequently, what is happening in your cup may be rather opposite to your expectations. The antioxidants get to work, reducing the oxygen molecules by giving them electrons. And when an oxygen molecule acquires an electron, it forms, - oh, dear… $O_2^{\cdot-}$, superoxide, an ROS! A reactive oxygen species. But is this secret double life of antioxidants bad? Perhaps it is antibacterial? Or hormetic (*Box 4*)?

Box 4. ON HORMESIS, OR ADAPTIVE RESPONSE

"Was mich nicht umbringt, macht mich stärker."
— Friedrich Nietzsche

It gets uncomfortably close to a debunked (save for placebos, see *Box 12*) concept of homeopathy, a brainchild of the 18th century German doctor Samuel Hahnemann, of "like cures like", such that infinitesimal (often diluting it to less than a molecule per dose, like resolvins, see *footnote(**), p.232*) quantities of say arsenic would cure the very symptoms it would cause at larger doses. But comfortably close to the key body-building principle, that your muscles grow by repairing the damage sustained through vigorous training. Some molecules may cosplay like antioxidants, yet they are anything but. They come in and do damage, but not enough to kill[(1)]. Affected cells mount a protective response, which somewhat exceeds the amount of sustained harm. So, when the damage is cleaned up, there will still some excess of "goodness" left. Different compounds may elicit different responses from the enzymes. For example, caffeine may upregulate cytochrome CYP26 so that it becomes more active in mopping up and incinerating other toxins. Coffee elevates hydrogen peroxide in urine yet has proven health benefits. Mild stress can also be caused by training, heat shock, irradiation, caloric restriction, - which, on the molecular level, may all involve LPO. For example, oxPL can increase stress tolerance of endothelial cells. HNE is cardiotoxic at high levels, while low levels induce endogenous antioxidant systems, through Nrf2. Plant flavonoids and various other compounds are often called antioxidants, and plants do need a lot of them, being exposed to sun irradiation and oxygen all day long. However, plants also have sophisticated chemical warfare to keep the pests, from amphid to zebra, away. Yet they still want their seeds dispersed, so the toxic weapons must not be deadly, - this topic could make a fruitful if intoxicating discussion over a cup of tea!

A seemingly simple term "antioxidant" requires clarification*. An antioxidant prevents damage when present at low concentrations compared to substrate. An

---

* An example of non-antioxidant, sulphoraphane, an isothiocyanate present in cabbage, broccoli and other brassicas, activates proteasome and the antioxidant and anti-inflammatory responses, by inducing Nrf2 pathway and inhibiting NF-kB.

antioxidant reduces an oxidant but gets oxidized in the process. For the threat of downstream oxidative damage to recede, the two newly formed species, - an oxidized antioxidant, and a reduced oxidant, - should both have low reactivity, unable to damage other molecules, which is rarely the case*. While oxidized flavonoids in the teacup storm above are often inert, a superoxide radical can be a pretty dangerous actor, stirring up a brain storm (see *ch.2.1-2.4*). In the LPO chain reaction, a lipid peroxyl radical $LOO^{\bullet}$, an oxidizer, rips a hydrogen atom away from a fellow PUFA next to it, turning itself into a non-radical molecule, LOOH. The newly "radicalized" PUFA, which thus far played a role of an antioxidant, has now been converted into a new oxidant, and attacks the adjacent PUFA in the next step of the chain (*ch.2.6*). PUFA molecules, therefore, are not antioxidants but relay stations for transferring the damage down the LPO chain. Should $LOO^{\bullet}$ bump into a membrane protein however, the chain may stop because the newly formed species is unable to abstract a hydrogen atom from the next PUFA. In this case, the membrane protein is an antioxidant. In chemical terms, antioxidant is a molecule that reacts readily with an ROS species but is much slower at transferring that damage to other molecules.

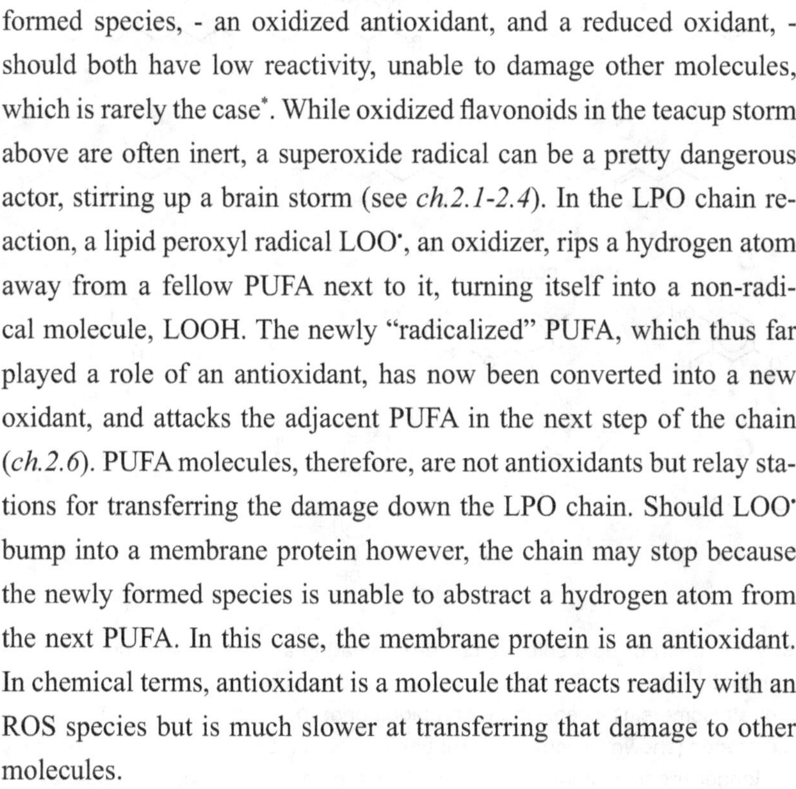

Flavonoids from tea are water-soluble, and so are not directly involved in LPO that happens inside of the lipid membranes. Of the two major groups of antioxidants, - fat-soluble (hydrophobic) and water-soluble (hydrophilic), only the fat-soluble (residing in lipid bilayers or lipoproteins) can directly inhibit the chain reaction. Let's meet some of them in person.

Vitamin E (vit E).

A big-ticket player, vit E comprises a mixture of 8 different compounds: 4 tocopherols (α, β, γ and δ) and 4 corresponding tocotrienols

---

* Fitting this bill still does not guarantee a usefulness of an antioxidant. Cyanide ion and hydrogen sulphide are great antioxidants, yet a little niggling issue of their lethal toxicity somewhat limits their attraction.

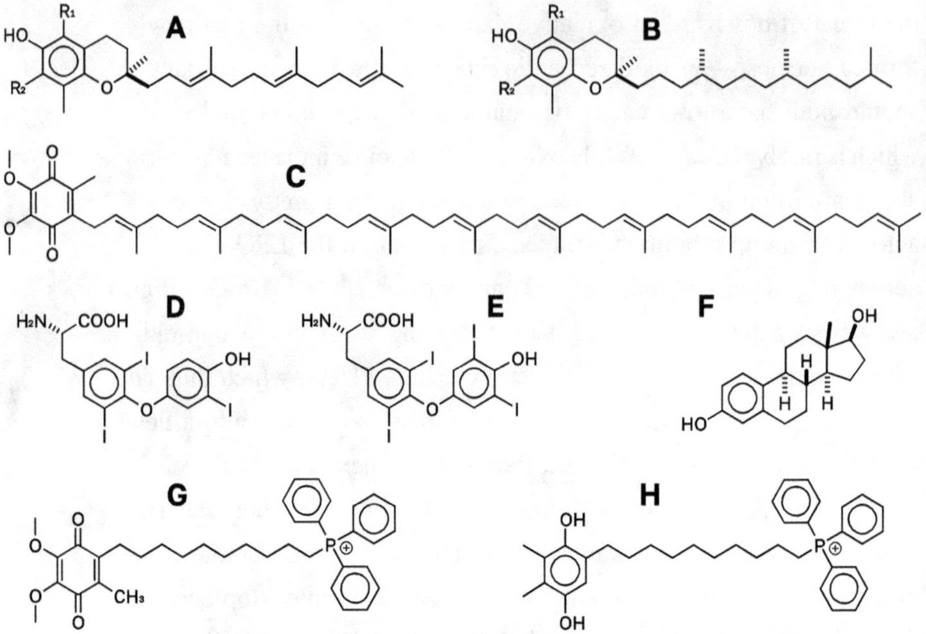

Figure 15. A selection of lipid-soluble, chain-terminating antioxidants all bearing a redox--activate phenolic hydroxyl moiety. **A**, tocotrienols; **B**, tocopherols (R1,R2=H; CH3). Vit E is a mixture of all isomers. **C**, coenzyme Q10 (ubiquinone). **D**, triiodothyronine (T3); **E**, thyroxine (T4). **F**, estrogen (shown as estradiol). Though present at low levels, it may be partially responsible for longer life expectancy in women. **G**, mitoquinone (mitoQ); **H**, Skulachev ion (SkQ).

(from Greek, *tokos*, birth, as it was found to be essential for live birth in rats), which vary in R1 and R2 groups, and hence in activity, with α-tocopherol (R1 = R2 = methyl) being the predominant form in humans (*Fig.15*). The aromatic chromane part is an antioxidant, and the hydrophobic side chain keeps this fat-soluble vitamin inside of lipid membranes. Recommended daily dose for adults is 15 mg/day, while the upper limit of intake is 1000 mg/day. Seed and olive oils as well as nuts are a good source of vit E, while fish, meat and dairy products have lower levels. Vit E prevents propagation, but not the initiation of the chain reaction by inhibiting $L^{\bullet}$ formation, because vit E reacts with $LOO^{\bullet}$ 1000 times faster than $LOO^{\bullet}$ reacts with PUFAs. Once a hydrogen is abstracted off a hydroxyl group of chromane, the newly formed chromane (tocopheroxyl) radical is much less reactive than $LOO^{\bullet}$ and so normally (see *ch.4.6*) does not propagate the LPO chain.

Instead, it is reduced back to the vit E form by a reaction with ascorbate (vit C), which takes place at the lipid-water interface. However, inhibition of LPO propagation also depends on how close, or far away, the nearest vit E is to the chain-carrying LOO. (and the intact PUFAs are always positioned right next to the lipid peroxyl, giving them an advantage). Vit E operates at the water-oil surface, intercepting PUFA peroxides which are less hydrophobic than PUFAs and so well up to the aqueous interface[2]. Vit E increases membrane fluidity, but decreases the barrier function, so its concentration in bilayers is tightly controlled and does not exceed a certain level (1:2000-3000[3], increasing tenfold, to 1:200 in the high oxidative stress compartments such as lysosomes and photoreceptors)* This is maintained by a strict vit E transport system (α-tocopherol transfer protein, TTPA). The vit E - TTPA complex is exported from liver to plasma, where vit E incorporates into lipoproteins for delivery to peripheral tissues. In brain, the neurons are subject to a lot of LPO as part of their normal activities (see *ch.4.2, 4.5*). The household maintenance team of astrocytes manages the TTP machinery, providing the neurons with a constant supply of vit E[4]. Excess vit E is oxidised by cytochromes into water soluble products for excretion in the urine and faeces. Vit E cannot stop the LPO for stoichiometric reasons, yet insufficient levels lead to increased LPO and neurological diseases including neuropathies, myopathies and retinopathies, as well as compromised immune function[5] (see *ch.4.9*). However, clinical trials of vit E supplements were either inefficient or worse ,possibly elevating CVD and cancer risks. This may have to do with a discovery, in 1992, of a two-face nature of tocopherol[6]. Apart from lipid bilayers, vit E also resides in various lipoprotein particles, like LDL. And this portends trouble. Roland Stocker turned the field upside down (or rather, put it solidly on its feet) by discovering that in the LDL environment, particularly on the background of CoQ deficiency (see below),

---

* But even at this level, looking from an antioxidant perspective, trying to intercept the squiggly LPO chains randomly scattering and branching in all directions is a bit of the “whack-a-mole” game.

vit E does not terminate*, but rather propagates the LPO chain[(7)].

Ubiquinone, coenzyme Q (CoQ):

Unlike vit E, CoQ (*Fig.15*) is made by the body and does not require supplementation because its inborn deficiency is rare. Still, for some unaccountable reason, CoQ supplements are extremely popular. It is enormously hydrophobic (making the delivery of the supplements questionable), resides in membranes, and is a major redox-active cofactor, shuttling electrons through the mitochondrial electron transport chain. It is also found in lysosomal membranes and LDL particles[(8)]. An ability to accept or donate one or two electrons at a time is at the core of its dual role as pro- and antioxidant. The reduced form, - CoQH2 - donates electrons to free radicals acting as an antioxidant, helping recycle vitamins E and C back to their active state. It was suggested in the 1960s that CoQ could break the propagating LPO chains, by directly quenching two lipid peroxyls as follows:

$$CoQH_2 + LOO^{\cdot} \rightarrow CoQ^{\cdot-} + LOOH$$
$$CoQ^{\cdot-} + LOO^{\cdot} \rightarrow CoQ + LOOH$$

CoQ might be able to quench $L^{\cdot}$ too, nipping the chain reaction in the bud. While the jury is still out on this, it actually seems rather unlikely, considering how clumsy and sluggish this hulk of a molecule would be, lying supinely between the two leaflets of a lipid bilayer trying in vain to reach out about 10 Angstroms away to intercept the radical, positioned at (roughly) the middle of a PUFA molecule, while those nippy oxygen molecules would be all over the place to react with $L^{\cdot}$. There is no chance! Come to think of it, nor is there much chance for CoQ to intercept a $LOO^{\cdot}$ neither, - once a peroxile is formed, it tends to migrate closer to the water interface, leaving the CoQ even further behind.

And the final nail. CoQ is only about 1/10 as efficient as alpha-tocopherol in $LOO^{\cdot}$ scavenging. This, in combination with the fact that concentration of CoQ

---

* A member of the antioxidant trinity of vit A, vit C and vit E, vit C (ascorbic acid) is not a fat-soluble chain-terminating antioxidant, so is not covered here. However, Victor Herbert's work of the 1990s reveals its dark side: when iron is present, vit C turns into an oxidant, helping recycle it, spinning the wheels of the Fenton process.

in lipid membranes is typically lower than those of tocopherol, it just does not seem reasonable to pin much hope on to it…

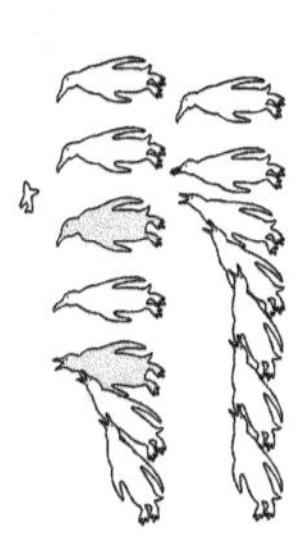

CoQ levels decline with age, and there is limited evidence that supplementing CoQ, particularly in combination with selenium (an element vital to the glutathione-making machinery, actively employed to neutralize LPO – derived toxic products), extends lifespan[9]. This is a chicken and egg question, which surfaces so often in biology. It is often impossible to use a metric without some background context. If an antioxidant is low, is it good or bad? Is it low because the cellular ROS is down, so the defence ministry saves resources by reducing the antioxidant production? Or perhaps the cellular ROS is elevated, so the antioxidant ammo gets used up? LPO increases with age, so the life-extending CoQ – selenium combo might in fact be interpreted as a booster to the chain-inhibiting and post-LPO-repair systems. Of course, CoQ is also helpful when wearing its electron shuttle hat, as it boosts the mitochondrial efficiency.

Hydroquinone CoQH2 is the core motif of many lipid-soluble antioxidants. Its substantial reducing potential has been used to a good effect in photographic development for film and paper to reduce silver halides to elemental silver. Being good at repurposing, nature found multiple other uses for hydroquinones. One of the most spectacular is based on the targeted release of explosive energy by bombardier beetles, shooting at trespassers from their multiple (up to 20 shots when fully loaded) rocket launchers. The ammo for this complex chemistry is stored in two separate sacs, and when the valves open, 25% aqueous hydrogen peroxide flows through channels lined up with catalases which partially decompose the peroxide into water and molecular oxygen. Another sac contains 10% hydroquinone, which upon release gets oxidized by peroxidases to a mixture of semiquinones and quinones, perhaps involving hydrogen gas as an intermediary. The exothermic reaction heats the mixture up to 100°C, and the vapour formed shuts down the valves protecting the beetle's innards, while killing, injuring or scaring away interlopers big or small[10].

Thyroid hormones

Thyroid hormones are also produced naturally but require a supply of iodine. Similar to vit E and CoQ, the two major thyroid hormones, triiodothyronine (T3) and thyroxine (T4), have a redox active phenolic hydroxyl group that is key to their properties (*Fig.15*). In the past, as a land-locked country, Switzerland was getting much less of it than its fair share. Many villages were severely affected by goitre, and when the theory was put forward connecting the disease to an iodine deficiency, Switzerland was the first country to start salt iodization, in 1922. Since then, the big data reveals significantly increased graduation rates and decreased incidence of intellectual disability and schizophrenia. Interestingly, LPO has been shown to be involved in both. Now, thyroid hormones act on almost every cell in the body. Essential to proper development, they regulate protein, carbohydrate and fat metabolism as well as heat generation (see *ch.4.3*). A less known but most pertinent to this book, thyroid hormones such as T4 may play a role of chain-terminating antioxidants. In red blood cells, at least half of T3 and T4 are associated with membranes. T4 is present in membranes at 1 molecule per 1000 phospholipids, comparable to tocopherol levels. Quantitatively, T4 is a more potent antioxidant than vit E[(11)]. However, the 1:1000 ratio is still way too low to stop the chain, and some of the other multiple roles the thyroid hormones play are pro-oxidant, with elevated levels leading to hyperthyroidism.

Multiple other types of antioxidants play a secondary role in LPO inhibition. Even estrogen can be a terminator. Carotenoids can reduce the levels of light-induced singlet oxygen and are used to a good effect in the eye. Other classes of antioxidants, produced by the body or supplied with food, play an important role in recycling the LPO-inhibiting antioxidants*. The turnover chains can have multiple units, coupled to one another through redox interactions.

Over time, bold attempts were made to chemically improve antioxidants[(12)].

---

* When vit E or CoQ terminate LPO chains by quenching LOO• radicals, they "get radicalised" themselves (see CoQ section above). This jeopardizes the peace and quiet and is unacceptable, so the law enforcement systems kick in to reduce the tension. While a CoQ radical can be "defused" by a mito ETC cycle, a water-soluble vit C (ascorbic acid) can "reset" the vit E radical back to its normal state, turning itself into a vit C radical in the process. Ascorbic acid itself can be recycled back to its normal state by GSH or lipoic acid, which can get discharged by the NADPH cycle. The net process therefore can be described as NADPH quenching the peroxyl radical.

Some of the strategies were remarkably inventive. Derek Pratt at Ottawa is at the forefront of the global rational design effort, employing the tour de force approach based on cutting edge kinetic, redox and physical chemistry principles to design, synthesize and evaluate multiple new candidates for radical trapping and LPO inhibition, with an eye on preventing cell death. Capitalizing on his intimate understanding of hydrogen atom and proton-coupled electron transfer processes, he and his team described a smorgasbord of novel compounds, often with delightfully unexpected properties and unprecedented efficacies, in model systems(13).

Another example of a cute rational antioxidant design was a contribution from two groups, over a couple of decades. In 1969, Vladimir Skulachev, a renowned Russian bioenergeticist at Moscow State University, showed how a triphenyl phosphonium (TPP) ion, a cation with a positive charge on a phosphorus atom stabilized by the adjacent phenyl rings, could get carried across the mitochondrial membrane riding on a trans-membrane potential(14) (see *ch.4.3*). In the early 1970s, several stabilized, mitochondria-penetrating cations such as TPP were named "Skulachev ions". As mitochondria are one of the major sources of ROS and LPO in cells, on account of their carbon fuel burning activities, in the mid-1990s Michael Murphy at Cambridge, UK, back then working in New Zealand connected the dots by combining a free radical-intercepting antioxidant moiety with a TPP cationic vehicle into a single humdinger device, to make sure these targeted antioxidants would get to where they are most needed, - in the inner mitochondrial membranes, overcoming the natural delivery limitations (*Fig.15*). Whilst there, with a charged TPP group being held at the lipid-water interface amongst the phosphate groups of PLs, the hydrophobic spacer arm of the molecule would reach into the lipid layer placing the antioxidant quinone moiety right in front of the PUFA's bis-allylic sites, at the ready to intercept the radicals, inhibiting the LPO chain. While the results of these efforts, the MitoQ compounds from Murphy(15) and the SkQ derivatives from Moscow(16) are widely used in various models and available as drugs (www.mitotechpharma.com) or supplements (www.mitoq.com), the jury is still out regarding their usefulness as medications. A widely broadcast ability of SkQ, at nanomolar concentrations, to help mitigate the dry eye condition seems fitting given the eye-watering price tag of the supplement (pun intended), even though a second mortgage may be needed to afford it. As the delivery is topical, it is far from clear to what extent if any the effect of SkQ, particularly at such low concentrations, may involve the LPO inhibition as its mechanism of action. Perhaps it may act through another component of the drug formulation, benzalkonium, - widely used as biocide and antiseptic?

When delivered systemically, the TPP derivatives concentrate in mitochondria and some other compartments which have transmembrane potential, but sadly this is not the only location where ROS are elevated. Moreover, cations crossing mitochondrial membranes en masse lead to decreased transmembrane potential, which could compromise mitochondrial performance. Besides, while low levels are not enough to stop the LPO chain, if concentrated to too high a level, these compounds may become pro-oxidants…

Quinone based antioxidants are a popular development concept even when not attached to a positively charged targeting appendage. Edison Pharma put a lot of effort into the idea, making various compounds for Rett syndrome, Friedreich's ataxia and other diseases. Quinone derivatives looked promising in cell cultures but failed in trials, so eventually the company gave up (the ghost).

Numerous observational studies and intervention trials, as well as meta-analyses (studies of studies) mostly failed to show benefits of taking antioxidant supplements. While there are rare exceptions, SUVIMAX study which revealed decreased cancer levels in men (but not in women, which may have been due to lower baseline levels of some antioxidants in men) on a low level antioxidant supplementation(17), sometimes people taking high doses were worse off than people that did not(18). How could that be? Antioxidants perform well in preserving various bulk materials such as oils, plastics, and rubbers. However, in lipid membrane bilayers the PL arrangement is different compared to the unstructured, bulk oils. Even though according to some estimates a PL molecule can circumnavigate a red blood cell in under a minute, still the lateral, side-ways movement of lipid molecules in membranes is more restricted compared to liquid oils. PL molecules in a lipid membrane are not unlike trees in a forest, or soldiers in the ranks. The chain reaction, once in gets going, moves laterally faster than the PL species, or membrane antioxidants, can diffuse.

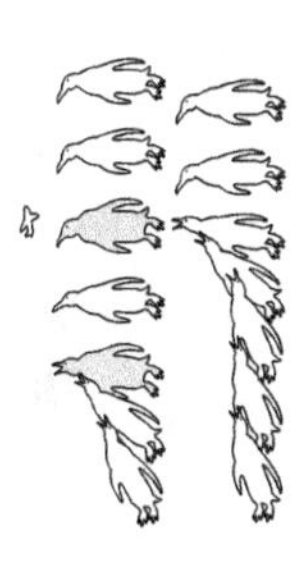

Another important reason is a very high degree of biological control exerted by organisms over transportation of antioxidants across the body. Such a control happens at multiple levels including organ and tissue, and accordingly experiments with simple models such as liposomes or cell cultures, which lack these control mechanisms and so can accumulate high levels of antioxidants, can show deceptively efficient antioxidant activity. Eating large doses of vitamin E would not increase its natural ratio in bilayers, which is approximately one vitamin E molecule per 2000-3000 PUFA residues(3). In fact, it would be dangerous if it did, as higher than normal levels disturb important membrane domains called lipid rafts in undesired ways. With this low a fraction, the chain reaction meandering laterally across a membrane may proceed for quite a few steps before bumping into a chain-terminator. Such a low antioxidant load in PUFA-rich lipid bilayers makes sense considering how important the structural integrity of membranes is. Typical membrane antioxidants, such as tocopherols or CoQ, are bulkier compared to PUFAs, and have aromatic ring systems which wedge themselves into bilayers, disturbing their integrity. Consider a zip fastener, where a slider interlocks two opposing rows of metal or plastic teeth, thus joining the two halves of the material to which the rows are attached. It works smoothly if the teeth fully match each other in size and spacing. Now imagine inserting different sized elements instead of the original teeth, messing up the spacing and creating protrusions or indentations, which will likely keep the barn door open. In non-living systems, it is possible to mix PLs with antioxidants at any ratio imaginable. Sadly, the lipid bilayers with that composition cannot be sustained in vivo…

The kinetic parameters carefully measured in solutions for various antioxidants in the LPO process are not really useful for the bilayer setting, where the molecule's mobility, like the sideways, lateral movement across membranes, can be rather restricted, due to the unwieldy, bulky shape. Certainly, while developing the next generation of the LPO-inhibiting, chain-terminating antioxidants, a little more focus on the mobility of these redox compounds wouldn't go amiss.

But as of today, antioxidants, synthetic or natural, fail to eradicate the radical chain reaction (pun intended). To summarize why this effort is largely a wild goose chase:

— There are near-saturating amounts of antioxidants already present in living cells. Typical ratios of around one tocopherol molecule per 2000 fatty acid residues in membranes, increasing to one per 200 in particularly oxidation-prone bilayers in Golgi, lysosomes, and retina cannot be exceeded by tightly controlled delivery mechanisms, for incorporating more would compromise structural integrity and fluidity of membranes(19). However, even at this high level of antioxidants, given the stochastic, random nature of ROS generation within the membranes, and the two to three orders of magnitude difference in the molar ratio of antioxidants to PUFAs, LPO propagation cannot be completely suppressed, particularly in PUFA-rich neuronal, retinal and mitochondrial bilayers.

— Attempts to provide extra antioxidant(s) are not just challenging from the transport systems perspective. Excessive levels of one would typically instruct the body to dial down on other antioxidants, which can be fraught with side-effects considering the intricate, in unison interplay of various redox molecules.

— Fat-soluble antioxidants in certain conditions can do a U-turn, becoming chain-propagating oxidants in non-bilayer environment, such as in apolipoproteins.

— Fully eradicating ROS could be harmful to some people because they modulate cell signalling(20), apoptosis and other processes, while human population is highly heterogeneous in the ROS levels(21). For instance, many LPO products at subtoxic concentration can exert hormetic response (*Box 4*), through upregulation of various defence mechanisms(22). Toxicity of oxidized forms of some antioxidants may be a significant factor(23). It is worth bearing in mind: long-living species often have less antioxidants, not more(24).

— Toxicity of LPO products themselves is a huge issue. They continue to degrade yielding various classes of toxic non-radical products such as reactive

α,β-unsaturated carbonyls, which also alter the cellular redox status by inducing electrophilic stress and reacting with thiol-containing biomolecules such as glutathione. Such secondary products of LPO are of non-radical nature and so cannot be neutralized by radical-trapping antioxidants. The less hydrophobic ones like MDA and HHE readily escape from the lipid bilayers, looking to do their mischief in the aqueous, while the more hydrophobic continue to make trouble in membranes

— Water-soluble antioxidants or antioxidants prone to forming hydrogen-bonding interactions with phospholipid head groups (such as polyphenols) are not efficient inhibitors of LPO[(25)], and at best can help recycle the LPO-inhibiting antioxidants.

Consequently, antioxidants provide a far from perfect defence against PUFA autoxidation[(26)].The listed pitfalls may explain why Denham Harman's hypothesis of oxidative stress and aging[(27)], proposed more than six decades ago to explain the pivotal role of ROS in the aging process, is still perceived as a conjecture.

Virtues of antioxidants have for decades been inculcated into the minds of health-conscious customers, with many a hagiographic paean extolling their goodness. However, you should not put your money on the "LPO-terminating antioxidants", because none of them can. Even though a rough eyeballing suggests the collection of antioxidants is currently more than a 1000 deep. Unfortunate antioxidants go out of the way (literally!) to catch the evasive peroxyl radicals. But be it tocopherol, CoQ (or insert whichever "interceptor" you want to consider), - the chain reaction is having none of it and proceeds ineluctably, capable of quickly putting organelles, cells and tissues out of commission, and leaving behind deposits of debris, superglued together by reactive carbonyl species. So, sadly, the new, improved and advanced, more "powerful" antioxidants are just hot air. Plucked out of thin air. They do not affect the LPO in vivo. The dog barks, but the caravan moves on.

It is unfortunate that the chain non-enzymatic chain reaction is so rare in biology, lipid peroxidation being the only example. Biologists are simply not well versed in a non-stoichiometric chain process. Borrowing from Jack London, stoichiometric damage compares to it like asteroids compare with the sun. It could have been beneficial to initiate a dialogue between biologists and experts working on different incarnations of the chain reaction, in chemistry and physics. However, because of the nature of the applications of the processes in chemistry and physics, the key opinion leaders may not necessarily be at the liberty to discuss the intimate details…

2.9. Peering into the tiny secret chambers of the LPO pyramid

"You can't manage what you don't measure."
— Peter Drucker

"I did not have three thousand pairs of shoes;
I only had one thousand and sixty."
— Imelda Marcos

To measure the progress of a disease, or the efficiency of treatments, an easily detectable "biomarker", that correlates with disease severity, is a very useful (and potentially lucrative) thing to have*. A fertile ground for research, the biomarker field has churned out thousands of candidates, but only a couple hundred made it into the clinic(1). Some diseases, such as neurological conditions, are particularly poorly represented. Suppose, for example, there is a small brain compartment, like the melatonin-producing pineal gland, in which a pathology slowly develops over

---

* As noted by Steven Austad using cholesterol and blood pressure as examples, individual biomarkers consistent with good health may not necessarily mean good health. High TG in blood is a CVD risk, while too low a level may indicate malnutrition or hyperthyroidism. Is zero CRP indicative of low inflammation – or of liver failure? High or low antioxidant levels (see CoQ, *ch.2.8*) are also difficult to interpret individually. The best approach is holistic, looking at multiple biomarker panels.

decades. To the nearest order of magnitude, a human brain, with 10^{11} neurons, weighs 10^3 g. Accordingly, this grain of rice - sized gland, at 10^{-1}g, would have 10^7 neurons. If a pathology becomes noticeable at say the level of 20-25% function loss, then 2 x 10^6 neurons will have to stop functioning for the telltale disease signs to appear. If that level of damage takes 60 years (2 x 10^4 days) to develop, then the daily loss of neurons would amount to just 100 cells. They may not be dead, bursting open to release their contents for us to measure, - they may be just feeling unwell, perhaps becoming senescent. Whichever molecules (if any) manage to leak out of those distressed cells would then need to, if they at all can, cross the blood brain barrier (for we cannot take brain samples), and dissolve themselves in 5 litres of our blood. Detecting those would be a tall order to say the least, because a comfortable current detection range for biomolecules by mass-spec-based techniques is in femtomoles (10^{-15} of a mole, or 6.02 x 10^{23} x 10^{-15} = 6 x 10^8 molecules per litre). As LPO is increasingly recognized as the common denominator for all sorts of pathologies, it is one of the legitimate biomarker candidates. Are the LPO biomarkers easy to detect?

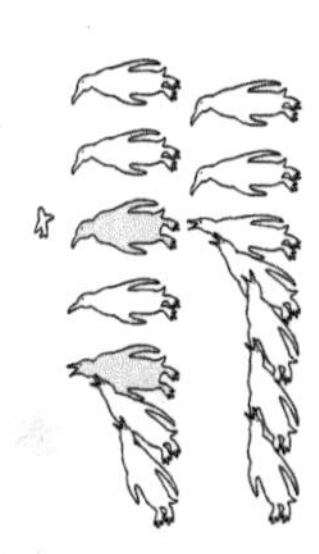

Cycling in a peloton formation may take only 5-10 % of the energy required for cycling solo[(2)]. Even according to more conservative estimates the savings are huge. Nature had hit on this peloton trick way back, when it genetically pre-programmed fish to assemble into huge schools. Ironically, back then it also helped fish to keep predators at bay. The schooling worked for a while, but lately this swimming strategy has led fish straight into dire straits. For when the apex predator is listening, the poor fish need schooling like, well, a fish needs a bicycle. Because sound waves are nicely reflected off the gas-liquid interface of their swimming bladders. In small anchovies, one of the main targets of the fish oil industry (see *ch.3.2*), the swim bladders are smaller still, invisible to sonars. But, amplifying by a mile-long school size, numbering millions of fish, strengthens the detected peak on the sonar screen, helping the apex predator lock on target. If only the poor fish could learn to scatter (*Fig.16*). The sound reflections off their bladders

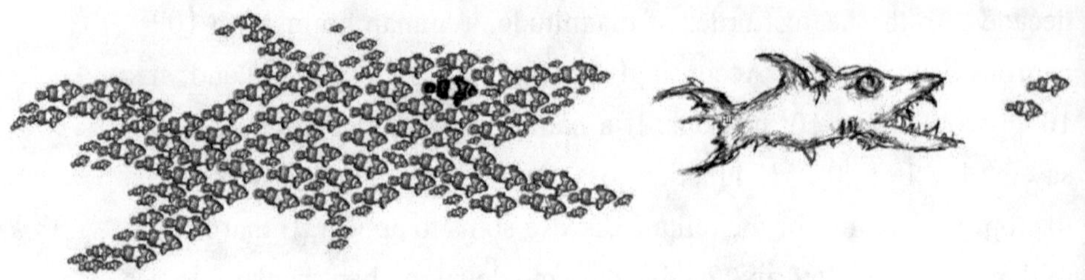

Figure 16. When small fry gather in a dense group to intimidate predators, the resulting school of fish can be visible from space. And certainly detectable by fishermen's sonars. Small numbers of individuals are much more difficult to spot.

would then dissipate into background noise, making them invisible…

The situation is very similar when fishing for LPO products in a sea of PUFAs. Once the LPO zips through lipid membranes or lipoprotein particles, there is initially a relatively small number of different oxidized PUFA species that can form. In LIN, a peroxide group can attach to one of the three carbons (*Fig.17*) Each position can be in one of the two optical isomer forms, giving the total number of LOOH derivatives as 6. This numbers will be correspondingly larger for LNN (12), ARA (18), EPA (24) and DHA (30), as with every double bond, six more potential isomers are added. There will be lower levels of each of the bis-allylic OOH derivatives, but all in all the detection of each species should be within the detection limit.

As a case in point, suppose we have 100 ARA residues in a lipid membrane that undergoes LPO, as 100 would be a reasonable number of steps a chain reaction would take before stopping. Let's assume, for simplicity, that all possible ARA-OOH derivatives are detectable with the same sensitivity. To be measured, they have to be converted first into stable alcohol derivatives, by reducing the peroxides. This reduction also happens naturally, when glutathione peroxidase (GPx) reduces LOOH to corresponding lipid alcohols (LOH), which are inert, whereas LOOH species can react with metals to yield more LPO-instigating radicals. It would also double the number of different compounds from 18 (LOOH) to 36 (18 LOOH + 18 LOH).

Isoprostanes (isoP) are the non-enzymatic relatives of the COX and LOX – produced eicosanoids (*Fig.9*). 64 different ARA-derived isoPs can be produced in vivo(3). Because of their structural resemblance to prostaglandins (PG), they have

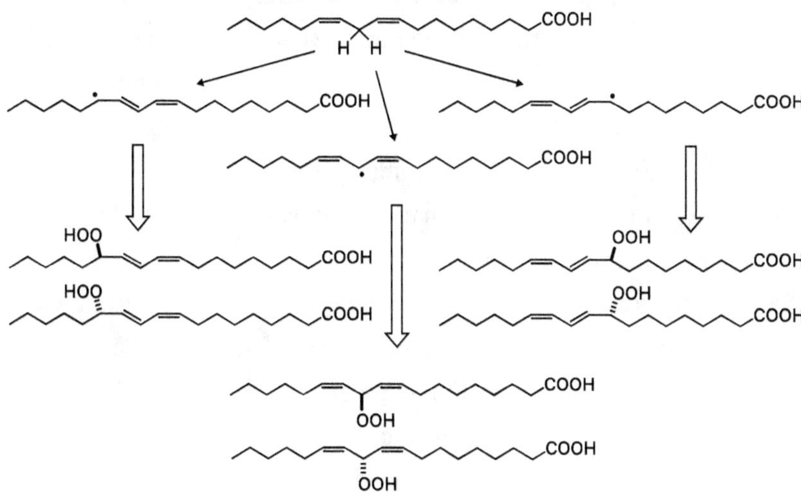

Figure 17. Even for the simple PUFAs such as LIN, with only one oxidisable bis-allylic position, three radicals (L.) can form upon hydrogen abstraction. Each can attach an O2 in a stereospecific way, giving six different LOOH derivatives. The number of decomposition products is larger still.

similar biological activity, although as non-enzymatic products, stereochemically they are racemic mixtures, having both possible chiral variants (unlike PGs which have one), thus multiplying the number of possible isomers*. As only some of the optical isomers are biologically active, isoPs can be said to be "weaker" versions of PG. Still, they are proinflammatory and augment the chronic pain sensation (see *ch.4.4*). They are also great markers** of LPO in our "precious bodily fluids".

* Chirality (Greek, *chiro*, hand) is when two isomers look like a mirror image, or a pair of gloves. Enzymes like COX and LOX produce one possible optical isomer (only one of the two gloves would fit), while nonenzymatic reactions produce both gloves (racemic mixture). Another dimension, ignored in this chapter, is the cis to trans isomerisation. In ARA, one trans double bond gives 4 positional isomers; two would give 6, and three – another 4. Adding an ARA with 4 cis bonds and one with 4 trans brings the total number of possible cis-trans isomers to 16.

** What do biomarkers mark? Care should be taken in collecting and interpreting the LPO markers. Neurological diseases are notoriously poorly associated with biomarkers, so about 25 years ago, a finding reporting telltale

Besides forming the isoPs, LOOH decomposition can take a different path. Through multiple mechanisms, peroxides can non-enzymatically transform into a large class of downstream LPO products known as the reactive carbonyl species (*Fig.9*). Hermann Esterbauer at Graz discovered HNE in the 1960s[(4)], and the RCS family that can irreversibly cross-link important molecules has been growing ever since. At least three different formation pathways for HNE alone were described[(5)], each involving multiple intermediary compounds. On the last count, there are at least 20 RCS (each with its own precursors) that can form from ARA, all too reactive to be found in a free form in the body[(6)]. The way to detect them uses fluorescently labelled antibodies that recognize the proteins to which these RCS covalently attach, the so-called conjugates, - a method often less reliable than mass-spectrometric techniques. Also, antibodies exist only for a small number of the most prominent RCS, like HNE and HHE. Our body deliberately makes some highly reactive peptides such as GSH trying to intercept the toxic RCS. The non-toxic conjugates are then washed out[*] and can be detected, for instance in urine, - as we have seen for mercapturic acid (see *ch.3.1*). Proteomics methods, involving chopping proteins down into small fragments and then detecting covalently attached LPO fragments (known as post-translational modifications, PTM) sound promising, but are not there yet. There are also techniques that indiscriminately detect RCS by using thiobarbituric acid to form coloured products (TBARS). Used widely in the food industry, they are not terribly quantitative. A more refined method looks out for LOOH products using a BODIPY-11 dye, but again it cannot discriminate between various isomers, nor different PUFAs.

Nature knows how reactive the RCS are, so it evolved several tricks to detoxify them. Both alcohols and acids are much less reactive than the corresponding aldehydes or ketones. Accordingly, two groups of enzymes are hard at work trying to neutralize the aldehydes. Aldehyde reductases reduce RCS to alcohols, while our good friends aldehyde dehydrogenases (ALDH) oxidize RCS to non-reactive

---

signs of elevated oxidative stress in urine of Friedreich ataxia (FA) patients caused quite a stir. Sadly, it turned out to be a dud: due to muscle weakness, the poor FA patients would take much longer to collect the urine sample, - enough time for the atmospheric oxygen to oxidise things during the collection process…

* If lysosomes and peroxysomes fail to digest the garbage, exosomes can eject it outside of cells.

acids. Obviously, the group of 20 RCS will theoretically give at least 20 new alcohols and at least 20 new acids, tripling the number of tell-tale molecules sloshing about in the peroxidized membrane to at least 60. Taken together with the stubs still left attached to the PL core after an RCS species has cleaved off brings the number to at least 70. There are also various "boutique" LPO products that require special detection techniques.

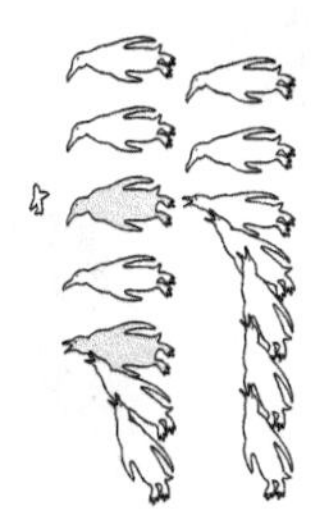

Box 5. LPO fragments as a superglue for bad things.

"It is much easier to make measurements than to know exactly what you are measuring."
— J. W. N. Sullivan

Different chemistries are involved in the downstream LPO processes, multiplying the number of LPO-derived compounds.

Maillard reaction, the darling of food afficionados (see *ch.3.3*), is responsible for browning of foods and praised for associated flavours[7]. This happens when amino groups of proteins and carbonyl groups of sugars react, accelerated by the high baking temperature, forming the distinct lipofuscin-like brownish tint on our bread[8]. Amines reversibly react with carbonyls to yield Schiff bases. However, the Amadori reaction makes the process irreversible by converting the Schiff bases into keto-amines, a point of no-return[9]. At the risk of irritating the gourmands among you, a very similar process, driven by elevated sugar, is responsible for the diabetes-associated clogging of blood capillaries, blocking the flow and sometimes resulting in gangrene. So much for the delicious smell, and yummy colour, of your brown toast. As with lipofuscin, RCS are never too far away and provide additional cross-linking of the Maillard products. The processes are so interrelated in biology, they almost always run together[10], posing additional analytical detection challenges. Indeed, what exactly is there to be measured in a monolith formed by heavily cross-linked proteins, lipid fragments and sugars, to make a call regarding the extent of the LPO?

More generally, a hodgepodge of proteins, sugars and LPO fragments all cross-linked together by a superglue is known as advanced glycation end-products (AGE)[11]. Purists may disagree, but there are no well-defined borders between the compote related to AGE, atherosclerotic plaques, various types of lipofuscins, drusen, or say a lysosome trying to digest a lysosome which choked trying to digest a lysosome which choked on a faulty mitochondrion. Perhaps even Lewy plaques, the deposits characteristic of Lewy body diseases such as PD, AD, HD etc, would belong to this group. The particular types may be enriched in a particular component, like oligomeric alpha-synuclein in PD[12] versus Amyloid-beta fragments in AD[13], yet the glue holding the tutti-frutti together would contain the LPO species (see *ch.4.2*).

An earlier mentioned decomposition of some types of LOOH results in release of pentane from n-6 PUFAs, while the n-3s yield ethane and some ethylene.

Gas chromatography techniques can be used to assay the hydrocarbons. Detecting the gases in exhaled breath as a hallmark of LPO may sound like a good idea. Sadly, this only looks great on paper, as the reality is much less rosy. All sorts of LPO markers have been detected in breath, including isoprostanes and various RCS, with elevated levels correlated to several diseases. As n-6 PUFA predominate in the body, pentane would be produced in greater quantity*. Yet this type of breakup is a minor pathway, which depends on high levels of transition metals in combination with relatively low O_2 levels. Accordingly, detecting elevated hydrocarbons may correlate more with increased levels of metals rather than with the LPO. Or with inhaled, environmental ROS species, which would bombard the PUFA-rich surfactant layer covering the lung alveoli, generating bursts of LPO and hydrocarbons[15]. From a personal experience, some geographic locations, like UC Davies in Sacramento CA, have atmospheric levels of hydrocarbons comparable to the expected exhale levels, and a known fix, - to flush the lungs with hydrocarbon-free air prior to collecting the exhaled air samples - is very difficult to implement in mice. This generated such an overwhelming background signal that exhaled breath of mice treated with a powerful systemic LPO-initiator bromotrichloromethane had ethane and pentane levels similar to the untreated group as the potential difference drowned in the sea of noise (see *ch.2.6.4*).

In healthy humans, the concentration of exhaled pentane is in the picomolar range, so another analytical challenge is separation of the pentane peak from

* Another reason for the pentane predominance is the extra mile that the human biochemistry (as compared to other mammals), is willing to go to preserve the precious n-3 PUFAs. The ratio of ethane exhaled to oxygen consumed is the lowest for humans as compared to other mammals[14].

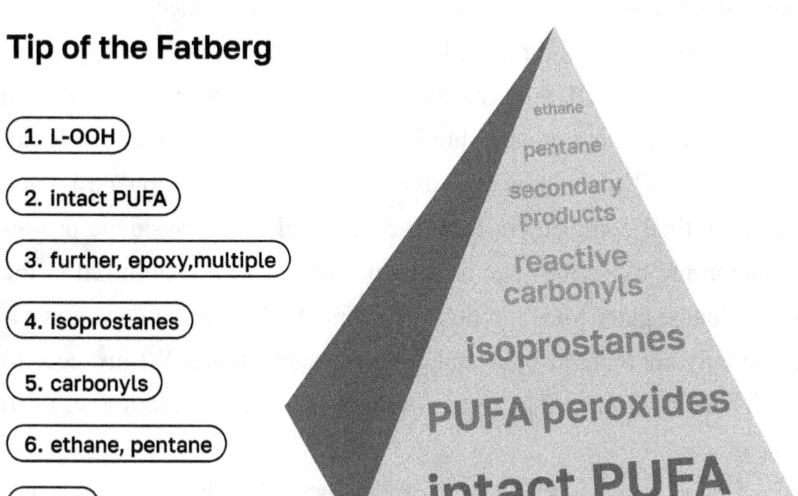

Figure 18. Oxidised PUFA pyramid: tip of the fatberg. While the species present at the lowest concentrations may be the most biologically important, they are also the most challenging to detect.

cholesterol synthesis-associated isoprene signal. Liver cytochromes further muddy the waters by oxidizing pentane to alcohols, so that the measured (reduced) levels may have to do with liver metabolism rather than LPO. Quantifying ethane has its own challenges, linked to contamination from skin and mouth bacteria. So even though elevated ethane levels were associated with other LPO markers such as CRP (see below) as well as aging and diseases in human clinical trials[16], the current focus in the field is on isoprostanes.

To summarise, there may be a good two hundred different molecules that could form from the initial pool of 100 intact ARA residues. And looking for "half a molecule" would not be easy, would it (*Fig. 18*). Important though they are, detection of LPO fragments is therefore a challenge*. As described in other chapters, the LPO-derived RCS

* Compounds formed at the lowest levels may well have more important biological effects, or higher reactivity compared to the more abundant species, emphasizing the need to improve our analytical capabilities. For example, the level of HNE is usually about 2 orders of magnitude below that of

can be found in every nook and cranny affecting mitochondria, sleep, pain, brain, eye and just about every cell and tissue. For instance, CEP, whom we shall meet in *ch.4.1*, - a very important DHA oxidation product, which forms conjugates with amino groups of proteins and phospholipids (*Fig.9*). It plays the major role not just in driving the AMD pathology, growth of blood vessels (angiogenesis) and in triggering the retinal complement response[(17)], but also as a systemic pro-inflammatory factor and oxidation-specific epitope (see *ch.4.6*), recognized in the bloodstream by innate pattern-recognition receptors[(18)]. The current detection capabilities are simply not there just yet. But there are roundabouts. We just need to send those fish back to schools. The signal strength can increase enough for us to rise above the noise.

Two things happen when PUFAs get peroxidised. A polar (water-soluble) peroxy group gets attached to a PUFA backbone. And, quite often, bits of that backbone fall off, leaving behind a truncated stub. Way more hydrophilic than the intact PUFAs, this cigarette butt is no longer comfortable sitting buried within the hydrophobic membrane. It turns away from it, twisting itself into the water layer above. And it is not alone, as multiple oxidised stubs reorient themselves away from the lipid environment into the aqueous compartment. Meet the lipid whiskers[(19)]! While each separate stub can be unique, just like individual bristles on a stubbly chin, they can still be recognized indiscriminately, as an unshaved whole. And the size of that lesion correlates with the LPO extent. Our old friends the pattern recognition receptors, like CD36, then kick in, and the recognition process is of the utmost importance in senescence, inflammation and apoptosis.

The body can certainly recognize the elevated LPO and mount a response. But can this translate into our analytical capabilities? Naturally occurring immunoglobulins G (mostly pro-inflammatory) and M (mostly anti-inflammatory) recognize oxidation-specific epitopes created during LPO and call for a complement cascade response. This is a very important mechanism, as LPO can affect both the adaptive (triggered by environment, such as T and B cells) and innate (genetically encoded, macrophages) immunity[(20)]. However, immunoglobulins recognize many different targets, so measuring their exact levels will not translate directly into the LPO extent.

MDA, yet MDA is much less reactive than HNE, ONE or acrolein.

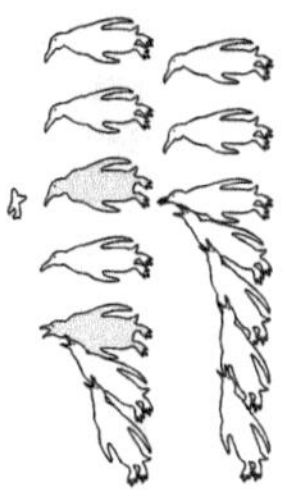

Meet C-reactive protein (CRP, the name reflecting its interaction with a cell wall polysaccharide of pneumococcus). This pattern-recognition receptor, made by the liver and other parts of the body used to dealing with PUFAs, such as smooth muscle cells, macrophages, endothelial cells and adipocytes, is sent on a patrol mission to inspect cell surfaces looking for LPO-associated bulges, because for membranes, thick means sick. So important is this lad for body-wide LPO patrolling and assessment that I will give you not one but two review references to it[21]. Even though the CRP family includes at least three members, - (1) the native CRP (nCRP), a pentamer which can fall apart giving (2) five monomers (mCRP), has been divided into several, and (3) a somewhat in-between state, for the sake of simplicity I will be politically incorrect and reduce the diversity, describing an old-fashioned, "combined" CRP, even though different forms may have different, sometimes even opposing effects[22]. From the LPO detection standpoint, that should be OK.

CRP recognizes lipid whisker patches, or similar LPO – related blebs and bulges in cell membranes and latches on to these[23]. The CRP team are like a police squad, patrolling the neighbourhood looking to apprehend those rowdy hirsute hippie cells. Oxidized membranes are almost always associated with inflammatory events, so the downstream effects of that docking call for pro-inflammatory cytokines to be produced sustaining the inflammation cascades, although some anti-inflammatory activities are also sustained. Regardless of these nuances, the take home message from the LPO detection perspective is simple: more LPO – more CRP. CRP is often used in various human RCT in all sorts of trials – in cancer, inflammatory and neurological conditions[24], implicating not just the inflammation, but LPO as the common denominator in all these, and many other diseases. CRP levels are measured in mg/L and any reading above one (say, 8-10) usually indicates inflammation, although acute infections may lead to the incredibly high, g/L levels (a thousand-fold increase). Lower levels (below 10) of LPO as measured by CRP are associated with numerous non-infectious conditions including CVD, vascular aging, autoimmune diseases, obesity, Type 2 Diabetes, AD, PD, chronic inflammation and inflammaging. A more precise, high-sensitivity CRP (hs-CRP) test may be better suitable when looking at the lower levels of the protein*.

Detecting rare LPO species individually is currently challenging, particularly in some small, "bottle-neck" locations, like single cells, synaptic contacts or organelles. But for practical purposes, just getting a holistic view of the LPO damage in available tissues such as blood (or better yet, in cerebrospinal fluid (CSF, sometimes referred to as "concentrated blood" which sits closer to the brain), may be enough to alarm us to the need to act. When a cute-looking Great spruce bark beetle munches its way through thousands of hectares of pristine boreal forest, or an island in Indochina is shaved free of its rainforest (and orangutans), to free more land for "vegetable" oil adored by the Western food-frying industry, these beetle- or human-made events are easier to spot from a satellite than looking for a handful of affected trees. Yes, it would have been better to spot the first affected tree, in good time. But short of this, developing the rapid LPO detection techniques for the bulk measurement of the proxy signs such as CRP, the harbinger of doom, should continue to be developed as the top priority.

---

* There are several transcription factors, controlled by oxidative stress events, which activate the expression of genes responsible for production of protective proteins. One of the well-knows systems, with particular relevance to LPO, is charmingly called "Nuclear factor erythroid 2-related factor 2" (NRF2). This transcription factor is activated by inputs such as elevated LPO and turns on the production of multiple repair and antioxidant proteins, such as glutathione machinery. NRF2 activation correlates with the LPO levels.

3. Protecting the dietary PUFAs

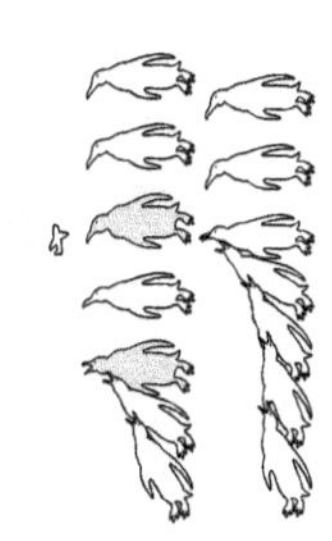

"Imprisoned in every fat man a thin one is wildly signalling to be let out."
— Cyril Connolly

"You can have your cake and eat it: the only trouble is you get fat."
— Julian Barnes

3.1. Chinese paradox: oil on fire

"There are no bad foods, only bad food habits."
— Alton Brown

"God made food; the devil the cooks."
— James Joyce, Ulysses

Kung Pao Chicken from Sichuan and Hunan's Sweet and Sour Pork are coveted for many reasons. Having withstood the test of time with flying colours, the dishes are discombobulatingly delicious, bodacious, nutritious, atrociously efficacious, capacious to share and precious to remember. Loved the World over, they are a great short-term solution to the cravings of many a hungry customer. The long-term consequences for the service providers, however, may be less rosy(1).

With about 300 million smokers, every third cigarette produced in the world is smoked in China. No surprize then that lung cancer is big in the country (more than 1 000 000 cases in 2020). Almost half of the Chinese men smoke, but less than 2% women. However, the lung cancer rates in males (50 per 100 000) are only twice as high as for females, at 24 per 100 000(2). Why would Chinese women have such high incidence despite of a low smoking prevalence? And only a quarter of lung cancer cases in women could be attributed to smoking… The Korean population is substantially smaller than the Chinese population, yet the male to female lung cancer trend is very similar, with 30% of men, and 5% of women smoking, though the numbers are currently declining. And in Taiwan. And in Singapore. Vietnam, too. How could that be?

The following picture conveys the entire drama. Stir fry cooking often involves setting a hearty quantity of overheated PUFA-rich "vegetable" oil* in a wok on fire. Flames engulf the bowl from both sides, helping to sustain a runaway LPO process. So massive and brutal the ensuing PUFA damage is that some of you may find it disturbing. While the lady of the house is busy sustaining the chain reaction run at scales incompatible with the inadequate kitchen ventilation systems, the man of the house is likely reclining on a sofa in a living room, ciggy in hand, glued to the TV screen, catching up on the latest geopolitics or sports developments. He therefore does not inhale. The LPO products, that is. But his wife does, and all the plentiful LPO-derived nasties in fumes and smoke get right into her lungs… Looking from the poor lady's perspective, a recent adage of eating a portion of French fries being equal to smoking 25 cigarettes (toxic aldehydes-wise) blossoms up with new hues, even though the exact number of cigarettes is not fully agreed upon(3). Indeed, "vegetable" oil in food is the new smoking! And it gets worse. As we know from (*ch.2.6.3*), people of South Asian descent have high levels of ALDH2 deficiency. And this is the very enzyme that could have helped to mop up the nasty lipid carbonyls. Wok cuisine, so popular in these parts, happens to be particularly dangerous for the very people who rely on it for their staple cooking…

Mercapturic acid is a garbage truck that helps excrete toxic electrophilic species (*Fig.9*), including the LPO-derived compounds, in the urine(4). Sure enough, elevated levels of some major LPO markers were detected in the urine of women frequently stir-frying food, hitching their rides on mercapturic acid on the way out of the body. The more wok cooking per week, the higher the levels detected(5).

---

* "Vegetable" oils consumed in China annually: soybean oil (50% LIN, 7% LNN), 20M tonnes; Peanut oil (PUFA 32%), 2M tonnes; Rapeseed oil (20% LIN, 8% LNN) 3M tonnes; Sunflower oil (60% PUFAs) low consumption in China; Corn oil (60% PUFA) 0.5M tonnes. This works out as about 16 kg oil per person per year. Hardly the right set of numbers to keep one's omega balance in good shape…

Who knows what fraction of the inhaled toxins does the excreted portion represent, and how many of these toxic electrophiles remained in the body, forming mutagenic DNA adducts (*Fig.19*), irreversibly cross-linking vital proteins, etc? A human body is like a cigarette filter here, with only a relatively small amount of the toxic intake finding the way out, having been filtered through lungs, circulatory systems, liver and kidney. All in all, the cancer-inducing power of inhaled LPO derived electrophiles seems to be on par with asbestos. And geographically, the more popular the stir fry in a given province (like Sichuan and Hunan), the more lung cancer incidence(6).

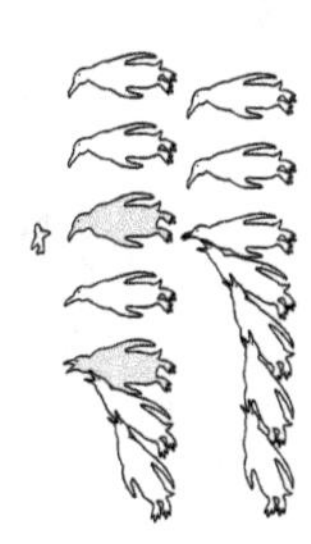

Things can get even worse when moving from private family kitchens to restaurants. Perhaps ventilation systems would be more powerful, - but the quantities of LPO products would scale up, too. The LPO product-containing fumes would be soaked up by the Chef's closings, just like linseed oil on canvas, turning his robes into a state-of-the-art fire hazard, - no kidding(7)! The fumes not intercepted by Chef's clothes rise to the ceiling, covering the walls with the linoleum (*ch.1.3*) type material impervious to all but the most powerful cleaning solvents(8).

There is a simple test to check if a restaurant is neglecting the LPO issue, to the detriment of customers and cooks alike. Sneak in and sniff the kitchen. Smells fishy or rancid? Rest assured: oils not suitable for heating are heated to too high a temperature and reused. Steer clear of this mortal breath of RCS.

Given the poor health and safety regulation track record as well as total disregard for expiry dates and ethical approvals in the food sector throughout the rise of vertebrates' period, particularly during the evolution of opportunistic mammalian below-the-top-of-the-food-chain-predator or scavenger types, the gastrointestinal tract could not help but evolve to withstand the daily brunt of toxic LPO products, picked with half-decomposed carrion. Perhaps this may partially explain why the intestinal epithelium cells have the fastest turnover(9). However, the respiratory system was obviously under lesser pressure and so the nature did not have to sacrifice generations of species, families and classes to

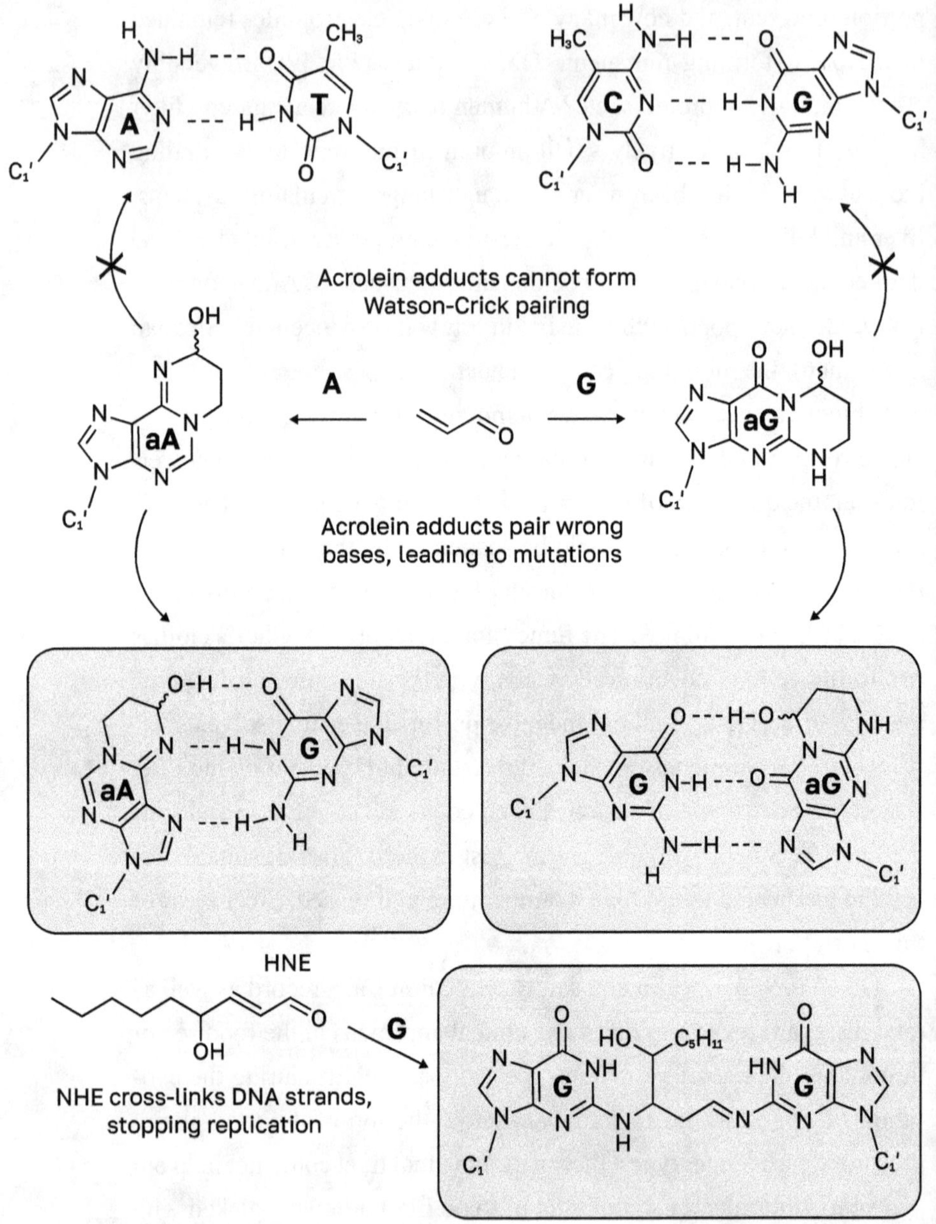

Acrolein adducts cannot form Watson-Crick pairing

Acrolein adducts pair wrong bases, leading to mutations

HNE

NHE cross-links DNA strands, stopping replication

Figure 19. Another vicious feature of RCS is their willingness to react with DNA or RNA bases, changing their complementarity. Such modified bases (as shown for A and G) would now pair non-canonically, with wrong partners, causing transversions (mutations). RCS can cross-link two DNA strands, causing major disruptions. Up to 2000 dG adducts are formed per 1 Bn DNA bases.

Table 1A

Oil or fat type	LIN, %
Safflower; Grape seed	>70%
Sunflower	65-70%
Corn; Cottonseed; Soybean	50-60%
Rice bran; Peanut	30-35%
Canola; High oleic sunflower	15-20%
Olive; Avocado; Palm; Lard	10-15%
Ghee; Butter; Coconut; Tallow	2-3%
Tallow; Butter, grass fed	1%

evolve the protection technique for lungs. But as "Nothing in Biology Makes Sense Except in the Light of Evolution" (Dobzhansky), one could also hypothesize that perhaps the economic considerations were important too, and the cheapest solution turned out to not reinforce the whole respiratory and gastrointestinal tracts, but to slightly reprogram a couple of olfactory receptors' signalling protocols to have the newly created mutants instinctively run away in disgust, without even realising what exactly were so unappetizing about that rancid piece of fatty flesh. Try testing the efficacy of this defence mechanism by sticking your nose in a bag of walnuts four years past the expiry date.

Pernicious as the above may sound, the fix might be easy. If not frying is not an option, then simply switching to oxidation-stable high oleic oils (Table 1A), like refined olive oil or high oleic sunflower oil* would substantially reduce the toxicity of the process, because as you have seen in *ch.2.6.1*, fatty acids need to contain more than one double

---

* In a rare example of a positive change in nutrition, there is a silent revolution currently ongoing in agriculture, with more and more "vegetable" oils being converted into high oleic varieties. Hope they would be coming to the wok lands soon.

bond in a molecule to make it vulnerable to LPO. And oleic acid (OLE, *Table 1*), with only one double bond, is much more stable to oxidation than PUFAs.

3.2. Fishy fish oils: a poisoned chalice?

"Fish oil is a great protein for your hair."
— Molly Sims

"Nothing is without poison."
— Paracelsus

As the Cat in the Hat, a big connoisseur of all things fishy, would have put it, if the smell isn't swell, well, that's a tell, that the fish oil's not necessarily well[(1)]. For the broad range benefits, real and fictional, ascribed to the fish oil supplements, the fish oil industry could be rather tatty with its quality controls. Fish oil can be bad in several ways. Mercury in the ocean mostly comes from two sources, precipitation from the coal-burning powerplant exhaust and leaching off the seabed. A lot of that mercury exists in a form of dimethyl mercury, an ultra-toxic compound with an unfortunate property of being able to penetrate through skin. Karen Wetterhahn, a chemistry professor at Dartmouth, died in 1997, when a tiny dose of the compound seeped through latex, as she tragically picked the wrong type of gloves, - and then through her skin. Many chemicals containing mercury are highly volatile, making the detection challenging. The levels of the metal increase with the increasing ranking of fish in the marine food chain. The issue is so serious, that some top of the food chain fish should not be consumed more than once a month, or not at all during pregnancy. To be high in mercury, a fish has to be large in size, in addition to its high status in the food web. So a large shark, swordfish (highest mercury level, at 0.995 ppm), tuna and marlin should not be regulars on your menu. Seafood at the bottom of the food chain, including the species from which fish oil is squeezed, like anchovies, herring, sardines (lowest mercury level, at 0.013 ppm), salmon (0.022 ppm) and krill, is OK to consume. The good news is, advanced purification and tough heavy metal standards (up to 0.1 ppm) result in the mercury levels being either undetectable in commercial fish oil samples, or being similar to the basal concentration normally present in human blood[(2)]. This is down to a combination of the marine species used in the industry being at the bottom of the food chain, and strict quality control and purification measures. In addition to the marine sources, some high-end DHA and ARA products used, for example, in the baby formula industry, are produced on land using genetically modified single cell organisms and so do not have any elevated levels of mercury.

Looking from the LPO perspective, mercury's d-shell is full, so it cannot initiate the LPO process (see *ch.2.4*). However, other transition metals are present in fish. In herring fillets, there is 7.4 mg/kg (or 7.4 ppm) of iron, and 0.9 mg/kg (0.9 ppm) of copper (copper is more active in Fenton than iron)*. In herring oil, there is less metals (iron, 0.03 ppm; copper, 0.1 ppm)[(3)]. But because of the constant recycling, given enough time, even one transition metal atom can see to it that every PUFA is peroxidised, so antioxidants and chelating agents (which, paradoxically, can sometimes make metals more active in Fenton (see *ch.2.4*) are a must to have. And for a good measure, as the van't Hoff's rule states that the reaction rate rises at least 2-fold with the temperature increase by 10°C, elevated temperature can be bad, too.

As fish oils are used to restore the skewed n-6 to n-3 balance and replenish the membrane PUFAs lost to oxidative damage, it is important to keep them in a good shape. Quality control relies on two standardized methods used together. Recall (*ch.2.6.1*) that LPO initially produces lipid peroxides. These are rather reactive, and to satisfy their hunger for electrons, they can oxidize iodine anions, which are colourless as iodide salts. This will yield molecular iodine I_2 which, depend-

---

* For comparison, humans, on average, contain 38-50 mg iron per kg bodyweight, and 1.4-2.1 mg copper per kg body weight. See *Back-of-the-Envelope-3.*

ing on solvent would have an intense yellowish-brownish-purple colour. Titration can therefore produce quantitative peroxide values (PoV, in milliequivalents/kg, mEq/kg) of LOOH impurities in bulk oils containing PUFAs, including fish oils[(4)]. Freshly squeezed herring oil would have a PoV of 6. The Global Organization for EPA and DHA (GOED) recommends the PoV for commercial fish oils of <5, while the industry accepts <10. When the PoV exceeds 30-40, rancid odour becomes perceptible.

However, with LOOH formation, the bad things are just getting started. *Fig.9* shows that once formed, the peroxides continue falling apart through multiple pathways giving toxic decomposition products. Just looking at the PoV value may therefore be misleading, as it may underestimate things, by not revealing other bad actors lurking in there. The end stage of LOOH decomposition yields highly toxic reactive carbonyls, discussed at length in *ch.2.6.2*. We don't want to be ingesting many of those for sure! A method of detecting carbonyls relies on reacting carbonyls (aldehydes and ketones) with para-anisidine to yield highly coloured products which are quantified by measuring light absorbance at 350 nm. Accordingly, what is measured is known as anisidine value (AV, mEq/kg). GOED's recommended AV is <20, while the accepted industry standard is <30.

Sometimes a combined metric called total oxidation value (TOTOX) is used as a single quality control measure, calculated as AV + 2PoV. Fish oils with TOTOX not exceeding 26 are considered good.

Analysis of South African commercial n-3 fatty acid supplements in 2009 revealed that more than half of the 45 products tested contained less than 90% of the claimed EPA/DHA. But more worryingly, the majority of capsules contained elevated levels of conjugated diene peroxides (LOOH)[(5)]. This report was noticed by the industry, in South Africa at least. When the authors had repeated the study in 2013, they found that the product quality improved[(6)].

It was similar in New Zealand. Again, that only less than 10% of the fish oils tested contained as much EPA and DHA (the rest contained on average < 70% of the claimed levels) is a smaller concern compared to the oxidation status. Only 8% of products did not exceed the PoV and AV thresholds. Perhaps surprisingly, the expiry dates, costs, and country of origin were not predictive of the fish oil quality[(7)].

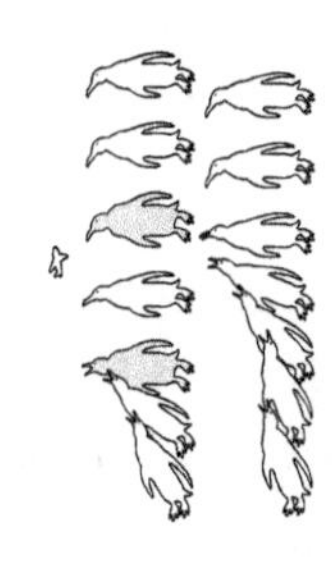

No country is immune to the problem of low-quality fish oil. Analysis of dietary supplements available in the US, again irrespective of the price range, revealed a very significant level of fatty acids other than the expected EPA and DHA. More than 30 different fatty acids were detected including various saturated species, comprising up to 36 % of the total fatty acid content, with levels of n-3s varying widely (33%-79%). The presence of so many different fatty acid types may not always be OK and depends on their nature. For example, for a long time, rapeseed oil was not good for human consumption owning to high levels of the toxic erucic acid (cis-22:1n-9), so a large effort went into producing cultivars with reduced content. The result is a well-known canola oil, with erucic acid not exceeding 2%. So, who knows what unwelcome fatty acids might be lurking in fish oils with such a poor n-3 content. Yet LPO is still a bigger issue here. If an extra effort is put into purifying n-3s out of fish oils or other sources, the resulting material is more expensive, but has guaranteed high levels of DHA and/or EPA, and is substantially less oxidized if stored correctly. Some of those formulations are so well standardized that they are FDA-approved. Oxidation of sdLDL, an important subtype of LDL particles, was shown to be fully (> 95%) inhibited by the high-quality prescription grade n-3 composition but not by fish oil dietary supplements, suggesting we may be overestimating the health value of fish oil supplements, while underestimating the importance of the ever-present damaging LPO processes(8).

Oxidation of PUFAs can be really fast. A UCLA professor John Edmond, who sadly passed away in 2022, recalled that in response to his scepticism about the speed of PUFA peroxidation in air, his PhD supervisor had him smear a small glob of fish oil onto a microscope slide surface. Having placed the slide on an analytical balance, John then watched in amusement as the significant numbers kept creeping up in real time, reporting on the oxygen atoms attaching themselves to the prostrated PUFAs.

Fish oil is big business ($2.6Bn annually in the US alone), naturally replete with hype and controversy, so fishing for answers is not straightforward. On the background of exaggerated claims, let's look at some facts. In a large group of patients with cardiovascular risk factors, daily dosing with fish oil did not reduce cardiovascular mortality and morbidity[(9)]. Fish oil failed in all measures of CVD prevention, both primary and secondary. Further studies were more nuanced, suggesting that while for a general population taking fish oil might be a risk for atrial fibrillation and stroke, taking the supplement could be beneficial helping against progression from atrial fibrillation to major adverse cardiovascular events and death[(10)].

A more worrying finding linked higher levels of EPA and DHA intake with increased prostate cancer risk[(11)], confirming similar observations in the earlier studies. However, other work suggests there is rationale to employ n-3 against prostate cancer[(12)]. The status of fish oil supplements is an equally moot point in neurology, with results spanning the range from not being beneficial, to elevated levels of DHA and EPA helpfully being inversely related to the risk of dementia[(13)]. Age-related macular degeneration is also in the moot category. Some trials reveal no risk reduction in people with AMD on long term fish oil supplements[(14)]. One of the most successful AMD-targeting supplement studies to date, the Age-related eye disease study (AREDS), did not show any additional benefits of adding fish oil to its composition of antioxidants and microelements. However, the latest recommendation is to consume omega-3 containing foods together with the AREDS-2 supplement. While human studies suggest that fish oil supplements improve biomarkers of aging and reduce the risk of all-cause mortality, studies in short-lived mice (see footnote (*) on page 247) show the opposite trend. Could inflammation be less important in short-lived strains?

This mostly sounds like a mixed bag. One common denominator of the above studies, however, is the use of fish oils. And we have seen earlier that they are often of sub-standard quality, in both composition and unacceptably high levels of oxidized PUFA impurities. The impurities which are very reactive and highly toxic. Could this low quality be the reason for fish oils repeatedly underperforming in trials? After all, as described in *ch.4.6*, feeding rancid fats to rabbits increases their atherosclerosis…

Tellingly, a highly purified product of defined composition (pure EPA) with carefully monitored levels of impurities as well as PoV and AV values performed well in REDUCE IT trials, particularly in comparison with several negative fish oil cardiovascular trials[15]. So, could the uncertainties and luck of success simply be down to inferior quality and unacceptably high levels of toxic LPO-derived impurities in fish oil supplements, where the goodness of n-3 (see *ch.4.9*) is simply overwhelmed by the badness of oxidized products?

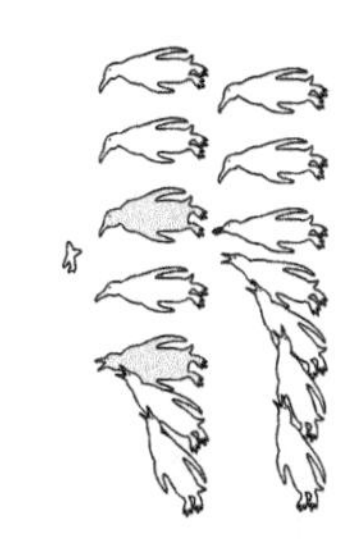

While the jury is still out, do not stop eating fish! And if you want to stick with nutraceuticals, try to find the supplements known for under-the-limit LPO content (by PoV and AV), and store them tightly closed, in cold.

3.3. Eat to keep your PUFAs happy

"When it's smokin', it's cookin'; when it's burnin', it's done."
— Norman Mailer

"Everything I like is either illegal, immoral or fattening."
— Alexander Woollcott

The pressure of natural selection has always been rather simple. Mature fast to reach reproductive stage before the predators get you (and if you are big, can fly, or are poisonous and smart, the pressure is reduced). Quickly build strength and weight to survive childhood, then outcompete others to reproduce (the Goal). Our food preferences evolved to support this, and anything not serving the Goal was discarded, for there was no selection for the traits that would only be beneficial post-reproduction. Traits appearing later in life would not affect reproductive success, as proposed by Peter Medawar in 1952. Some adaptations that would help to reach the Goal early in life could actually be detrimental later in life,*[(1)]. Apres moi, le deluge ("moi" is "reproduction" in French, n'est-ce pas?). Longevity unfurls way past the reproductive peak, so there was no evolutionary pressure for human longevity. In fact, species may be "discouraged" from hanging around for too long as they will start competing for resources with their own descendants, jeopardizing the Goal. But it is not just the genes.

Longevity depends on genetics, environment, lifestyle and good luck. The estimates vary widely, but our genes are probably responsible for about half of the outcome. Despite our best efforts, many afflictions are down to sheer chance. Say, a billion years ago, a faraway supernova explosion had ejected a 1 EeV proton (Exa, 10^{18}; one EeV would be about 1/20th of the energy of a penny dropped off the Empire State building). As destiny would have it, a billion years later two earthlings, having just had the most salubrious breakfast, would be on the wrong outdoors scenic route at the wrong time, jogging and chatting about their future. Man proposes, but God disposes. One of the joggers (and it could have been the other one) would get pierced through by that rogue proton. Had it been a falling brick, the outcome would have been instant, and too fast to initiate LPO. But stricken by the cosmic particle (which will convert its energy mostly into splitting water molecules to generate ROS), the jogger will not feel a thing, not for the next week or year, until… (see *ch.2.3*).

As most of us have neglected to pick parents with perfect genes (and good habits), we have no power over our (epi)genetics. Environment is also largely beyond our control. Our lifestyle, however, we can fine-tune. Hygiene, light exposure (Try to reduce it. Have food with high vit D instead, like fish, beef or lamb

---

* According to antagonistic pleiotropy theory put forward by George Williams in 1957, an ability to increase blood pressure may help reach peak performance but wears out blood vessels in the longer term, elevating the risk of cardiovascular conditions. A high-octane neuronal performance may give advantage early in life while overtaxing the neurons faster, - something like that may be the case with HD patients, who seem to have more kids. A selected for, favourable trait may also be genetically linked to a different characteristic, which might be detrimental but would take longer to develop. Less "restricted" calcium processing would help stronger bones, while over time increasing the chances of cardiovascular and peripheral artery disease. Antioxidants improve vasodilation in 70-year-olds while worsening it in 25-year-olds. What is good for reproduction is often bad for longevity!

liver dishes), exercise (not too much, just to trigger the hormetic response) and sleep aside, the main factor here is the food we eat.

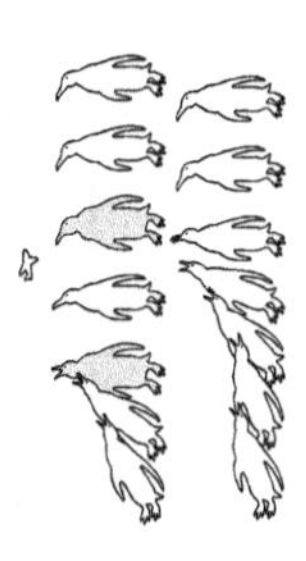

With regards the food preferences, "all's fair in love and war". Resources were scarce, time was short, any food was fair game. Only the most toxic stuff was programmed to be avoided, such as cadaverine in badly decomposed meat, or really rancid fats. Detection relied on olfactory systems, providing a DO NOT EAT warning at a distance. Everything else was programmed to be welcome and craved as long as there were short-term gains, - regardless of the long-term effect, – in pursuit of the Goal. It could be that our attraction to the smell of the burnt meat, reeking of Maillard-Amadori products (see *Box 5*), was the result of the evolution "realising" that cooked food would yield quick energy, on account of nutrients being more digestible. Hence smaller teeth, hence smaller face muscles, hence more room for the brain (*Back-of-the-envelope-8*). It must be a recent human trait, as wild animals do not seem to be attracted by the cooking odours. Therefore, the instruction was, eat, - good stuff or bad, - as long as you grow fast, beating others to the Goal. Eat all you can get, like there is no life after reproduction. The five basic tastes, - salty, sweet, sour, umami (all good) and bitter (beware, keep off) evolved to support the Goal, too. Test yourself: mix fat ("vegetable", suet, lard or tallow), sugar, a drop of vinegar, and some soy sauce (salt plus umami). Taste a bit of this mixture. Delectably delicious, isn't it? Q.E.D. No need for social status-boosting things like foie gras, oysters, samphire or goji berries-stuffed snails with a glass of vintage-1815 Champaigne to deceive our palate. For full satisfaction, just titillate your main taste receptors.

We have a hard-wired reaction to certain irritants. If a noise is too loud, temperature too high, smell too evil, or taste too pungent, the automatic pre-programmed response would be AVOID. Oddly, the visual cues for "unappetizing" colours seem to be overruled, maybe as the result of the selective pressure to cook food to help get quick energy and materials. Otherwise, why would looking at a half-charred, pitch-black, carcinogenic smoke-reeking, smouldering barbequed steak, straight from the medieval holy inquisition cook books, not trigger the avoidance response? The char part is not that dissimilar to the material that would trigger scrotum cancer in chimney sweeps – and the sweeps were not even eating it... In addition to evolutionary pressure, recent industrialization of food production also threw a spoon into the works of our longevity mechanism. Hugh Sinclair's famous letter to Lancet in 1956, controversial back then, was vindicated over years: a great number of the Western diseases are to do with highly unnatural (wrong) combinations of dietary PUFAs(2). Yet, 70 years on, the distorted ratios of dietary PUFAs are still everywhere... The emergence of "vegetable" oils and non-existent-in-nature refined sugars, low prices and the popular novel concept of non-stop snacking are largely responsible for a pretty low, and still declining life expectancy in the USA, the richest country on Earth (at 76 years in 2021, the lowest since 1996; it is currently 84 years in France). Light snacks take heavy toll. "Vegetable" oils are not distinguishable by our palate from other fats and over-consumption ruins both the BMI and the finely tuned omega balance. High blood pressure, elevated cholesterol, inflammation, diabetes type 2 and other metabolic diseases are on the rise, but only in the countries which have adopted the Western dietary protocols, not in the rest of the world.

Wrong types of oils and overeating join forces to give us grief. We are not genetically programmed for long term maintenance of our bodies. Even the top public educators like New Scientist and Scientific American often extol the deliciousness of baked cakes and suchlike, citing the wonderful gamut of fine smells provided by the Maillard chemistry, whereas Maillard, oxPUFAs and other waste materials are just plain toxic, - but in a longer term. "Baking turns it into a gorgeous gold brown, with delicious crunchy crust..." – Oh, really? With a little help from Maillard and the whole smorgasbord of LPO products, acrolein and acrylamide (the latter is a product of aspartic acid decomposition, but forms faster when LPO is elevated) and other nasties? No thank you! Even hormesis (see *Box 4* and *ch.2.8*) is unable to compensate for this flood of nasties*. Remember what happened to rabbits fed on oxidized lipids (see *ch.4.6*)? Garbage in, garbage out.

---

* Steven Austad sites oxidation and "browning" as the main detrimental chemical processes associated with aging. Browning is a mixture of Maillard-Amadori and RCS cross-linking processes. Why would one do to our food what aging does to our bodies?

The longevity trait in many senses may be the opposite of the Goal. Follow your cravings, - at your peril. Otherwise, here is a set of suggestions to mitigate the indifference, ruthlessness and cruelty of the natural selection*. As dietary advice is a matter on which there is sure to be profound disagreement, the following is not a list of recipes. The following pontifications only pertain to the well-being and the healthy environment for your PUFAs, and can thus be overlapped with existing or popular approaches to nutrition, and I am not at all pernickety about your favourite dietary schemes. Squeeze your preferred cuisine or food preferences into the boundaries of the list below (see *ch.3.4*), and chop off and discard whatever does not fit and sticks out**.

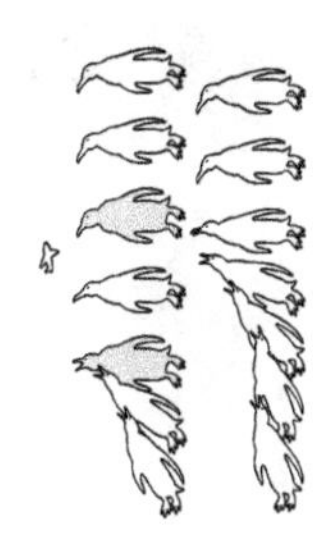

---

* Just like the beer that students in Oxford pubs are informed they never actually own - merely borrow it for a while - the only thing nature cares about is that we return the atoms that make us up. Human recycling, a la Aldous Huxley.

** As a tribute to Eastern European cuisine, no discussion about dietary fats can be complete without mentioning salo (slanina), a type of cured pork fat. Fat-wise, salo to Slavonic people is what surströmming is to the Vikings, or Casu Marzu (the one your little friends share with you) to the Sardinian centenarians. However, in a clash with the Global Safety Officer, the Highest-ranking and most powerful officer on this planet, such traditional foods are on the way out. Recently, the Officer even banned unpasteurized fermented products - renowned for their health benefits and beneficial microflora, such as sauerkraut, - and insisted on pasteurizing them, thereby killing all the good bacteria in the process. Why wouldn't he scrutinize the "vegetable" oils instead? Anyways, if there is ever gonna be a Russian edition of this book, make sure you check out the Salo section in there.

3.4. PUFA-friendly menu: let's put our best food forward

"Good advice is always certain to be ignored,
but that's no reason not to give it."
— Agatha Christie

"In giving advice seek to help, not to please, your friend."
— Solon

As you are working your way through your coffee, a bowl of porridge, and this chapter, - here is a disclaimer. The suggestions below, based strictly on metabolic input, ignoring the social aspects, are not claiming any medical, nutritional or health benefits. Just the well-being of your precious PUFA stock, as the traditional cooking advice regularly misses the LPO plot.

1. Keep your PUFAs well oiled, well omega-balanced(3). Make sure you get enough n-3 PUFAs. Adequate supply will not stop LPO, but will help replenish the damaged PUFAs.

2. Olive oil, Camellia oil, Macadamia oil, Avocado oil, other modern GM varieties with high oleic content are all good. Opt for at least 75% oleic in your olive oil. Store your oils (this relates to fish oil, too) tightly closed, cold, in the dark. A rule of thumb, just like in other organisms: olives from down south (Tunisia) would have substantially less n-6 LIN (the one to be avoided) than the northern varieties. Do not heat high PUFA oils.

3. Remember: the "vegetable" oil lobby is resourceful, is watching what you eat, and is plotting to have you eat even more "vegetable" oils. When buying a pot of olives, pitted or not, you might be assuming, wrongly, that the oil they are in is olive oil. Unless you go an extra mile to find the right stuff, the default option would always be some "vegetable" oil. Same applies to cooked wild mushrooms, and sadly to my favourite, sun-dried tomatoes, too.

4. Minimize not just the high LIN "vegetable" oils(4) but the food-chains based on corn feed. If your chicken was raised on the n-6 diet, you would keep

getting excessive LIN even when using the best olive oils on the side. Looking from the omega-balance perspective, grass-fed lamb, beef and dairy products are not bad[3], although some of the grass PUFAs may be converted into saturated FAs by the ruminants.

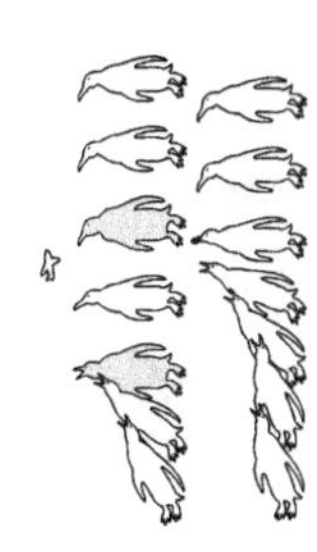

5. As mentioned in *ch.2.5*, for reasons not understood plants as a rule have plenty of n-3 in the green, and a lot of n-6 in the grain. Make sure you have an adequate intake of the former, at the expense of the latter. Cereals (rice, buckwheat, quinoa) and legumes such as lentils are desirable for reasons other than PUFAs. While their total fat level is low (around 1 % of dry weight), the LIN to LNN ratio in red lentils is a respectable 4 to 1, highly recommended as a side to your salmon and greens.

6. *"Avoid fried meats which angry up the blood".*
— Leroy Satchel Paige

"No one owns life, but anyone who can pick up a frying pan owns death."
— William S. Burroughs

Do not fry anything, particularly oily fish, - just stop using this stone age recipe. Heaven forfend you grill your food, cut out this barbaric medieval practice. Do not toast. Do not bake beyond ever so slightly yellow. Totally cut out deep fried food, - French fries or vegetable crisps/chips, - anything fried in "vegetable" oils. Boil, stew, simmer or steam. Cook in a mixture of oil and water, not in pure oil, to keep it not much above 100°C. If you must cook at higher temperature, use saturated fats like dairy butter or OLE-rich oils, not PUFAs. The more double bonds in a PUFA, the faster it oxidizes. That "beautiful" golden crust batter, or the "delicious looking brown crust" on your cake, is poison. If you see a queue to a popular barbeque restaurant, queue the other way!

7. Keep an adequate intake of n-3s. This includes pregnancy and breast-feeding terms. Go for bottom of food-chain seafood -

anchovies, sardines, mackerel, herring, salmon - to get your n-3 PUFAs while avoiding mercury. When buying caviar, make sure it is NOT in "vegetable" oil, otherwise all the benefits will be ruined, just like with olives. The idea of ruining noble caviar with oil developed to lubricate machines leaves me blinking in an utter bafflement. The machinations of the dark market forces must be at play.

8. Reduce smoked products. Even before the LPO kicks in in earnest, cis to trans isomerisation in PUFAs tacitly takes place, and trans PUFAs are bad. The connection is usually multifactorial, but the elevated incidence of stomach cancer in East European and Baltic countries is to some extent linked to high consumption of smoked oily fish, particularly sprats, and other cured products. Chew on cheese but eschew smoked cheeses. Apart from oxidized PUFAs, there are multiple other potential carcinogens arising from this cooking method, not to mention the salt content.

9. Even though they cannot stop the LPO, having not enough antioxidants makes LPO even worse. So do care about your antioxidant intake, and make sure you get different ones*.

10. *"How plump and how tender she looks,*
just as if she'd been fattened on nuts!"
— Old robber woman (H.C. Andersen, "Snow Queen")

"Too much of a good thing can be wonderful."

— Mae West

Do not take this poetic licence too literally. Eat nuts, but don't go them over

---

* This book is not about antioxidants, so I will just mention some rank and file briefly (*ch.2.8*). Vit A, vit C, vit E, alpha-lipoic acid, selenium (needed for GSH machinery, Brazil nuts being the best source). Carotenoids, including carotenes, lutein, lycopene and zeaxanthin. Some indoles. Some lignans. Flavonoids (isoflavonoids, flavons, anthocyanins, catechins) and other polyphenols, particularly from herbs like rosemary. Some garlic - (such as alliin) and onion-related sulfoxides. Pomegranate seeds have punicic acid, a type of PUFA with protective properties. The same is often said about pinolenic acid in pine nuts. Mushrooms are a good source of ergothioneine. Ellagic acid, relatives of resveratrol and curcumin, many hundreds of compounds derived from spices, cocoa, coffee and teas, chocolate beans, fruits, vegetables and berries... We can barely scratch the surface in this footnote. Many members of these groups may work through hormesis (mild toxins), not as antioxidants. (*Box 4*).

Table 2

Nuts	Total fat	Saturates	OLE	PUFA	LIN	LNN	Notes
Almonds	50.6	3.9	32.2	12.2	12.2	0.0	
Brasil nuts	66.4	15.1	24.5	20.6	20.5	0.05	selenium
Cashews	46.4	9.2	27.3	7.8	7.7	0.15	
Hazelnuts	60.8	4.5	45.7	7.9	7.8	0.09	
Macadamia	75.8	12.1	58.9	1.5	1.3	0.21	
Peanuts	49.2	6.8	24.4	15.6	15.6	0.0	pulses
Pecans	72.0	6.2	40.8	21.6	20.6	1.00	
Pine nuts	68.4	4.9	18.8	34.1	33.2	0.16	pinolenic acid
Pistachios	44.4	5.4	23.3	13.5	13.2	0.25	
Walnuts	65.2	6.1	8.9	47.2	38.1	9.08	

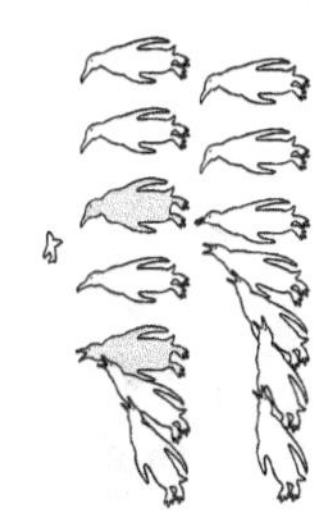

them*. No nut should be fried, roasted, or salted. Particularly since roasted and salted often come together. Visual test is simple: any colour darker than their natural colour means they are not good. Check out the expiry date. Walnuts have the highest level of the good LNN at 9%, but as we are talking about omega balance, this is offset by the highest (38.1%) level of LIN, of all nuts. See for yourself what fatty acids are in your nuts of choice (*Table 2*).

Go for macadamia nuts (they are like little chewable gelcaps of olive oil, really, on account of their high OLE content), hazelnuts and a little bit of the rest. There is a daily required dose of selenium in just one Brazil nut(5). Macadamia nuts have the best ratio of OLE to PUFAs of all foods inclusive of olive oils, and hazelnuts are not bad either.

11. Diary butter and dairy products are welcome** (particularly

---

* Translating from pidgin: don't go nuts over nuts.

** "Low fat" yoghurts deprive the consumer of one of their most valuable components. The fats removed will find their way back into your stomach, piggybacking on a knob of diary butter on your bread. But the sugar, for children: after all, milk evolved to help mammalian babies grow*), unlike hydrogenated "vegetable" oils. Cholesterol is not as bad as they say, while trans-fats are probably worse than they say (see *ch.2.5.1, footnote on p.45*).

12. Avocadoes should be an important part of PUFA-conscious diet. They are nick-named "the big olive" in the nutraceutical world, a nod to the composition similarities, particularly on account of OLE[(6)]. We should thank millennials, the main "**advocadoes**" for catapulting avocadoes into the social networks – supported mainstream. Many an Instagram post features photogenic avocado toasts, cementing their cultural zeitgeist status. Well covered by the popular literature, they may all look the same, green with bumpy leathery rind, but different varieties have different PUFA composition. Avocadoes have about 15 g total fat in 100 g flesh. All varieties can be recommended, but if there is a choice, go for the highest OLE and lowest LIN, such as in Zutano (best) and Bacon, and go easy on the Maluma Haas variety[(7)]. Apart from FAs, avocados are a good source of vitamins C, E and K, as well as lutein and beta-carotene. They even have some LNN**.

13. A fitting topic for a vigilant reader. About 4 million tons of pesticides and 2 million tons of herbicides and fungicides are used by global agriculture annually for sprinkling plants to boost the yield. Almost all of these are highly toxic. Rotenone, paraquat, DDT and many other agricultural chemicals were linked to elevated risk of PD[(8)], AD[(9)], ALS and numerous other neurological and non-neurological conditions. LPO is often the connection between a pesticide and a disease (see *ch.4.2*). The following suggestion is probably the easiest to implement. WASH YOUR FOOD before eating. Fruit, veg, berries, everything from apple

---

added copiously to make the "low fat" palatable by giving it back its fullness, will stay. And these mutilated products are called "healthy diet"... Beware.

* This is less relevant for us grown-ups, content with the way we are, and certainly unwilling for anything else to grow in our bodies as we age. Fermented products are less of a concern as microorganisms, to an extent, degrade the mammalian growth factors. Back to the PUFAs: sheep milk has the highest n-3 content, followed by goat milk and then cow milk.

** Known by various aliases, Chayote fruit, a distant relative of pumpkins from the Caribbean, has substantially less total fat than avocadoes. However, its FA profile is so good (18 % LNN, 11.2 % OLE, 8.9 % LIN) that it might make sense to freeze-dry the fruit for human consumption, or use it as an animal feed, breeding rabbits or pigs with the most favourable PUFA profile.

to zucchini. Fruit and veg are often covered in a wax-like material in which the industrial chemicals accumulate, so tap water may not be enough. Perhaps a little soap in warm water. A tipsy tip: in winemaking, grapes are not washed prior to fermentation, while pesticides and fungicides are used abundantly in vineyards*. A handful of molecules, particularly consumed regularly, may be enough to trigger a pathology over time, - while being well below the current detection limits. There are now organic varieties of wine, which presumably do not use pesticides, although this is not always the case. Wine makers: do not get affronted, just wash the grapes please. Or, as a fallback, one can switch to liquors, known as spirits in the Old country. They are distilled and so will reduce the inadvertent pesticide intake**. Go down well with n-3-rich caviar, BTW. Make sure the caviar is "vegetable" oil - free, and the drink is ice-cold. Consume in moderation. Cheers!

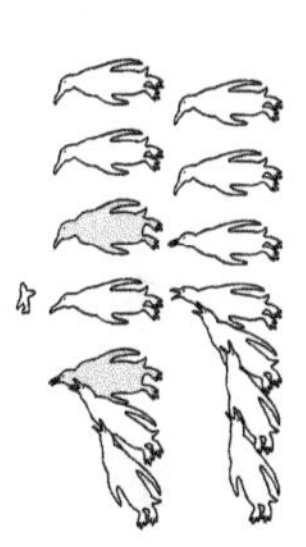

A distantly related group of toxic compounds with proclivity to stay in lipid membranes are plasticizers leaching out of food containers and wrappers, such as phthalates and bisphenols. Best to avoid them altogether.

14. PUFAs make membranes that maintain ion gradients in cells, keeping different ions separated. Neuronal membranes rely on special pumps to keep sodium and potassium ions parted. In mammals, maintaining such membrane gradients across the body consumes about

---

* To make it more interesting. Do you know this zesty taste (sorry, I mean „note of") of red wines the purveyors of fermented grapes products are normally tight-lipped about? The grape skins are not removed when making the reds. So, this tangy astringent feeling comes from some crushed ladybirds. Cheers! :)

** Another point on wines. Dating back to antiquity, a lump of sulphur would be lit up inside an empty wine barrel, or an amphora, to extend the expiry date by eliminating microorganisms, giving rise to the practice of spiking wines and beers with sulphites. Tragically, sulphites are mutagenic – as any epigeneticist who has ever used the bisulphite sequencing to read the methylation patterns can confirm. But despair not, just go for sulphur-free wines (exactly how the Sardinian centenarians do it. Their wines, to be consumed within a year, would not store – and the next year's harvest would produce the next batch), or for spirits.

50 % of the total energy output. The daily intake of sodium chloride and potassium chloride should be approximately equal, yet it is seriously distorted. The 5 g per day requirement for NaCl comes from just one portion of ramen noodles, while potassium is often underdosed (a banana has 10 % of the daily norm). Membrane pumps are overstretched dealing with excessive sodium while trying to squeeze enough potassium from large volumes of water (elevating the blood pressure in the process), stressing themselves and their PUFA environment. Without potassium, table salt is not worth its salt. Make the life of your precious membrane PUFAs easier, provide enough potassium. Go for potassium supplements, and use Lo Salt (66 % less sodium). See footnote* on p.143.

15. Dietary or caloric restriction (DR, CR) has been shown to favourably re-form the PUFA composition in membranes, resulting in reduced membrane peroxidation index[(10)], translating into lower levels of toxic RCS[(11)], essentially being a modest rejuvenating agent. But if just thinking about permanently reducing your daily caloric intake makes you feel peckish, despair not. An alternative approach, called intermittent fasting, could be easier to implement. And do it to give your body a chance to carry out repairs, not for the body-mass sake: there is almost no people with low BMI among Sardinia's centenarians.

16. A few words on metals. Try not to overdose on these, double-faced as fentanyl, actors which pretend to be the good guys, like vitamins, yet drive the Fenton reaction. Your PUFAs will not appreciate it. Usually, iron and copper are safely locked up in proteins, but excess is as bad as deficiency (see *ch.4.2 and 5*). Excessive iron can easily come from foods rich in haem, such as red meat.

17. *"Europe's the mayonnaise, but America supplies the good old lobster."*
— D. H. Lawrence

"Mayonnaise is basically the only condiment there is. Anything tastes better with lots of mayo... even a spoon."
— Tinsley Mortimer

Shopping for mayonnaise is a good exercise, for one has to walk about a mile along the condiment shelves looking for the desired brand. And that's where the good part ends. Appallingly, all mayo products would contain "vegetable" oils as the main ingredient. Even the "Olive Oil mayonnaise" would typically have only 5 % – 10 % of olive oil, with the rest being LIN. While the conspiracies behind this fact are discussed in *ch.2.5*, it is easy to remedy the injustice. Get a bottle of olive oil with at least 80 % OLE. The wonderfully delicious, slightly bitter kick of cold pressed virgin oils comes from flavonoids that add a green tinge to the oil. Tragically, it does not work well in mayonnaise, so make sure that the bulk (at least 75 %) of the oil you use is the lighter, non-green version. Paradoxically, however, the pungency of mustard complements mayonnaise really well. For about 600 ml of oil, you will need four-five egg yolks. Chicken egg yolks are **eggcellent**, yet if one of them is a duck egg yolk, so much the better. Add a spoonful of mustard to the yolks in a one litre jug and, while whisking with a hand-held electric mixer, slowly drizzle in some oil. Once you add about 100-150 ml, the dense emulsion would form – at this point you can increase the rate of pouring. Having added about 0.5 l, stop mixing and squeeze some lemon juice into the jug. Add Lo Salt*, a hint of pepper, a few drops of soya sauce, and a couple drops of Balsamic vinegar, then continue with stirring till you add all the oil. Should you decide to combine the best the US and Europe have to offer (as per the D.H. Lawrence epigraph), opt for rock lobster or Mexican spiny lobster as both are high in n-3 PUFA, and duck the pro-inflammatory, ARA-rich European spiny lobster variety(12). Finally, be warned: the result may be so finger-licking good,

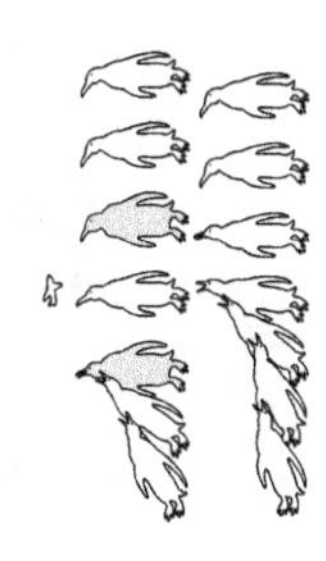

---

* As already mentioned, a lot of body energy is expended maintaining the sodium-potassium gradients across cellular lipid membranes in various tissues. Sweat contains 0.9 g/L sodium and 0.2 g/L potassium, and a somewhat similar ratio of Na to K is excreted with urine. Yet table salt has 100% Na and 0% K, wrecking the normal balance. And only a few of us consume 15-20 bananas per day to get enough potassium. This sends the vascular apparatus into overdrive, in a futile attempt to compensate for the mismatch between the income and the outcome. Water is pumped in, blood pressure is increases, CVD is guaranteed. Whatever it is called in your country – Microsalt, Lo Salt, reduced sodium, whatever, - do start using that, give your cardiovascular system, and your PUFAs, a break.

exercise restraint or else you may end up in the Mayo clinic.

18. Vegetarian and vegan diets have many advantages for health, environment and animal welfare. Looking from the PUFA perspective, an extra effort should be made to get an adequate supply as the key ones only come from animals. There are now biotech sources of PUFAs which should be used as supplements. This is important, as vegans compared with meat eaters may have elevated levels of COX – and LOX – produced eicosanoids (see *ch.4.9*) resulting, for example, in increased platelet aggregation and platelet volume[13]. Their levels of harmful homocysteine were elevated, too[14].

19. *"Seaweed of the finest hues, from deep purple to soft green, entwined her hair."*
— Hans Christian Andersen

Millenia of rains had washed microelements out of soils into oceans, where they combined with elements leached off the seabed, to provide a plentiful supply for the ocean flora. Microalgae flourishing in this **vitamins sea** are the starting point of the marine food chain and a rich source of DHA, - but plankton is somewhat impractical as a food source. However various seaweeds have plenty of healthy stuff, ready to lend a kelping hand. Just ask the Okinawan centenarians. Dry seaweed is high in fibre, microelements, vitamins and protein (10-50% of dry weight). Some types of seaweed have decent levels of n-3 PUFA[15]. EPA is high in brown seaweed, while particularly high DHA levels are found in Gracilaria sp (Irish moss) [16].

As mentioned (*ch.1.1*), to keep our PUFAs well stocked and healthy, we are relying on sea food, a finite source. It is about time we used the advances of DNA technology to solve that problem. CRISPR, gene editing and biotechnology are the only way to go. They are so well advanced that even the regulators are quick with their approvals. Do not be scared of GM foods. It is just a faster way, compared to doing traditional agricultural selection by hand. PUFAs can be made safer and cheaper, this already happens but needs to accelerate. With global warming and ocean depletion, GM can provide nitrogen-utilising plants, heat and drought – resistant plants, PUFA-super-producing organisms, and more. There are already good precedents. Koninklijke DSM N.V., a Dutch multinational corporation, does a great job producing non-seafood-based DHA and ARA used in baby formula the world over. But more needs to be done.

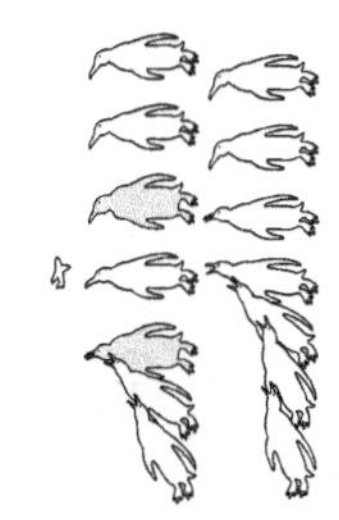

4. The all-pervasive LPOctopus tentacles

"I believe the common denominator of the universeis not harmony; but chaos, hostility and murder.."
— Werner Herzog

"Careful! The kraken is wagging its tentacle, again."

— Efrat Cybulkiewicz

4.1. There's more to LPO than meets the eye

"Each day a few more lies eat into the seed with which we are born."
— Norman Mailer

"The eye altering, alters all."
— William Blake

There is no need to extol the significance of vision. 80% of all memories come from our eyes. Just close them, and see for yourself the importance of seeing, for truly, "you know it when you don't see it". Half the brain is dedicated to vision. The eye is a brain's 500-megapixel camera, with some preliminary visual signal processing and the contrasts-recognizing system that can detect up to 500 shades of grey.

Charles Darwin famously confessed that the idea that the eye could have evolved through natural selection seemed absurd at first. He later

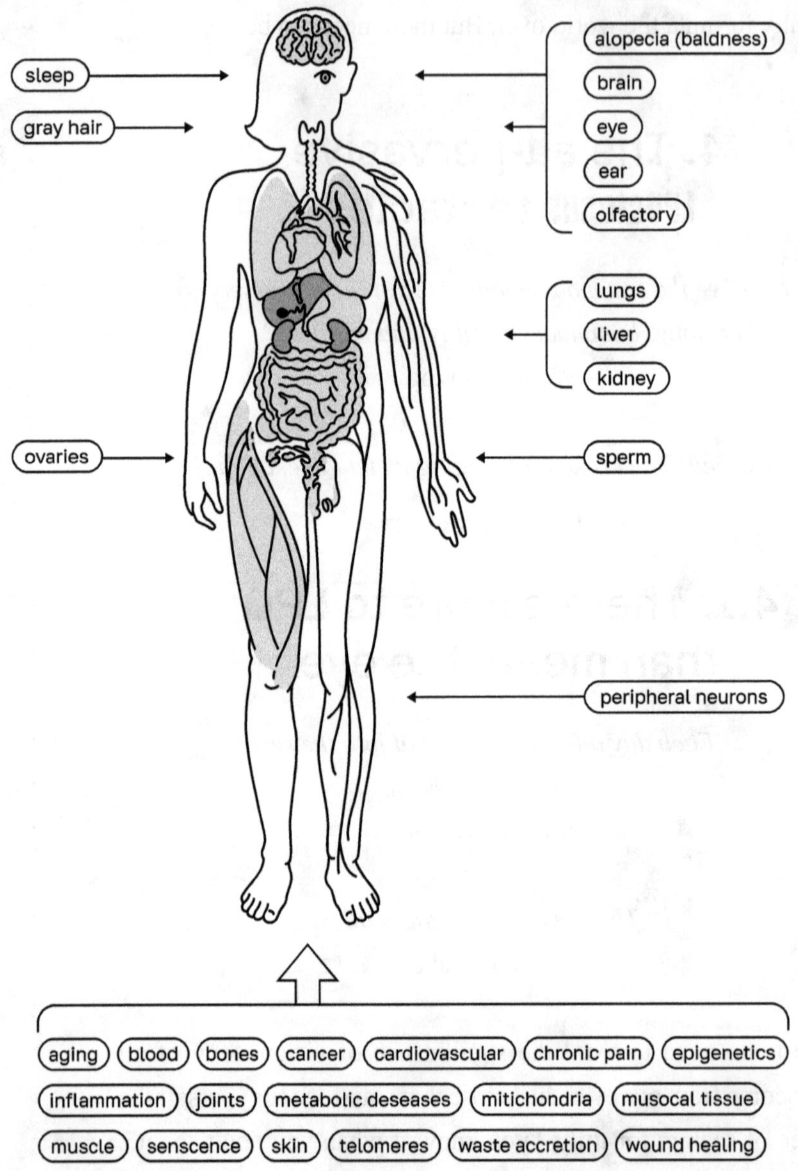

Figure 20. LPOctopus tentacles reach to every nook and cranny: body processes, organs and systems are all in the LPO crosshairs. Many body parts are interconnected through LPO. For example, LPO-dependent olfactory decline is often a harbinger of dementia and PD, as the initial LPO damage in the olfactory neurons, perhaps initiated by the environmental triggers, can travel into the brain. Cancer, the most multifaceted of all diseases, can be affected by LPO negatively or positively, depending on the stage, and type.

came around, hoping to find examples of eye design with intermediate complexity. It might have been more difficult for him to change his mind had he known, back then, about finer details of the mechanics of vision. Richard Feynman, in the optics part of his Lectures on Physics, elaborates in some detail on the visual signal transmission, the mind-boggling complexity of which defies belief. Or, if you prefer, imagination. Before the signal even leaves the ping-pong ball-size eyeball en-route to the visual cortex, a lot of sophisticated processing takes place in the light-detecting retina, amplifying and converting an analogous event to a digital spike(1).

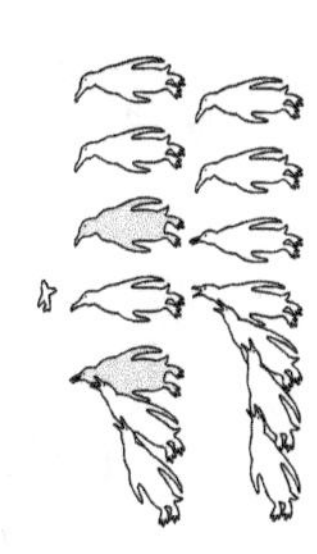

Yet while being a marvel of a device, the eye, upon a closer look, would hardly stand up to scrutiny of the intelligent designer. The rods can detect incoming light at the level of single photons (a candle from 14 miles away), yet these photons have to first cross multiple non-photosensitive layers before hitting a photosensitive part. This awkward layout points instead to a random process of repurposing, ever since a single cell evolved a light-sensitive patch to follow the sun or moon.

Compound eyes of insects are even more disastrously designed, with tens of thousands of separate units needed to give a very coarse, pixellated view. Insect eyes are worth a little PUFA-related aside. A trade-off between pixel size and colours seen dictates that insects' vision is shifted towards shorter wavelengths, as this allows for narrower conical units (called ommatidia) to be used, hence larger numbers can be employed. But the shorter the wavelength, the more energetic the light, hence more LPO. Curiously, this is much less of an issue for insects, for they do not have DHA. Locusts feed on green so their eyes and brains mostly get LNN, while cockroaches have LIN and some ARA from their preferred food sources. Only a small group of aquatic amphibious insects have some EPA in their neurons(2).

Another peculiarity is the way rods and cones signal the detection of light. In any physical measurement, it is always more reliable to detect an increase in signal, not a decrease. Yet retinas produce the potential in the dark and only stop when a photon is detected, so a signal would be a dip in the background level. A quirk of natural selection? Or a chosen way the meaning of which we do not grasp? The jury is still out on this one.

Our eye is not built to last. It is the only organ where some cells survive the entire lifespan from the time we are born. As retinal cells are mostly formed by about age one (a bit later for cones), continuous body growth (including the eyeball) is associated with stretching and distortion of the retina which may result in retinal diseases later in life. The evolution might be too optimistic when pinning a lot of hope on the resilience of eye tissues, which are exposed to light, have the highest oxygen levels after lungs, and are packed with PUFAs and metals, so LPO in our retinas is always a foregone conclusion[3]. If that decision were made when we, as troglodytes, lived in caves, it would have made sense as light exposure, and the related photodamage, would have been low. Or perhaps the anticipated at-birth longevity, slightly past the reproductive age, were compatible with the expected short lifespan of various eye compartments. The non-renewable lens turns cloudy and opaque with time, due to light exposure, LPO and a slow but ineluctable process of amino acid racemisation. Photosensitive rods and cones, which make up but a monolayer of cells, cannot divide (are post-mitotic) and so must last the entire lifespan. Of course, there are wonderful redundant neuroprotective mechanisms that evolved for the retina to defend itself from LPO. There are high levels of antioxidant enzymes, high levels of GSH and vitamins C and E*. There is an insulin receptor that activates various metabolic pathways. The garbage disposal systems are amazingly efficient as a very important part of the retina, - Retinal Pigment Epithelium (RPE) - works non-stop to maintain rods and cones in a good working order. That is why, in some (rare and lucky) cases, post-mitotic photoreceptor cells could last over 100 years. But however good the defences, damage repair is a losing game, as retinal tissue is post-mitotic. Inspecting human skin, you see the relentless onset of age spots. And skin is a renewable, multi-layered tissue (see *ch.4.7*), while light exposure and oxygen levels are similar between eye and skin. Imagine, then, what kind of damage is accumulating inexorably in

---

* Unlike other membranes which have 1 vit E per 2000 PUFA residues, the OS have 10 times more. As such a high level of vit E might be making bilayers less stable, perhaps one of the reasons VLC-PUFA (see below) are abundant in OS is to additionally stabilize membranes whose integrity is reduced owning to their high vit E content.

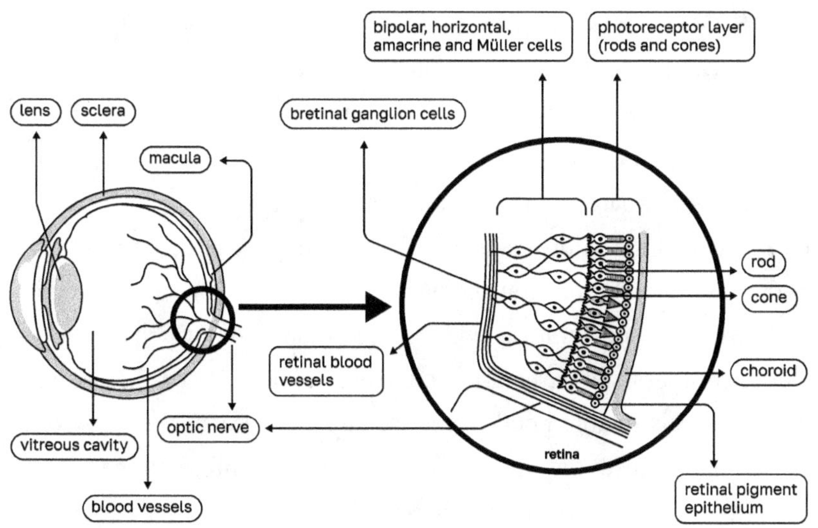

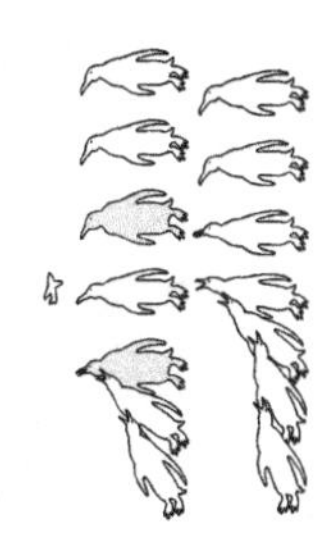

Figure 21. Eye has the highest levels of PUFAs, oxygen, iron and light. So what could possibly go wrong? Any eye compartment can be hit by LPO, but neuronal parts (circled) are particularly badly affected.

a monolayer of post-mitotic retinal cells, unable to renew themselves. Eventually, the waste accumulates and spills out, smothering the biochemical pathways, and the cells, similar to other tissues. And, taking a little poetic license here, the LPO is at the helm of the downfall.

Let's eyeball some eyeball basics. Light is focused by the lens on the monolayer of photosensitive neurons lining up the back of the eyeball. These cells come in two flavours*. Rods do not discriminate colours and rely on rhodopsin to provide sensitive twilight vision. At 100 -130 million per eye, they make up most of the retina, outnumbering the colour-sensitive cone cells by 20:1. Three types of cones detect red, blue and green colours, using three different opsins. Detection is not "digital", and absorbance spectra overlap, leaving it to the brain to reconstruct the colours. Green and red cones are concentrated in macula

* There are more photoreceptor types. Melanopsin-containing ones react to blue light, affecting the go-to-bed time. They continue operating even in blind eyes, and only the surgical removal of an eyeball would disturb the sleep patterns.

(and its centre, the fovea) the small part of retina responsible for the highest visual resolution, which we use for reading. Curiously, blue cones (the rarest, at 6-7% of all the cones) are not in fovea but scattered around mingling with rods. This is because blue light, on account of its short wavelength, is always scattered by the lens a bit out of focus compared to other colours, and so has to be "collected" from a wider area*. Rhodopsin and opsins are G-Protein-coupled Receptors (GPCR) which sit in the lipid membranes translating received signals into physiological effects. (Rhod)opsins form complexes with retinal (vitamin A), which has one of its double bonds in a less stable, pent-up, cis configuration. When the complex catches a photon, its cis double bond isomerises into a more stable trans form, changing shape, which causes the (rhod)opsins to undergo conformational changes. This triggers the phototransduction process, which reports a captured photon as an electric pulse. Roughly, a rod or a cone represents a pixel. To maximize the detection, (rhod)opsins in photoreceptors are arranged in a dense multi-layer, populating both sides of flattened lipid membrane structures known as outer segments (OS) that form stacks of around 1000 discs (rods have discs; cones have membranous invaginations resembling the teeth of a comb) placed on top of each other (*Fig.21*). This increases the chance for a photon, piercing through the stack, to be intercepted. The electrochemical signal then passes from a disc to the plasma membrane of that rod or cone and travels down to the synaptic terminal, an electric socket that connects the cell with the rest of the bioelectronics, including the brain**.

The main circuitry is placed on top of the photosensitive cells, so that the light must cross these layers first. The layers are made of several types of specialized neurons including horizontal, bipolar and amacrine cells, all involved in the initial signal processing***. Crowned by the innermost layer of ganglion cells, the tip of the

---

* Many issues pertaining to blue, green and red cones remain a grey area. Why are they shaped as cones, as opposed to rods? A speculative hypothesis (which does not seem to have been published) was put forward by John and Nancy Medeiros around 2014, suggesting the conical shape helps spectral dispersion, processing the incoming light according to its wavelength, with PUFA membranes making up an optical waveguide.

** A direct electric contact between neurons is also possible, through Connexin proteins that form gap junction channels. This type of rapid signal transmission is of particular importance in the eye and heart.

*** For example, one of around 40 different types of amacrine cells is responsible

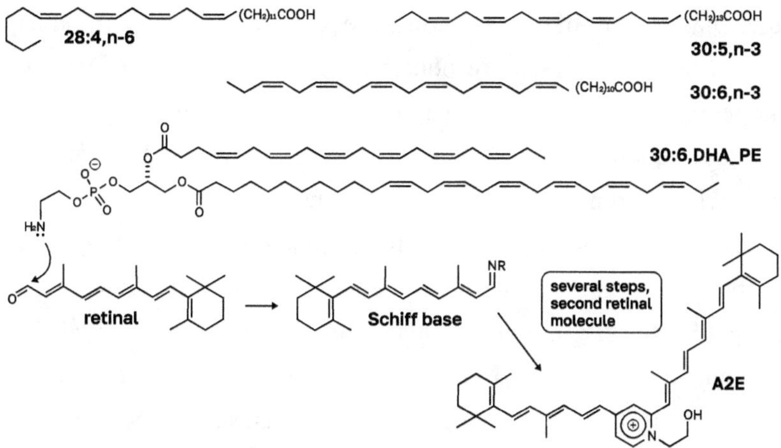

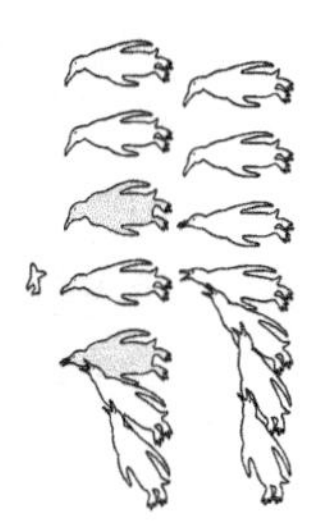

Figure 22. VLC-PUFA are extended versions of ARA, EPA and DHA up to 32-38 carbons long. Shown are 28:4,n-6 (from ARA), 30:5,n-3 (from EPA) and DHA-based 30:6,n-3. Found only in retinas (2% of total PLs, or less in DR and AMD patients), synapses, sperm and testes, they form choline PLs, whereby a VLC-PUFA is attached to a primary (sn1), and DHA – to a secondary (sn2) hydroxyl of a glycerine moiety, as shown. A2E, a component of retinal lipofuscin, forms in RPE lysosomes from ethanolamine PLs and two retinal molecules in disease- and age-specific manner.

---

eyesberg that directs the signal through the main cable called optic nerve to the brain for further processing. Pre-processing saves us the reaction time. For around 130 million photoreceptors, there are only about a million fibres in the optic nerve, so the eye has to "concentrate" the information. The importance of this cable for vision was investigated by Isaac Newton, who, lacking non-invasive equipment, inserted a bodkin (or a spoon - accounts vary) between his eyeball and eye socket until it reached the optic nerve at the back of his eye, causing unusual colour effects. This tight neuronal network is permeated by long Müller cells, which among other things provide structural and immune support and are somewhat similar to the brain glia cells.

---

for the initial automatic resolution of object contours, and when switched off, mice could see low contrast objects in the dark, with abnormally elevated contrast sensitivity[4]. Psychedelic image generation may be a related phenomenon of "closed-eye visuals" as amacrine cells have receptors for psylocybin and N,N-dimethyltriptamine (DMT), the hallucinatory "spirit molecules".

At the other ends, the tips of rods and cones are plugged into a major monolayer called RPE, a life support system responsible for photoreceptor maintenance. The OS stacks get renewed every 11 days, as RPE bites off (circadian rhythm-controlled phagocytosis) about 100 (10%) of OS for recycling daily, shortly after the onset of light[(5)], while 100 new discs are added at the other end*. Outside of the fovea, RPE cells are huge and house up to 30 embedded rod and cone tips each, meaning that about 3000 discs end up inside of an RPE cell every day. The discs get recycled as RPE cells sift through the materials, chopping up damaged DHA and other compounds, and recycling FAs back to rods and cones (a process termed "a short loop"). RPE rests on a supporting choroid layer, and together the two make up the major barrier and relay station for supplying blood-delivered materials and removing the waste. This description is ruthlessly simplified, to focus the tunnel vision on just one component of the eye. What relevance might PUFAs have to all this?

Well, direct! Visual processing takes place in the oceans of PUFA. DHA is higher in the OS (40-45 % by weight) than in any other organ bar tails of sperm cells. Just like in the brain, while much remains to be elucidated, DHA has some vital yet mysterious role to play in the eye mechanics, as speculated in *ch.2.5*** Rods have severalfold more DHA in the outer segments than in their plasma membranes. This enrichment, the mechanism of which is unidentified, is striking, considering the disk membranes are actually made from plasma membrane invaginations. DHA is concentrated in the discs' central parts, with the rims containing much less[(7)]. For reasons unknown, rods have more DHA in the outer segments than cones. Perhaps this could be linked to the observation that rods start dying first in AMD. As the discs gradually move from the bottom of rod outer seg-

---

* This cycle seems to be running even when eyes are not exposed to light. Perhaps light-induced LPO and other damage were found to be such a major problem that the cycle got hardwired and runs irrespective of the actual damage. Or, remembering that rods and cones counterintuitively keep firing in the dark and stop when photons are detected, this idle running produces enough LPO even without light exposure, to justify the regular renewal. The constant turnover may explain why, while being such a complex and active neuronal machine, eyes do not require sleep (see *ch.4.5*).

** One possibility is that only DHA - (and not DPA) – containing PLs allow the expansion of rhodopsin following the incident photon of light[(6)], and rhodopsin regeneration after the bleach.

ments towards the RPE, the content of DHA in them increases (there is no corresponding data on cones). The mechanism for this enrichment is also unknown, but one possibility might be the Miyashita paradox (*ch.2.7*).

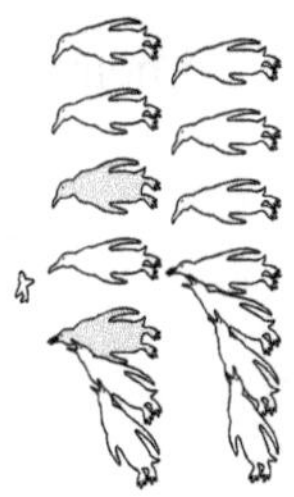

The troupe of retinal, "**l-PUFAs**" has some strange actors at play. A new set of length record holders called Very Long Chain PUFAs (VLC-PUFA, both n-3 and n-6)* was first reported in 1987[(8)]. Seven dedicated Elongation of Very Long Chain (ELOVL) enzymes, which vary by their preference to the PUFA substrate and cell location, build VLC-PUFAs by adding multiple CH2 groups to ARA, EPA and DHA[(9)]. With up to 36 carbons in their unbranched backbones (*Fig.22*), VLC-PUFA make up less than 2 % of retinal PUFAs (but up to 20 % of rod outer segments content is VLC-PUFA).

VLC-PUFA are highly preserved in vertebrate species as distant as fish, birds and mammals, underscoring their importance. Calling VLC-PUFA "fatty acids" is a bit of a misnomer as with that many carbons they are occupying their own niche. 36:6,n-3 is long enough to pierce through both leaflets of OS bilayers, presumably helping move retinals around and keep (rhod)opsin molecules in the right orientation, as light has to hit them at the exact angle for the detection to occur. Stiffening the bilayers fluffed up by high levels of vit E might be another reason. VLC-PUFA are also located in photoreceptor synapses, pointing to their role in neurotransduction. In PLs, VLC-PUFA are typically found at sn-1 position, with the sn-2 position occupied by DHA. Two PUFAs so close to each other makes such PLs highly oxidizable. The levels of VLC-PUFA are depleted in Stargardt, RP, AMD and some non-retinal diseases, likely due to elevated LPO and/or malfunctioning ELOVL[(10)]. A fine balance of VLC-PUFA has to be maintained, as too much of VLC-DHA is also bad. This happens when peroxisomes, organelles responsible for digesting VLC-PUFA (oxidized or intact) are dysfunctional. The ensuing abundance causes the Zellweger Spectrum of Disorders, affecting multiple tissues. VLC-PUFA are also found in

---

* VLC-PUFA are present in sperm.

brain, meibomian gland (which sits in eye lids secreting lipids to "oil" the eyeball, helping the tear gland protect cornea from drying) and gonads of mammals.

So, there is a lot of DHA and other PUFAs, oxygen (second highest after lungs), light (do not take light lightly: 94% of visible premature aging comes from UV, even though cornea and lens absorb a lot of it), and transition metals in the eye, all conniving with each other. Sometimes there are also environmental factors, such as an LPO-inducing smoking, a strong factor in AMD development*. And facing it, there is a monolayer of highly metabolically active, PUFA-rich, non-renewable photoreceptors, which, once lost, will not be replenished. What could possibly go wrong? So much in fact, with DHA sitting right in harm's way, that most of retinal dystrophies, including Age-related Macular Degeneration (AMD), Diabetic Retinopathy (AMD and DR are the main causes of blindness in the West) and LHON are linked to LPO. Some of the subtypes of Retinitis Pigmentosa (a notoriously multifaceted disease with at least 250 different mutations resulting in the phenotype) also involve LPO, as do two main subtypes of Stargardt disease, the AD (mutation in ELOVL4) and AR (retinal lipofuscin accumulation). Even cataracts are associated with elevated LPO[11]. Eyes also simply get older, gradually deteriorating and accumulating LPO-containing muck as they age…

As ever, antioxidants cannot prevent LPO (*ch.2.8*)**. Could supplying fresh DHA to replenish the oxidized PUFAs be helpful? Or would it be like putting proverbial oil on the fire? This is controversial. While several clinical trials failed to find improvements or in some cases (prostate cancer) reported worsening, some population studies suggest that high intake of dietary DHA helps patients with AMD, Stargardt, DR and RP[10]. It is also possible that helping one part of the body might make matters worse in another. Consider lipid-transporting apolipoprotein APOE that has three main variants. APOE4 predisposes White people to AD (but is protective for Black people), APOE3 is the most common variant, with no effect on AD, and APOE2 is protective against AD. The situation reverses itself for AMD, with APOE4 being protective, and APOE2 predisposing to AMD!

---

* Smoking also badly compromises night vision. As there is more DHA in night-active rods, it may have to do with smoking-induced LPO preferentially affecting rods.

** Singlet oxygen (see *ch.2.2*) is a real problem in the eye, as oxygen, photosensitizers and light all conspire to produce a lot of it. So, carotenoids, such as lycopene, lutein and zeaxanthin, can actually help the eye to keep singlet oxygen in check.

These correlations are true for Caucasian, but not Asian populations[(12)]. No one size fits all…

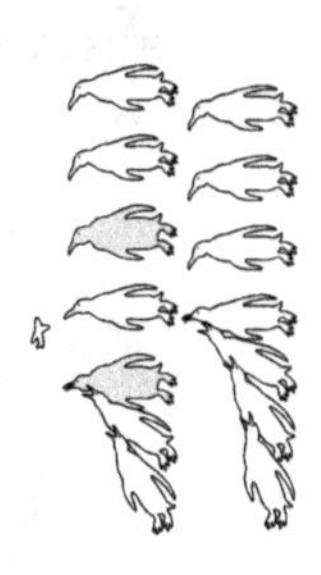

Imagine what kind of concoction is brewing in RPE, as it bites off the LPO-ravaged parts of rods and cones. Things easily get out of control. Two major types of unwelcome waste deposits can form in the retina. RPE-specific lipofuscin (see *ch.4.7*) accumulates in lysosomes blocking their capacity to recycle proteins, lipids and mitochondria originating in the RPE cells. Unlike in skin and other tissues, it has several particularly toxic ingredients, including retinal-derived A2E and iso-A2E ("A2" refers to two retinals, and "E" stands for ethanolamine, Fig. 22) and some eye-specific DHA oxidation products like immunogenic (through complement factor H) carboxyethyl pyrrole (CEP; *Fig.9*).

The abundance of fluorophores, oxygen, PUFAs and light makes this retinal lipofuscin proficient at generating new ROS species. So, the larger the deposits, the faster even more deposits can be generated. Another type of waste, - drusen, - builds up between RPE and choroid. While the two types of muck share some components, there are also differences. There is no A2E in drusen. Unlike lipofuscin, it has some lipoprotein particles, pushed out by RPE as a way of getting rid of excessive cholesterol (made by retinal glia), bitten off rods and cones. Another origin of PUFA- and cholesterol-containing LDL and HDL, is from blood, as lipoproteins enable transport of lipids across aqueous compartments of the body. So much cholesterol and PUFAs are exuded, that sometimes they even spill out of drusen. Perhaps lipoprotein particles can even travel through the eye-filling, water-based vitreous: an age-associated pale white ring at the front of the eye, around the outer edges of cornea, called arcus senilis, consists of cholesterol deposits. Drusen is also high in calcium. Both types of deposits are related to vision loss as the early signs of nascent AMD, DR, Stargardt, RP, LHON and other indications, as well as aging eye. As the ravaging LPO preferentially destroys PUFAs, there is a gradual accumulation of "rigid" lipids and PUFAs are decreasing with age. Since lipofuscin and drusen are both held together by a degradation-resistant oxidized PUFA "glue", LPO yet again can be named as the common denominator of various types of retinal diseases. Overwhelming quantities of garbage and ensuing inflammation lead to swelling in the RPE – choroid area, which is misinterpreted by the body as new "tissue" with missing blood supply. A network of new blood vessels is duly grown into the swelling, distorting everything. The network is pretty leaky. This is wet AMD. Dry AMD, also known as geographic atrophy, manifests as patches of photoreceptors and underlying RPE layer dying out. In the retinal real estate, it is the same mantra: location, location, location. Macula, the most metabolically active central part of retina, gets hit first, leading to the loss of central vision, - affecting reading, driving and face recognition.

Non-enzymatically oxidized PUFAs are pro-inflammatory, and PUFA also get oxidized enzymatically. Many ARA-derived eicosanoids are pro-inflammatory but some are anti-inflammatory, such as prostaglandins which can reduce intraocular pressure. Pro-inflammatory eicosanoids can be blocked by NSAIDs. n-3-PUFA-derived maresins, protectins and resolvins and VLC-PUFA derived elovanoids are claimed to have anti-inflammatory and neuroprotective properties. We will talk about them in *ch.4.9.*

Arnold Schwarzenegger would be devastated to learn that he may have wasted his prime time pumping all the wrong muscles. His iconic Terminator's "scratched eye" image could have been so different had his eye muscle been pumped strong enough to be able to dodge bullets when being shot at, at point blank, from a machine gun… If only Arnold knew that the most powerful muscle in the body, per unit volume strength, is in the eye's steering mechanism, - he could have been taking down the bad guys just by squinting hard. Always in motion, this group of muscles is responsible for constant eye movement. The muscle is 100 times stronger than it needs to be to do its job, so a redundancy was built in to guarantee uninterrupted performance. A nearby muscle that controls the eye lid movement may be affected by a muscular dystrophy condition, known as Drooping Eye. While not involving malfunction of retinal biomechanics, the condition, also known as ptosis, can severely affect vision. In this regard, various ataxias, muscular waste, dystrophy, sarcopenia and cachexia are all known to have an LPO play a role, through mitochondria (see *ch.4.3*) and other mechanisms.

The long-term prognosis for the eye is, sadly, not exactly rosy. It is packed with PUFAs, and you "can't live with them, you can't live without them"... UV light, oxygen and ever-increasing (see *ch.4.2*) levels of transition metals will all work as a team, if not to say a gang, to see to it that slowly but surely, LPO will take its toll. There is a one in forty chance that the ocular pressure will start increasing with age, resulting in glaucoma, an LPO – involving disease(13). Half of the people 80 or older would develop cataracts, an LPO-entailing condition(11). A three-in-ten chance of LPO-driven AMD. And a 100 % chance that as we age, the wear and tear would result in increased LPO-driven toxic deposits, photoreceptors would be lost, vision would deteriorate, time to adjust to light would increase and resolution would decrease. Visual acuity and even colour resolution suffer with age. In the absence of cataracts, perhaps the only reason is some decline in the cone performance, particularly along the green-magenta colours, while the blue-yellow axis stays relatively intact(14). Things get much worse when on top of this, cataracts develop which diminish blue light. This can be seen vividly on Monet's paintings as he was getting older, gradually losing his colour perception.

No exposure to bright light, no smoking, abundant carotenoids (put ketchup on your ketchup to boost your lycopene, as they do in Pittsburgh, and slap that on your spinach, kale and goji berries for a bit of lutein and zeaxanthin) and good genetics will only postpone the inevitable. If only there was a way to slow down LPO...

4.2. Brain LPO: the lard of the matter

"Behind every glorious façade there is always hidden something ugly."
— Stanislaw Lem

"I rob banks because that's where the money is."
— Willie Sutton

We have already mulled over the main elements of the evil that will befall the brain over its lifespans. The neurons are full of PUFAs (*ch.2.5*), oxygen (*ch.2.1*) and metals (*ch.2.4*), all the necessary prerequisites for the chain reaction, so the brain is the hot spot for LPO. The way the neurons store information, - through synaptic contacts, - means that most brain nerve cells last a lifetime and cannot renew themselves by dividing (are post-mitotic), as splitting in two would destroy the synaptic contacts and wipe out the stored information*. The synaptic contacts that allow you, at 103, to recall your grandma baking a cake for your 3rd birthday. How do neurons cope, unable to renew themselves by division, while facing a constant ROS onslaught and the LPO pressure? They actually cope remarkably well, but eventually, time will take its toll. Some parts of the brain shrink, neuronal communications slow down, brain inflammaging takes hold. Worse, as the garbage deposits, such as lipofuscin, accumulate, more garbage and stress, including LPO, is produced at ever-increasing rate, accelerating the decline. This accelerated garbage accumulation concept is easy to illustrate. Place a slice of fresh bread in a plastic bag with a few stale slices, and another fresh slice into a separate empty bag. Seal both. Leave for 24 h. Open and inspect. The outcome is a cinch to predict.

Aging is directly linked to LPO (see *ch.5*), but some diseases may see various

---

* This has to be taken with a pinch of salt. Some animals do renew parts of their brain. Memory is not that "digital", probably more like a "hologram"[(1)], so some degree of very slow turnover might be possible. But to the first degree of approximation, neurons do not divide.

aspects of the process accelerate, such as in particular types of neuronal malfunctioning, or with certain types of LPO – initiating deposits forming[2]. These conditions are known as neurodegenerative and psychiatric disorders. The role of LPO as the driver of CNS pathologies had been recognized a while ago[3].

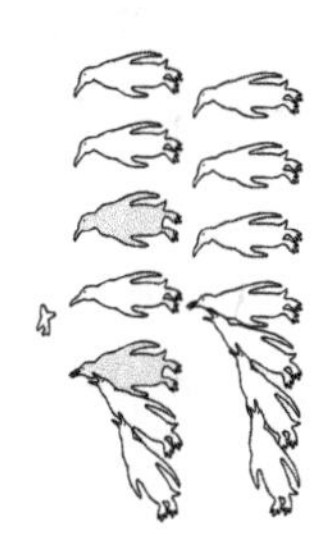

Alzheimer's Disease (AD) and Dementia

AD is the cause of 60-70% of all dementias, affecting 50 million people worldwide (2020). 10% of cases are early onset (before 65 years of age). Women are more likely to be affected. No treatments can stop or reverse its cognitive decline or memory loss progression, and the current global annual cost to the society is US$ 1 trillion. This seventh leading cause of death kills about 2 million people annually.

The hallmark feature of AD is the mis-cleavage of an Amyloid-beta Precursor Protein (APP), a membrane protein present in various tissues. In neurons, APP is important for synaptic plasticity and iron metabolism, but thorough understanding of its role remains elusive. Mice without the APP are fertile but show minor behavioural differences. The mostly 39-43 amino acid long cleavage fragments called beta amyloid (Aβ) peptides are initially formed within membranes, but then deposited outside of neurons, close to neuronal membranes, as Aβ plaques, having been famously observed by Dr Alois Alzheimer in 1906. These are the neuronal membranes, chockfull of ARA and DHA (*ch.2.5*). Beginning in 1994, D. Allan Butterfield and Mark Mattson carried out a series of elegant experiments trying to understand what makes the plaques toxic. The plaques, like spark plugs in an internal combustion engine submerged directly into the gasoline tank, would ignite the ROS generation in membranes, setting the PUFA fuel on fire in minutes. So efficient were the Aβ fragments at initiating the LPO that would blow the membrane components to smithereens that Butterfield and Mattson named the process "molecular shrapnel"[4]. The LPO onslaught would severely compromise various membrane-based enzymes, mediating the neurotoxicity of Aβ[5]. Zooming in on finer details, Mark Mattson further singled out our old friend 4-HNE as the main culprit behind the disruption of ion homeostasis and neuron degeneration(6).

There is also an Aβ-independent railway to AD, which involves a different protein, called TAU (tubulin-associated unit). TAU plays a role in transporting things along the sophisticated microtubulin railroad system inside of neurons (some neurons can be more than a metre long). Normally water-soluble, TAU can malfunction, turning into sticky insoluble neurofibrillary tangles inside of neurons, eventually suffocating them to death. And you guessed it: LPO plays the major role. As part of its function, TAU gets phosphorylated (phospho-TAU) and, equally importantly, dephosphorylated. Dephosphorylation is compromised when LPO products latch on to the phospho-TAU surface, triggering the tangle formation(7).

While frontotemporal dementia, dementia with Lewy bodies and vascular dementia are somewhat similar to AD and have a very strong presence of LPO in their aetiology(8), other diseases may seem to have a less obvious connection to this rubric.

Down syndrome patients have three copies of chromosome 21 (instead of the normal two). DS occurs in one of every 800/1000 live births and presents as accelerated aging*, cardiac and brain defects as well as cognitive disability. Relevant to LPO, two genes lie on chromosome 21: Superoxide dismutase 1 (SOD1) and APP. The 50% elevated level of SOD1 (3 copies instead of two) may generate 50% more H2O2 when SOD1 neutralises superoxide by dismutation (simultaneous oxidation and reduction):

$$2\,O_2^{\cdot-} + 2\,H^+ + SOD1 = O_2 + H_2O_2$$

---

* Several ultra-rare genetic disorders are often considered forms of accelerated aging, including progeria (Hutchinson-Gilford syndrome) and Werner syndrome. However, these conditions, at best, are caricatures of the aging process. For example, progeria patients typically do not develop cancers nor cataracts, rarely get high blood pressure or strokes, and have higher-than-average intelligence, as some aging-associated problems do not affect those patients. Werner syndrome patients are prone to cancer, but not to diabetes, high blood pressure, stroke, or AD. DS may actually be the closest to accelerated aging, even though while leukemia, CVD and AD all develop early, other important hallmarks of aging, like osteoporosis, prostate and breast cancer, high blood pressure, cataracts, skin wrinkles, etc,- are rare in the DS patients.

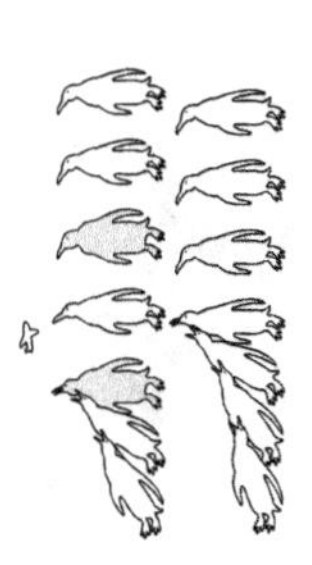

However, catalase is encoded by chromosome 11, so its level is normal in DS patients compared to SOD1, and it fails to mop up the excess H_2O_2 (dismutation again), leaving behind enough to

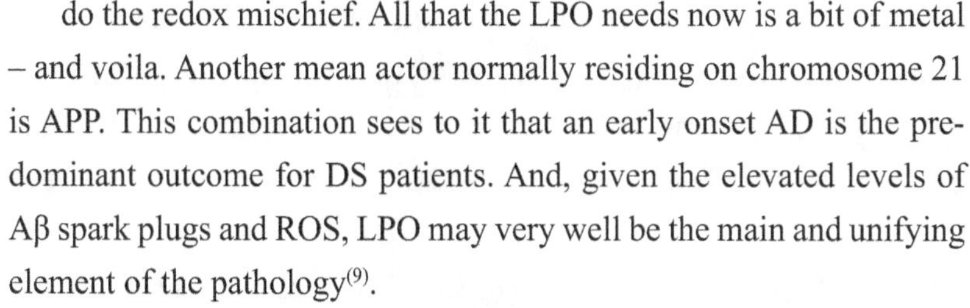

$$2\ H_2O_2 + \text{catalase} = O_2 + 2\ H_2O$$

do the redox mischief. All that the LPO needs now is a bit of metal – and voila. Another mean actor normally residing on chromosome 21 is APP. This combination sees to it that an early onset AD is the predominant outcome for DS patients. And, given the elevated levels of Aβ spark plugs and ROS, LPO may very well be the main and unifying element of the pathology[(9)].

Several diseases fall under the category of Tauopathies, or disturbances in the TAU processing. An example is an orphan (prevalence of around 10 cases per 100,000) condition called Steele-Richardson-Olszewski syndrome, also known as progressive supranuclear palsy (PSP). Often misdiagnosed at an early stage as PD, it involves a build-up of phospho-TAU in neurons. Again, LPO is a major player[(10)].

Traumatic Brain Injury (TBI) is a lifestyle- or occupation-related injury which arises when the brain, upon impact, hits the inside of the skull. It affects tens of millions of people, more often males, annually. Woodpeckers do not get TBI, as their brains are too small for centrifugal and concussive forces to reach the necessary threshold. So, when looking at your windscreen after a summer drive through the fields, you can be safe in the knowledge that the bugs did not die of concussion. After the initial impact, all bets are off as to the direction the pathology might take, and a latent period follows that could take decades. Then, when the original injury is long forgotten, the retribution may come back, with the vengeance. A bane of contact sports like American football, rugby and boxing, the original trauma can transform into one of several neurological conditions. Mohammed Ali famously developed PD (although a more frequent outcome for a boxer is an aptly named dementia pugilistica), while other crippling afflictions include dementia-like Chronic traumatic encephalopathy (CTE) and even ALS, particularly among football (soccer) players specializing in headers.

Animal studies, very limited human postmortem data and modern non-invasive methods[11] suggest that accumulation and spreading of abnormal phospho-TAU may be the very first pathological step[12]. And the dots-connecting glue is, again, LPO. In fact, as the centrifugal force jolts the PUFA-rich membranes, they momentarily lose the barrier function and lots of metal (including transition metal) ions* and debris collide, resulting in an LPO-initiating burst of ROS. The LPO is the first step, and the LPO products envelop everything inside of the neurons, including phospho-TAU busily helping shuttle cargoes inside damaged neurons[13].

Once again, LPO and its toxic descendants are a common factor that glues these different indications into one big problem, insoluble in every sense of the word.

Parkinson's Disease (PD)

Worldwide, about 9 million (1 million in the US) people have PD (double the number from 25 years ago), causing around 350,000 deaths every year. Compared to women, men have a 50% higher chance of developing the disease. Early onset (before 50) represents about 4% of all cases. PD is a progressive neurological condition that affects the brain's nerve cells, causing movement problems as well as cognitive and mood issues. The target part of the brain is the dopamine-producing substantia nigra.

PD belongs to a wide category of Lewy body diseases, characterised by pathological deposits of waste material inside or outside of neurons. It has been established 40 years ago that LPO is the driver of the substantia nigra's demise[14]. The exact perpetrators of this, however, were unknown. (*ch.2.6.4*)

Unlike the AD, one of the key culprit protein responsible for Lewy bodies in PD is alpha-Synuclein (aSyn). The jury is still out regarding its exact biochemical role, but one of the functions seems to provide insulation to lipid bilayers. Tiny liposome vesicles, that ferry signalling molecules like serotonin and dopamine between synapses, are miniscule aqueous droplets wrapped in lipid bilayers. Imag-

---

* For example, released through microbleeding.

ine a flat lipid bilayer. Once you start bending it, the tight arrangement of PLs will get distorted at folds, moving the lipid molecules further apart and opening crevices, like folding a slice of hard cheese. Taken to an extreme, think of a cube with 8 vertexes, where the membrane arrangement would be the least orderly. Taurine is the tool nature uses to plug those holes. This small molecule, often erroneously referred to as an antioxidant,

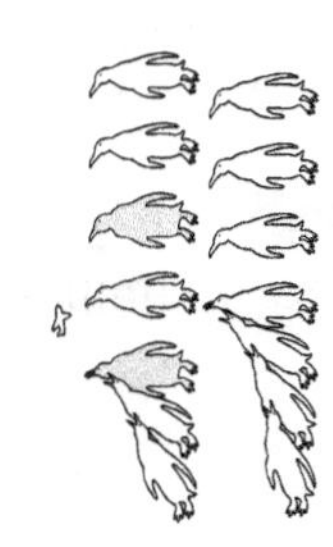

$$H_2N\text{-}CH_2\text{-}CH_2\text{-}SO_3H,$$

plays that role. A-Syn is a protein equivalent, which sits in membranes of vesicles helping them deliver the cargo without leakage. It naturally evolved the propensity of finding comfort when inside of lipid membranes(15). But then the things took a sinister turn.

Andrey Abramov (UCL, London), the top authority in the field at the overlap of neurological diseases, calcium signalling, bioenergetics and oxidative stress, in his youth was an accomplished fencer, winning high-ranking tournaments in the USSR. It might be a bit of a trite to state that the ability to stay on target, to overcome resistance, to take good aim at the key well-hidden target, to persevere and, finally, to go for the kill stood him in good stead ever since. Having fenced with the problem of aSyn toxicity for a while, he hit the bull's eye of the most unexpected vicious circle target with his foil. ASyn exists in several forms, including the functional monomeric, and the sticky oligomeric. The oligomeric version, it turns out, prefers getting into membranes that have been affected by LPO. And the nub? Whilst in there, it catalyses even more LPO(16)!

The machinery that pumps out Lewy body-forming oligomeric aSyn relies on LPO*. And the more monomeric aSyn is there in neu-

---

* This is true for the two main PD subtypes: "brain first", and "body first". The latter involves the oligomeric aSyn formation outside of the brain, perhaps in the guts of heart. The oligomer then travels to the brain through the vagus nerve, so the two subtypes converge, as the oligomer ignites the LPO in both.

rons, the more of a chance that the monomers would be targeted by reactive LPO products, driving the formation of oligomeric aSyn(17), which will then get into PUFA-rich membranes to sustain the circle. This is the reason why the unfortunate families that carry three copies of aSyn versus normal two have a substantially elevated risk of early onset PD(18).

While hallmark signs of LPO are elevated in postmortem brain samples of patients with neurological conditions, access to sample materials in alive patients is mostly limited to blood and CSF (*ch.2.9*). CSF sampling (particularly for isoprostanes) tends to yield stronger LPO product signal, but is more invasive compared to a blood draw(19), with many biotech companies currently active in this area.

Other Neurological Indications

Many more neurological diseases can be mentioned in this section. There is plentiful evidence that LPO plays a major role in ALS, Huntington's disease, multiple sclerosis, and a large number of various orphan diseases, rare diseases with loosely defined prevalence, typically less than one patient per 1,000 people. Numerous ataxias (such as Friedreich ataxia), Barth and Rett syndromes, Batten disease and many more are neurological diseases with proven LPO involvement. Even dividing them into categories is challenging, as idiopathic diseases (arising sporadically, for unknown reasons) would overlap with genetic conditions as well as environmentally inflicted (chemicals, lifestyle). Mitochondrial diseases are a large group taken separately, yet it has large presence in neurology. Even the dividing border between central (brain) and peripheral neuronal problems is often difficult to draw. Is eye a neuronal organ, being full of neurons? Complicating factors are galore. Smoking, while giving protection against PD, much exacerbates AMD… Genetics should also be taken in context. Same genetic traits can be protective for people with one genetic background, while detrimental for others (see *ch.4.6*).

In clinical neurology, numerous attempts were made to address various facets of the LPO curse with drugs. The results, so far, are disappointing. Replenishing damaged PUFAs with fresh ones, employing chain-terminating antioxidants (*ch.2.8*), encouraging autophagy to get rid of the damaged material, trying to counteract the effects of elevated levels of RCS in ALDH2 carriers (*ch.2.6.3*), reducing the levels of transition metals (*ch.2.4*), all failed. Ashley Bush of the Florey Institute in Melbourne has been spearheading the concept of the ever-accumulating transition metal levels in the brain[20]. Once upon a time, when men fought with swords and spears, there was a robust (if not healthy) and reliable mechanism of getting rid of some (too often, all) blood. The torch was then carried by the long-standing practice of bloodletting, a method purportedly curing all diseases and increasing wellness. The method, believe it or not, was only discredited in the late 19th century. And guess what? That date roughly coincided with Dr Alois Alzheimer's discovery... Prof Bush believes metals accumulate because currently in men, there is no protocol for shedding excessive metals with spilled blood*. Elevated incidence of AD in women may tie up with this: up to a certain age, some blood is removed regularly, and so are metals. So, while men can be gradually getting accustomed to high iron through hormesis or whatever, women rid themselves of the excess metals periodically, until one day, the metals are suddenly no longer removed. And then the Fenton reaction ensues, instigating LPO[21]... The concept, of course, could not have possibly been missed by the ferroptosis community[22] (see *ch.4.8*). Be it as it may, clinical trials of various iron-arresting chelating agents, sadly, drew a blank[23]. This, however, does not rule out the mechanism: it is a tall order to remove metals from brain**.

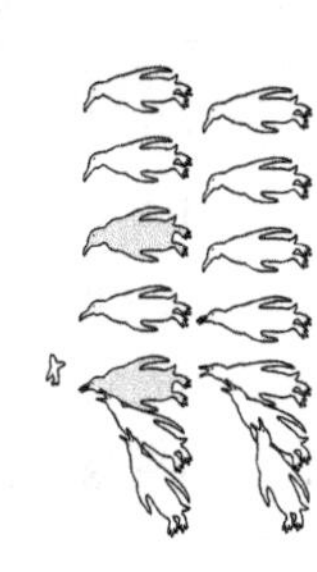

* I could not find references on prevalence of AD, or other neurological diseases, in serial blood donors. Perhaps Ashley Bush might have the numbers. Meanwhile, a couple of recent intriguing reports seem to point in the right direction. An Australian named James Harrison gave blood more than a thousand times between ages 18-81. He stayed free of neurological diseases and died aged 88, well exceeding the Australian men's life expectancy of 81.2 years. Separately, people who frequently donate blood see a boost in the growth of healthy blood cells[31]. Either, or both, observations might be linked to the benefit of decreased LPO, courtesy of reduced iron levels, - or might not be.

** Curiously, metals can be inhaled! BBB restricts access of manganese, a transition metal, into the brain, letting in just enough of this important cofactor. But nasal uptake can deliver it directly into the brain. Welders can absorb so much of airborne manganese that brain LPO can ensue, resulting in

For hundreds of neurological diseases, with thousands of genes implicated and thousands of chemicals known to trigger the pathology multiplying the number of possible combinations, there are only two towers that stand unshakably: lipid peroxidation and neuroinflammation*. And when put under pressure, they may very well morph into one: LPO. The common denominator of neuronal distress**. Let's see if this target can be druggable.

4.3. Mitochondria: dirty diesel engines with large ROS footprint

"Everything is energy and that's all there is to it."
— Albert Einstein

"Our terminal decline into old age and death stems from the small print of the contract that we signed with our mitochondria two billion years ago."
— Nick Lane

A biochemistry student walks into an English pub and says, "Please, may

---

cognitive impairment and PD-like symptoms. Staying close to fireworks can similarly elevate LPO, as strontium (vibrant red) and copper (blue) salts fall down and can get inhaled.

* Inflammation can induce LPO (see *ch.4.9*). But brain neurotransmitters can do that, too. Adrenaline (epinephrine) can both lead to LPO-initiating ROS, for example by activating the ROS-producing NADPH oxidase, - and turn into the ROS itself, particularly if manganese is present. Metabolism of dopamine in mito-membranes generates ROS which can kick-start the LPO. Serotonin, by binding to some receptors, can activate some ROS-producing signalling pathways. Yet a more accepted role for serotonin is that of an LPO-stopping antioxidant(24), similar to the LPO-chain-breaking potential of melatonin(25). Good job caffeine seems to be on our side – while capable of generating ROS, it seems to inhibit the LPO(26). In addition to boosting our productivity and output as we are thinking hard trying to figure out how to stop the LPO.

** LPO also plays an important role in psychiatric diseases including Schizophrenia(27), bipolar disorder(28), autism(29) and ADHD(30). This area is poorly explored. Current anti-psychosis drugs have horrible side-effects that may include mood changes, itching, diarrhoea, mass-murder and jumping off bridges. Perhaps it is time the LPO inhibitors were tested as anti-psychotics.

I have a pint of adenosine triphosphate". – "Sure, mate", - goes the barman, - "That's eighty p". Now, there are two types of people. Those to whom "ATP" means "Adenosine triphosphate", and those to whom it means "Association of Tennis Professionals". The second type are much better off financially. Which is ironic, considering the first type has been, and will be doing just fine without tennis, while tennis entirely depends on adenosine triphosphate, and on the ATP-making factories called mitochondria. While (a) there are some non-mitochondrial sources of ATP including aerobic and anaerobic glycolysis, and (b) apart from ATP, the energy "currency" also includes GTP, NADH and NADPH, mitochondria still stand out as the main "power stations". Although a lot of ATP is expended driving enzymatic transformations, one of the major consumers is the plasma membrane, which uses ATP to generate the sodium-potassium gradient, the driving force for various membrane activities. About half of the total energy output is expended maintaining the lipid membrane gradients[(1)], particularly in neurons. Mitochondria (mitos) were discovered by Richard Altmann in 1890. Decades later, in the 1960s, Peter Mitchell postulated their mechanism of action. Mitchell's chemiosmotic hypothesis explained the oxidative phosphorylation, the key internal combustion machinery overseeing conversion of food-derived fuel into the main energy currency of all things alive, the ATP. This fascinating organelle is the subject of several great popular books[(2)].

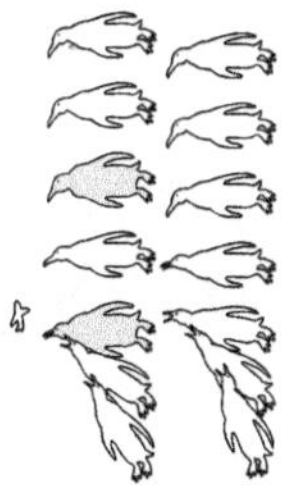

Mitos are an integral part of almost every cell in the body. Notable exceptions include red blood cells, which also lack nuclei, DNA and RNA, likely forever wiping the RBCs off the Nobel committee's radar screen (*ch.1.1*). Perhaps the absence of mitos is due to the inability of the nucleus-free RBCs to control mitogenesis through gene expression, as mitos, with their own tiny genomes, rely on the nuclear DNA machinery. Or maybe RBCs stay "sterile" to ensure that no oxygen is wasted unnecessarily, enroute from lungs to the periphery. So tightly are the mitos interwoven into various aspects of cell/tissue function (energy production, calcium signalling, cell death, aging and cancer,

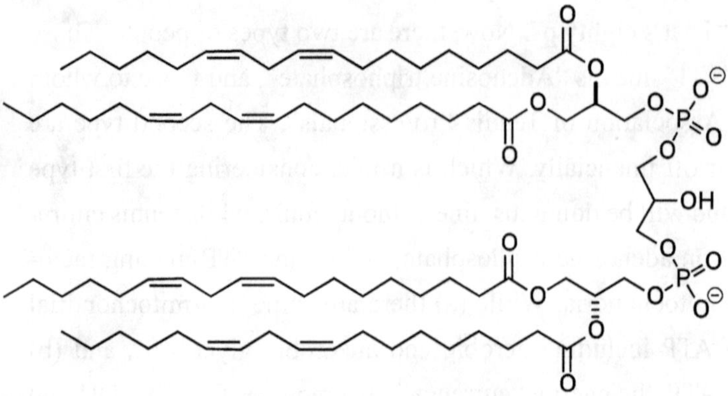

Figure 23. Structure of a mitochondria-specific PL, cardiolipin (CL). In humans, the predominant PUFA is LIN. CL is also found in bacteria, which is not surprising, considering mitos evolved from bacteria, engulfed by an ancient host cell, perhaps an archaeon.

to name a few) that it is impossible, when talking about them, to avoid repeating topics covered in other chapters.

According to Guy Brown at Cambridge, if not for haemoglobin in blood, muscle myoglobin and skin pigments, we would all be the colour of mitochondrial cytochromes, changing hues depending on how energized we are. Mito lipid bilayers are an impressionist painting of a dense stew of proteins, peptides and small molecules seemingly jumbled in chaos, but actually all working in a highly concerted way, as the respiratory chain places O_2 molecules on complex IV, having cytochromes feed electrons, with their energy depleted by prior interactions with Complex I and Complex III, to it (complexes I and III are where most of the ROS are produced). If the task were simply to burn the fuel to release the energy, the mito's layout could have been much simpler. In fact, that's one of the mechanisms of how our body temperature is maintained. Membranes separate the acidic and basic environments, and combining the two generates heat:

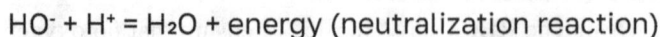

$$HO^- + H^+ = H_2O + \text{energy (neutralization reaction)}$$

Electrons release their energy gradually, allowing for better utilization of every last bit, as compared to an explosive discharge in one go, which would make it difficult to utilize the energy fully. The reduction of molecular oxygen also happens through the step-by-step reduction to water, but for a different reason. Four electrons are required in total, but they come one by one, not all at once. The reduction of oxygen takes place on Complex IV. Superoxide and hydrogen peroxide on Complex IV are under tight containment, and do not leak out to ignite the LPO,

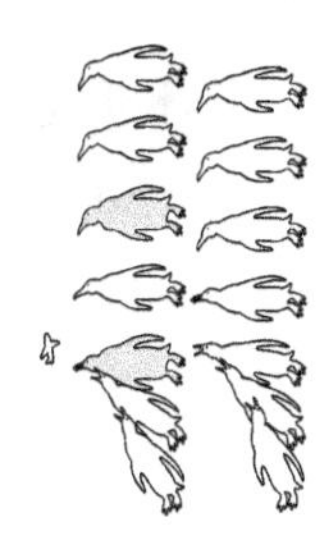

$$O_2 \rightarrow O_2^{\bullet-} \rightarrow H_2O_2 \rightarrow H_2O$$

as do the ROS species generated along the OxPhos pathway. Gradual release of the electron energy requires an extended, multistep process, involving intermediate proteins and small molecules like CoQ, iron-sulfur clusters, etc. This all takes place in inner membranes, and imperfections at every step (*Box 10*) allow for small leaks of electrons, leading to the ROS formation*.

The number and type of mitos varies hugely, depending on the cell type, from tens to thousands, although quantification may be challenging, as all mitos may be linked into huge "mitochondrial reticulum"-like body. The large numbers should not be too shocking considering there are peripheral neurons more than a metre long. Extending Nick Lane's epigraph, to a large degree the problems with mitos stem from the contract they themselves signed with PUFAs. As mitos pack a lot of PUFAs, iron and oxygen, we basically have a large depot of tiny live grenades that keep **exLPOding** in every cell.

The literal meaning of "mitochondrion" (singular of mitochondria; Greek, *mitos*, thread + *chondrion*, small grain, from their initial appearance) proved to be a bit of a bane for their purification, for even today, making pure preparations of mitochondria is challenging. Purified fractions tend to contain other organelles which co-purify with the mitos. It is therefore only with some uncertainty that the lipid composition of the organelles can be ascertained. Even though we focus

---

* Remember stirring sugar in that cup of tea (*ch.2.8*)? Excessive reducing species such as NADH are a source of ROS, too. This happens when an electron is passed on to an oxygen molecule: $NADH + O_2 \rightarrow NAD^+ + O_2^{\bullet-}$

here on PUFAs and not PLs, one type of a mitochondria-specific lipid must be mentioned. Cardiolipin (*Fig.23*) contains three glycerol moieties, two phosphoric acid residues linking them, and four PUFAs attached to the two flanking glycerols. It makes up 15-25% of PLs in the inner mitochondrial membrane, and is localized, synthesized and taken apart exclusively in the mitochondria. Both the deficiency of CL remodelling (Barth syndrome) and excess (Tangier syndrome) of CL are detrimental to human health. CL's unique functions include helping immobilize cytochromes, which hold to two of its four arms in their grip[(3)], while the other two PUFA strands are immersed into a bilayer. (Cyt is also positively charged, and holds on to negatively charged phosphates of PL). A PUFA arm of CL can get oxidized while sitting inside of its cytochrome c "glove", and the cytochrome can then slide off, initiating apoptosis, the programmed cell death protocol. This cytochrome c - CL interaction determines whether cyt C acts as an electron carrier in the OxPhos pathway, or as a PUFA peroxidase. Cyt c release is suppressed when mito's LPO-preventing hydroperoxide peroxidase (Gpx4) is elevated[(4)], cementing the role of LPO in apoptosis. CLs, distributed in the bilayers non-symmetrically, are also vital for maintenance of the inner membrane-specific wrinkles called cristae (from Latin, *cristae*, crest), the distant relatives of the (rhod)opsin-holding discs in rods and cones (which rely on their own unique PUFA-based contraption, the VLC-PUFAs (*ch.4.1*)). Cristae make the surface area of the inner membrane huge, giving plenty of room to the OxPhos machinery. Vast surface, plenty of PUFAs and ROS and inefficient antioxidants lead to rampant LPO, with PUFA tortured corpses to be found everywhere in these compartments.

Russian botanist Konstantin Merezhkovsky suggested in 1905 that mitos were procaryotic bacteria entrapped by early eukaryotes 1.5 billion years ago. The symbiotic relationship led to mitos transferring most (but curiously, not all) of their genes to the cell nuclei. As mitos always retain some genes, but these genes are different between different species, a theory was put forward suggesting that the retained small genomes help cells to distinguish between thousands of mitos, which can rapidly signal individual distress or respond to environment by tuning their own gene expression. While human mtDNA codes for 13 OxPhos-related proteins and some local tRNA and ribosomal RNA (37 genes in total), mutations in these genes are associated with numerous downstream systemic effects, like

HDL, cholesterol and TG levels in blood (*ch.4.6*), so the mito genome punches way above its size[5]. Other peculiarities are worth mentioning as they make a very direct impact on aging and longevity. As mtDNA is basically bacterial, it is circular and has no introns (non-coding sequences), so that any ROS shrapnel would hit the gene-related DNA bases and not the irrelevant intron decoys. It also lacks effective repair systems, as well as histones. Sitting unprotected, mtDNA is constantly bombarded by ROS leaking from the inner membrane, as well as plentiful LPO products which form adducts with the DNA bases (*Fig.19*). A big safety issue in this PUFA-packed ROS factory is the exposure and insecurity of its own DNA. A bacterial descendant, mitochondrion was not concerned with longevity, continuing its lineage by constantly dividing, with no need to preserve its working parts. Things changed once it got swallowed up and an endosymbiont was born. Having swallowed the ancient ancestor of all mitochondria, the host continued stressing the hell out of it, in multiple ways. And one of the most vicious of those was the forced adoption of PUFAs. Bacteria use oxidation-stable branched fatty acids as the way of keeping their membranes fluid (*ch.2.5*), so they did not have a chance to get used to PUFAs. When, through whatever mechanisms, the host cell decided to adopt PUFAs, it did not really consult its prisoner. Forced upon the mitochondria by the host to boost membrane processes, PUFAs in combination with the constant flux of ROS created this tinderbox, which, as if this were not enough, also had its circular DNA fully exposed. The ensuing mutations, different for each individual organelle[6], would generate a very heterogeneous pool of mitochondria*. The combined result is an age-related decline in mito efficiency, associated with even more ROS production. MtDNA damage may be followed up by mitofusion, as mitos would form a single large structure.

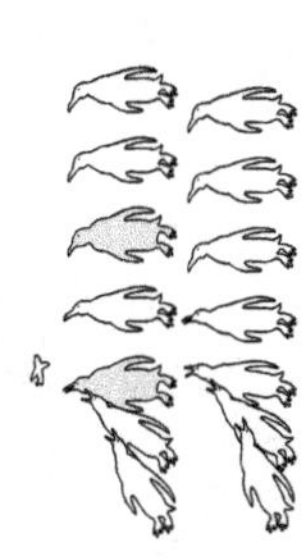

Nuclear DNA has more degrees of protection compared to mtD-

---

* Some parts of mtDNA are less exposed to the ROS attack compared to some other regions, presumably because they are buried inside of the spatial 3D-structure[7], hindering the access of damaging species.

NA. However, if LPO products manage (as they do) to sneak into a nucleus, which has its own PUFA-containing membrane, with a proton gradient so that the LPO is never too far away, LPO- derived carbonyls can "mutate" the nuclear DNA as well[8]. This has repercussions in inflammation, cancer, and beyond, but let's stay focused on mitos.

PUFAs smoulder in the inner mitochondrial membrane, forming reactive carbonyls. More fat-soluble carbonyls, such as n-6-derived 4-HNE, prefer to stay in the hydrophobic membrane environment, damaging membrane proteins. More hydrophilic carbonyls, like MDA and the n-3- derived 4-HHE, tend to leach out into the aqueous compartments, cross-linking water-soluble proteins, and mitochondrial DNA. An unprotected blob of mitochondrial genome, called nucleoid, with typically 5-10 copies of a 16,569 bp circular double-stranded mtDNA, is targeted by the LPO-derived reactive carbonyls (*Fig.19*). These vermin get down to work, forming covalent adducts with DNA bases and cross-linking DNA strands. Reactive carbonyl adducts with DNA nucleobases yield freak structures which either have difficulty, or are completely unable to form the canonical, Watson-Crick base pairs. Instead, they form hydrogen bonds with the "wrong" bases, generating mutations. These lesions are repaired by various pathways in both bacteria and mammals, but under elevated stress conditions, the damage can overwhelm the repair capacity. Another type of damage is even more detrimental: the carbonyls can covalently cross-link two opposite DNA strands (*Fig.19*), stopping DNA replication[9], just like the anticancer drug cisplatin does. The crosslinks can also form between DNA and proteins, generating a different set of challenges for the poor mitochondrion. Mito DNA damage, exacerbated by poor protection of its DNA and elevated levels of ROS and LPO products, is behind many theories linking this damage with aging. As is often the case with LPO (see *ch.5.6*), the vicious circle of damage eventually generates even more LPO. Various mechanisms were put forward, including the mitochondrial free radical theory of aging (MFRTA) and the "survival of the slowest (SOS)" hypothesis[10].

Mitos can reduce the level of oxidative damage by splitting into more and less active groups, such that the less active pool sustains less damage[11]. Damaged mitos could be renewed*. This is happening less efficiently as we age, suggesting

---

* Another approach to minimizing damage is linked to exercise. Why wouldn't in-

that another flaw, in addition to the vicious cycle of ever-increasing damage, is the defect in turnover mechanisms. Some organs, like substantia nigra and colon, have particularly high levels of mtDNA damage. The reason for that is unclear. Ultimately, preserving mitos in good shape requires resources. If cells are reluctant to expend the extra energy for repairs, content with the post-reproductive decline and aging, then perhaps a little help from the outside could go a long way, propping up the mitos and postponing the inevitable? Look for answers in *chapter 6.*

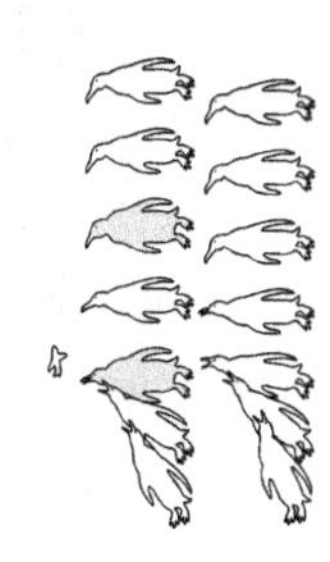

The Mitochondrial Free Radical Theory of Aging is based on several observations: (a) mitochondrial ROS production increases with age as mitochondrial function declines, (b) activity of several antioxidant enzymes declines with age, (c) mutations in mtDNA accumulate with age, and (d) a vicious cycle occurs because somatic mtDNA mutations impair respiratory chain (RC) function, which leads to further increase in ROS production and LPO, generating more damage still. Accordingly, mitochondrial innate imperfection drives the aging process.

The SOS hypothesis was put forward by Aubrey de Grey to explain the age-related respiratory decline, whereby respiratory chain-affecting mutations in mtDNA would lead to accumulation of less efficient mitochondria, as normally performing mitochondria would suffer more oxidative damage to their membranes as a direct result of their superior respiratory activity, and would accordingly be more often relegated to lysosomal utilisation*. The inferior, slower mitochondria would therefore accumulate, greatly reducing the overall cellular performance, re-

---

tense exercise dramatically elevate the levels of toxic mito exhaust, damaging all in and around them? Perhaps because in response to training, the number of mitos increases substantially, so that the energy demand per mito is actually lower in trained people[(12)]. An additional factor is a decreased oxygen level.

* Languid mitos can also produce ROS, for example when electrons are prevented from flowing through the OxPhos pathway. Instead, they sit idly on Fe-S clusters or semiquinone, increasing the chance of collisions with O_2, and consequent ROS generation. In general, there are at least two types of bad mitos: the ones which are good energy producers, but leak a lot of ROS, too. And the ones that are bad energy producers.

sulting in aging at the organismal level, which manifests through decreased mitophagy, chronic inflammation, cellular senescence, age-dependent decline in stem cell activity, and other key aspects of aging. All these calamities stemming from the upstream LPO…

OxPhos pathway-associated proteins (electron transport chain, ATP production and multiple other important processes, including beta-oxidation and its key player, Complex I – bound mitochondrial trifunctional protein[(13)] are located in the inner mitochondrial membrane. The outer mitochondrial membrane is also buzzing with biochemical activity. As discussed earlier, biochemical processes taking place inside of lipid bilayers require the bilayers to be fluid, so the mitochondrial membranes are highly enriched in PUFAs. And herein lies the problem. Many hubs of the OxPhos pathway are notoriously leaky. According to some estimates, up to 1-2% of all the oxygen we inhale slips out of the proper OxPhos channels and goes astray. As it bumps into one of many electron carriers within the OxPhos chain, for example an iron-sulphur cofactor used by various enzymes, an electron can jump on to that oxygen molecule. In other words, the iron-sulphur cluster gets oxidized by oxygen, while oxygen gets reduced by that electron, forming a superoxide. Electron transport chain (ETC), particularly complexes I, II and III, can easily get out of balance, squirting the ROS in all directions. This happens inside of the bilayers, and the newly-born ROS go straight to work to initiate LPO in the surrounding dense forest of membrane PUFAs. We already know what happens next. And the nearest help, in the form of vitamin E or CoQ plodding languidly through the membrane, may be hundreds of PUFA residues away… We do not have to wait long for the chain reaction to get started.

Mitochondria should not pin too much hope on the double-faced antioxidants, as these perfidious actors are either inefficient or, because the tightly controlled delivery mechanism will not deliver more antioxidants* than it is programmed

---

* And for a good reason, too! Too much of a good thing with regards antioxidants is known as Reductive Stress. This imbalance of redox homeostasis leads to an excess of reducing molecules such as NADH and GSH, compared to their oxidized counterparts NAD^+ and GSSG, for instance when NADH precursor supplements are taken. This can disrupt various cellular processes. In mitos, it leads to an increase in the ROS production through the OxPhos pathway, because a high $NADH/NAD^+$ ration impairs ETC efficiency, promoting electron leakage and ROS generation (*see ch. 2.2, 2.4*). This immediately increases LPO in the inner membranes.

to do, make things worse (*ch.2.8*). Vitamin C is often used as an iron recycler in generating constant flux of ROS in the lab setting. In 2001, Ian Blair at UPenn explained how excessive vit C, in the absence of transition metals, induces lipid peroxide decomposition, actively helping to increase the levels of mutagenic carbonyls[(14)].

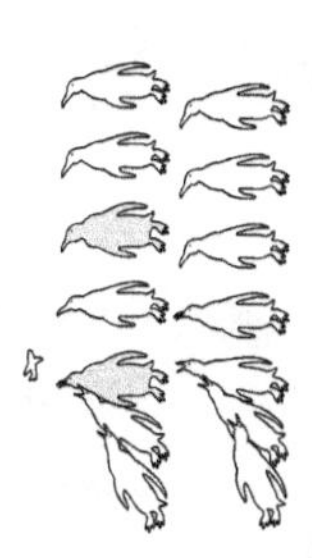

As the mitochondrial furnace constantly pumps ROS into the PUFA-rich lipid bilayers, bullying the defenceless PUFAs into the chain reaction, it can only bode ill for the PUFA-rich mitos. Both mitochondrial structure and function are highly sensitive to the LPO – derived RCS[(15)].

How could this be mitigated? In an example of a biological adaptation (see *ch,5)*, mitos in longer living species reduce the amount of LC-PUFAs in their membranes. Mito PUFA analysis across all classes of phospholipids, and the peroxidation index (PI) (a metric reflecting the amount of PUFAs, and their level of unsaturation, in a tissue) of skeletal muscle, liver and brain in mice (maximum lifespan (MLS) 4 years), pigs (MLS 27 years) and humans (MLS 122 years) revealed that mouse mitos had the highest level of PUFAs, and PI. The levels were the lowest in humans, while pigs fell in between[(16)]. Whether this selection happens at the cellular or mitochondrial level, and through which mechanisms, is still largely unknown.

There is a beautiful red-coloured Mill steam engine on the ground floor of the Science Museum in London. Built in 1903, it is now almost as old as Jeanne Calment, yet still fully operational, at 700 hp, taking up an area of 30 m^2. The well-oiled massive parts moving in unison are remarkable, but my personal favourite is a humble spinning feedback contraption called centrifugal governor, that keeps the pressure from building up. It is made of two hinged rods with metal weights at the ends, which hang listlessly down when stationary, "impaled" on the drive shaft, keeping the fuel feeding valve fully open. As the balls rotate and rise, the valve stem is forced downward, closing the valve. Invented by one of the key figures of the Scientific Revolution, the Dutch polymath genius Christiaan Huygens 350 years ago, the contraption is so important it still finds use in some internal combustion engines and turbines.

Mito performance and output are also controlled by many ingenious mechanisms and feedback loops, some relying on LPO. A biochemical equivalent of this mechanism uses LPO as a proxy for the speed of rotation. What if mitos get a little over-worked producing the trans-membrane proton gradient? Suppose there is no need for so much ATP, yet the mitos continue burning fuel to pump protons across the inner membrane. The transmembrane potential is huge (*Back-of-the-envelope 10*), and the overdrive results in elevated levels of ROS, and the LPO galore. There is a group of mito membrane proteins called uncoupler proteins (UCP). They are membrane channels, and one of their functions is a negative feedback loop: they release some excessive proton gradient when activators, such as HNE, attach to them covalently[(17)]. This opens the UCP pores, increasing proton leak and lowering the proton motive force, effectively "letting the steam out". As a result, the production of ROS decreases, leading to reduced LPO.

Mild uncoupling dramatically decreases $O_2^{\cdot -}$ production. A 10 mV (6%) decrease in mitochondrial membrane potential results in a 70% decrease in superoxide[(18)]. The initial evolutionary role of UCP could have been to slightly decrease the output, with a huge protection from ROS. It could have been how the warm-blooded animals evolved*, - as a side-product of protection against ROS!

In addition to UCP, HNE also reacts with multiple other proteins. TRPA proteins involved in pain signalling will be mentioned in *ch.4.4*. A large cluster of oxidative stress response mediators, such as an NRF2 transcription factor, are activated by low concentrations of LPO products[(19)] produced by mito tinderboxes, reminiscent of the hormesis effects (*Box 4*).

Mitos produce heat not just when needed, but as a side-effect of their activity (see *ch.2.3, footnote p.173*). While there are other heat-producing processes, mitos consume 85% of the inhaled oxygen and generate about 75% of the total body heat output. Mito metabolism and core body temperature decrease with age,

---

* There are some people, living at high latitudes close to the Arctic Circle, who, courtesy of the "leakier" UCP, have a higher body temperature as an environmental adaptation. These folks are known to never put on excess weight. Elevated LPO and RCS may be the side-effects this adaptation, perhaps explaining why these people are not particularly long-living. A similar effect can be achieved by a now banned OxPhos uncoupler, the weight loss remedy dinitrophenol (DNTP).

while their ROS production increases. However, the brain temperature, which is normally about 2°C warmer than the rest of the body, increases with age for not well-understood reasons, likely impairing brain function through increased protein denaturation (see *ch.5*). Glycolysis produces 50% less heat than OxPhos and can be preferred by tumours which, owning to poor vascularisation, have an inferior system of excess heat dissipation, predominantly carried out by blood flow. Mitos were reported to be substantially hotter (up to 50°C) than the rest of the cell[(20)], although exactly by how much is unclear as many proteins denature at above 43°C[(21)].

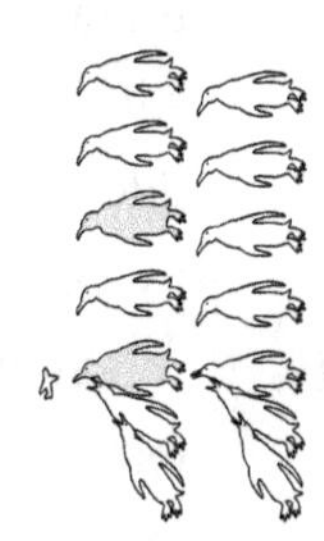

Mitos produce plentiful ROS but are not the main cellular ROS source, being responsible for 4% - 44% of all ROS in different cell types[(22)]. They also consume back a lot of superoxide and H_2O_2, further reducing the ROS load. Bigger sources of ROS in most cell types include peroxisomes, plasma membranes, ER and NADPH oxidases. However, as the ROS production happens inside a PUFA-rich inner membrane, there is plenty of LPO-igniting ROS on tap to constantly harass and stress the mito PUFAs. The resulting LPO products damage OxPhos complexes, often resulting in even more ROS produced, leading to more LPO, sustaining the vicious cycle directly linked to inflammation, neurodegeneration[(23)], CVD and metabolic diseases like diabetes[(24)]. Since the decline in glucose metabolism[(25)] and mito function and increased ROS and LPO often precedes clinical pathological symptoms by decades, ROS and LPO were suggested as the cause not the consequence[(26)].

CL may be undergoing LPO so fast that the CL making systems may be unable to keep up, compromising the inner membranes, somewhat similar to Barth syndrome-associated CL deficiency. In this case, boosting the transgenic expression of CL can attenuate mito dysfunction[(27)], somewhat similar to positive effect of supplemental PUFA on conditions known to involve LPO (*ch.4.1*). This, of course, is a rather drastic way to keep the CL pool in a good shape. So, when it was claimed that a series of mito-penetrating Szeto-Schiller peptides could protect CL against LPO through an unknown mechanism[28], that caused quite a stir. Sadly, the RCT results for SS-31 did not stand up to the expectation. Similar fate befell the much-acclaimed CoQ mimetic EPI-743, an antioxidant developed by Edison Pharma for treatment of mito diseases.

Similar to the eye (*ch.4.1*), there is plenty of carotenoids in mitos. AD patients exhibit low levels of lutein, lycopene and zeaxanthin, and the associated increase in LPO and neuroinflammation. This, in AD models at least, can be fixed by carotenoid supplementation, which also improves memory. Mito-targeting carotenoid compounds were developed accordingly[29].

Non-enzymatic oxidation of ARA yields LPO products, while enzymatic, stoichiometric (one on one) ARA oxidation by COX and LOX can produce ARA-peroxides which can jump-start the LPO upon contact with mito-abundant transition metals[30]. Separately, COX and LOX eicosanoid products such as PGs, TBs and leucotrienes are predominantly pro-inflammatory (see *ch.4.9*). In agreement with this, non-steroidal anti-inflammatory drugs (NSAIDs), through both mitigating the LPO (by dialling down on one of the LPO-initiating mechanisms) and reducing the levels of pro-inflammatory eicosanoid mediators, were reported to have beneficial effect on some neurological diseases[31].

PUFAs are involved in various mito processes including mitochondrial calcium homeostasis, gene expression, respiratory function, bystander effect, ROS production, mitochondrial apoptosis and many more, as LPO can affect them all. There is also a whole FDA recognized class of mitochondrial diseases. We will examine some of them in other chapters.

4.4. How oxidized PUFAs partake in ache

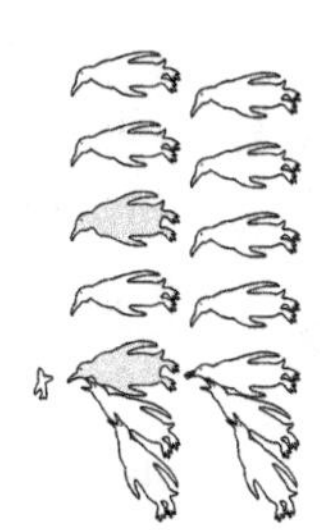

"We forget very easily what gives us pain"
— Graham Greene

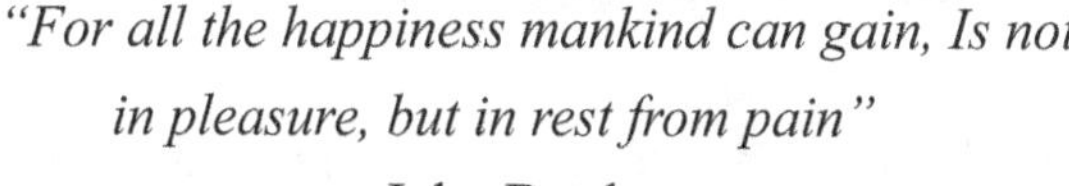

"For all the happiness mankind can gain, Is not in pleasure, but in rest from pain"
— John Dryden

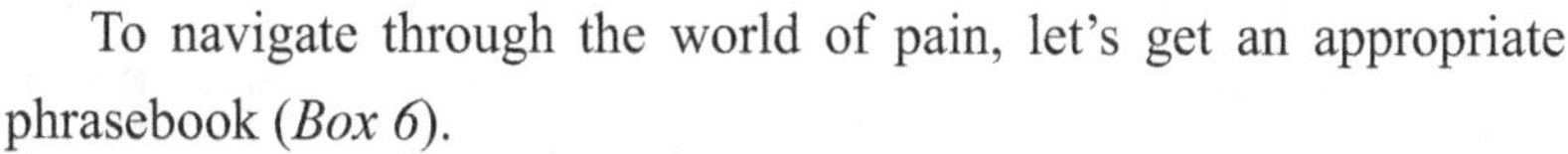

To navigate through the world of pain, let's get an appropriate phrasebook (*Box 6*).

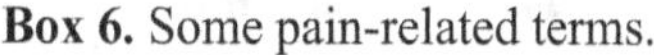

Box 6. Some pain-related terms.
Acute pain is a sudden or urgent pain caused by injury. It is time-limited and lasts up to 30 days.
Chronic pain persists beyond the recovery period or accompanies a chronic health condition.
Nociception (from Latin *nocere*, to harm), is the neuronal encoding of noxious stimuli (mechanical, cold, hot or chemical), which consists of receiving a painful input by a nociceptor (nerve ending), carrying this sensory information through afferent (going from periphery to CNS) nerve fibres, and reacting to this signal to trigger an appropriate defensive response. Nociceptive pain includes somatic and visceral.
Somatic pain typically comes from the skin or musculoskeletal system as a normal response to stimuli that harm these tissues, like skin cuts.
Visceral pain originates in internal organs like lungs, heart, pancreas, and blood vessels, or externally, like a sunburn. It may feel like a deep ache or pressure and may not be as well localized as somatic pain.
Migraine is a headache with unknown aetiology. Even though there are no nociceptors in brain, migraines are not imaginary, and involve electrical and chemical pain transmission processes.
Inflammatory pain occurs when inflammation activates nerves. As a subset of nociceptive pain, it can be somatic or visceral.
Neuropathic pain (neuralgia) may come from damaged or malfunctioning nerves, spinal cord or brain. Sometimes it can be chronic. Neuralgia includes allodynia (pain from a normally nonpainful stimulus like touch) and dysesthesia, a softer version of allodynia, which is unpleasant but not painful. Both may result from lesions in the nervous system.

The roller-coaster journey of a neuronal signal, sometimes referred to as a "spike", generated by a peripheral pain receptor in response to, say, a toe cut, proceeds through the spinal cord, to the thalamus sorting station for redirecting to various pain perception areas of the brain. It travels across multiple synaptic junctions and more than 1.5 metres of neuronal membranes. Owing to Mark Humphries' wonderfully detailed, visual description of how the signal travels through neurons, in his book "The Spike" (highly recommended!)[1], I can skip that part here and go straight to PUFAs

The electric spike, which is generated in, and then travels through, these main cables, must navigate the veritable sea of PUFAs, the key building blocks of neuronal membranes (*ch.2.5*). Many different ion channels, which convert cellular stimuli into electrochemical signals, help sustain the pain signal transduction, including calcium, potassium and sodium channels, as well as an important separate group of Transient Receptor Potential (TRP) channels. The latter are mostly calcium-permeable pores that sense ROS, triggering diverse responses such as cell death, chemokine production and pain transduction[2]. Their name reflects the fact that (in fruit flies) deactivation of TRP fails to generate the calcium dependent photoreceptor potential required for light adaptation. All these channels are attractive drug discovery targets, on the bet that blocking them would have painkilling effects. The molecular mechanics of the process is complex, but we are only focussing on PUFAs here. The channels, and many other auxiliary components, reside in lipid membranes, and so are surrounded by PUFAs, particularly ARA and DHA, the dominant fatty acid species of neuronal membranes. Neurons are metabolically active, consume plenty of oxygen, and since the elaborate network of antioxidants is not bulletproof (*ch.2.8*), neuronal oxidative stress sometimes gets out of control. At low levels, ROS are required for functional changes involved in normal synaptic plasticity, but crossing the fine line into higher concentrations is bad – and far from understood. And when there are ROS in the vicinity of lipid membranes, we do not have to wait too long for the LPO to start, - with direct effect on pain transmission.

4.4.1. ROS, LPO and pain

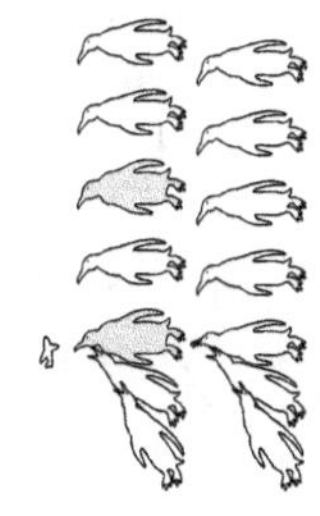

"The longer you have pain, the better your spinal cord gets at producing danger messages to the brain, even if there is no danger in the tissue."
— Dr. Lorimer Moseley

"Living with chronic pain is like trying to get comfortable on a cactus sofa."
— Anonymous

Several cation channels involved in pain transduction are affected by ROS and LPO products in different ways. Various non-selective cation channel TRP receptors on nociceptive neurons get activated differently by ROS. TRPV1, the vanilloid TRP receptor, which responds to the chilli pepper ingredient capsaicin and even to anandamide (*ch.2.5.2*), gets turned on by hydrogen peroxide, as well as by some lipids, all of which enhance its sensitivity to heat and promote thermal hyperalgesia. TRPM2 is likewise activated by H_2O_2, but non-directly, and a go-between mediator is required. TRP ankyrin 1 (TRPA1) receptor, responsive to wasabi and pungent ingredients of curcumin, mustard, garlic and cinnamon (they all react with its cysteine residue) is important in modulating nociceptor excitability and neurogenic inflammation in the setting of tissue injury. It is targeted by environmental irritants such as acrolein, which can provoke nociceptive and inflammatory action of smog and smoking. TRPA1 and TRPV1 work together to mediate neurogenic proinflammatory responses of pro-algesic agents such as bradykinin, which act upon the channels indirectly. But tissue injury also produces endogenous compounds that cause pain and inflammation by directly sensitizing TRPA1, which possesses the highest oxidation sensitivity of all TRPs. At sites of inflammation, ROS, generated by immunocytes, epithelial cells, and elevated levels of transition metals, get to work, inducing LPO. The strongest effect is elicited by 4-HNE, the LPO end-product of ARA oxidation (*Fig.9*). Formaldehyde, acrolein and HNE, LPO products generated in a lipid membrane in the vicinity of TRPA1, react with cysteine and lysine residues of the channel to elicit acute pain and neurogenic inflammation in response to oxidative stress, tissue injury and inflammation(3). Even a reactive prostaglandin 15d-PGJ2, a product of enzymatic oxidation of ARA (see *ch.4.9*), can activate TRPA1(4). While these LPO products form chemical bonds with the components of pain transduction pathway, other oxidized PUFAs bear no reactive groups, and so must interact with their targets non-covalently. Several LPO products of LIN, including 9- and 13-HODE and various DiHOMEs (*Fig.9*) have been shown to play a distinctive role in pain transmission and inhibition(5). The formation of these LPO products depended on the elevated inflammatory status(6)*. In agreement with this, reducing the extent of LPO led to reduced mechanical sensitivity in a chronic pain model in rat(7).

Other channels important in pain sensation and hyperalgesia can also be modulated by the LPO products. Methylglyoxal (*Fig.9*) has been shown to drive hyperalgesia by reacting with a sodium channel. In diabetic neuropathy patients, methylglyoxal above a certain threshold concentration in plasma was associated with pain, whereas individuals with levels below that stayed pain-free. Several strategies including a methylglyoxal scavenger are effective in reducing methylglyoxal – and diabetes-induced hyperalgesia(8).

Pain is one of the cardinal signs of inflammation. Pain-associated processes described above operate on the molecular level. A somewhat different mechanism of pain generation involves inflammation-induced swelling. As the body responds to harmful inputs like damaged tissue, it releases immune cells and fluid locally, close to the affected area. This leads to swelling and the resulting elevated pressure can physically affect the nerve endings, generating the sensation of pain. To mitigate this sensation, anti-inflammatory intervention may be employed, which would also provide the sought-after pain relief. For example, COX inhibitors, by dialling down on the pro-inflammatory eicosanoids, would reduce pain. It was also reported in the literature that enzymatic oxidation products of EPA and DHA,

---

* COX-2 upregulation in inflammatory conditions leads to increased production of PGE2. This prostaglandin amplifies the response of sensory neurons to painful stimuli and is involved in both acute and chronic pain.

called resolvins, can reduce ("resolve") inflammation and pain*, while an enzymatic oxidation product of ARA called leukotriene B4, which is often up-regulated in skin diseases, operates through TRPA1 and TRPV1 to produce a sensation of itch[9].

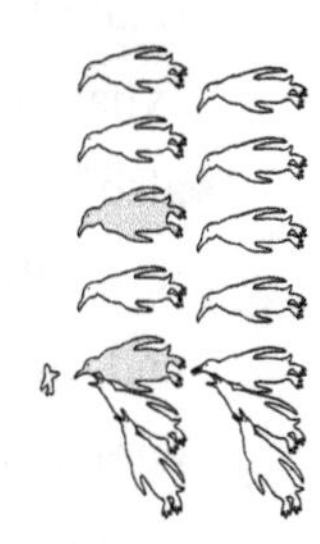

One would expect, theoretically, the antioxidants to show some pain relief effects. They could reduce LPO, and hence the levels of acrolein, HNE and methylglyoxal. They could theoretically reduce inflammation... Sadly, antioxidants were found to be only marginally effective at best, with a few trials reporting positive results counterbalanced by a large number of negative findings[10].

4.5. LPO's grim grip on kip

"I love sleep. My life has the tendency to fall apart when I'm awake, you know?"
— Ernest Hemingway

"A riddle wrapped in a mystery inside an enigma."
— Winston Churchill

The requirement for sleep, or at least for alternating active and resting states, exists across phyla. Even certain single cell organisms have some resemblance of a resting state, perhaps reflecting ancient lunar cycles and tides, yet these organisms need neither memory consolidation nor dreaming (or so we think). In higher animals, sleep disconnects the brain from the outer world, seriously endangering their lives**. The risk must therefore be offset by a very substantial benefit, the exact nature of which is still unclear. There is an agreement that sleep is needed for memory consolidation. Both potentiation and depression of synaptic transmission are induced during sleep, helping

---

* Specifically, resolvins E1 and D1.

** Dolphins sleep with half a brain awake, and one eye open.

retain the important bits of memory, and purge the rest*. Thus, it plays a role in synaptogenesis, synaptic pruning, neuronal myelination and neurogenesis. Sleep also seems to be the time for waste removal and housekeeping, thus having a restorative function. Sleep facilitates the synthesis of molecules that help repair and maintain the brain. Growth hormones, which drive the formation of synaptic contacts, are preferentially secreted during sleep, and that's also when the energy supplies, such as glycogen, are replenished. While the rest of the body can engage in maintenance during quiescent waking, brain restoration requires sleep. Indeed, "sleep is of the brain, by the brain and for the brain"[(1)].

As many aspects of sleep are still vague, naturally there are on-going disagreements and controversies in the field. One is a discovery of the glymphatic (glial cells + lymphatic) system, reported in 2013 by Maiken Nedergaard, - a network of channels in the brain which she had claimed served to eliminate toxins using cerebrospinal fluid (CSF). Purged toxins would include bits of Lewy bodies (A-betas, aSyn) as well as LPO debris, so the system could be considered a guardian sentinel and cleaner-in-chief. The Science magazine recognized this as one of the top 10 breakthroughs of 2013. Glymphatic system was reported to play the major role in restorative function of sleep, as metabolic waste products of neural activity were cleared out of the sleeping brain at a faster rate than during the awake state[2]. Similar claims of accelerated waste clearance during anaesthesia were also made (*ch.2.5.3*). Yet not everyone was on board with these claims, and disagreements are still simmering. Fast forward to 2024, and an article in Nature Neuroscience claimed exactly the opposite: fluorescent markers injected into the brain moved independent of sleep or awake states, while the clearance system was found to be actually running slower, not faster, during sleep and anesthesia[(3)]. Go figure! It seems both sides need to sleep on the issue a bit more to resolve it.

But let's zoom in on the possible role of lipid peroxidation in the mechanics of sleep. It has been known for a while that sleep deprivation increases oxidative damage to the brain[(4)]. Methylglyoxal was implicated (*ch.4.4*), but its provenance could be a bit duplicitous. While it does form from PUFAs during LPO, it could

---

* The question of what mechanism is responsible for making decisions on which memories to keep and which to dump might itself be related to PUFAs, particularly to DHA (*ch.2.5*).

also be a non-LPO product of both aerobic and anaerobic glycolysis (the former is employed in synaptic growth, while the latter is involved in synaptic transmission). Neurons often obtain energy by glycolysis (hence brain's cravings for sugar), which takes place in cytosol, not in mitochondria. But from the LPO clearance perspective, the exact source of methylglyoxal in neurons is not that important. Whatever the origin, mopping it up would take a toll on the resources normally spent neutralizing other LPO-derived reactive carbonyls, such as the glutathione-based machinery. Methylglyoxal would deplete these stocks, leaving fewer reserves available to defuse the other LPO products, thus exacerbating the LPO-associated damage.

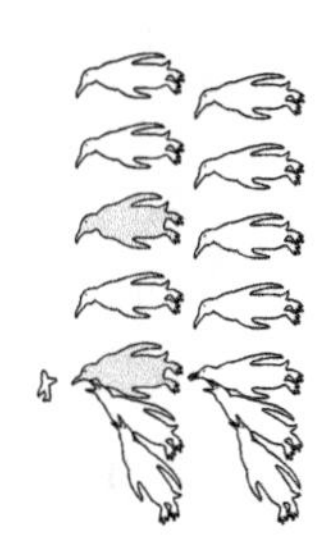

Neurons seem to never stop firing, generating spikes day in day out, but their workload changes at regular intervals. PUFA are mostly oxidized when the brain goes full bore through the day. In fruit flies, the caretakers of neurons called glia cells show a daily cycle of mitochondrial oxidation linked to an elevated LPO. The cycle depends on prior wake and involves an apolipoprotein E analogue, whom we shall meet in *ch.4.6*. This protein carries out neuron-glia lipid transfer, taking care to remove the LPO waste from neurons for further processing and disposal by glia. Compromising the transfer leads to accumulated LPO products and oxidative stress in neurons, but a full night's sleep is enough to remove this waste from glia. This mitochondrial oxidized PUFA removal cycle, operating between neurons and glia could be a fundamental, homeostasis-restoring function of sleep[(5)].

While honing his insect brain surgery skills, - human neurosurgery, in comparison, would seem like a piece of huge cake, - Gero Miesenböck, the father of a widely adopted optogenetics technique, now in charge of Centre for Neural Circuits and Behaviour at Oxford, must have gotten to know every one of approximately 123,456 Drosophila's neurons up close and personal, if not by name. Deep understanding of voltage-gated cationic channels in neuronal membranes helped him to make an astonishing observation. One of the subunits of the potassium channel known as Shaker, acted as a memory storage unit, operating through the most elegant mechanism. The NADPH-$NADP^+$ pair would play the role of tiny capacitors on a flash memory stick, and the memory storage would be switched on by reactive carbonyls such as 4-oxo-2-nonenal. This binary "one" would be kept until the membrane is depolarized*. The collective LPO memory of the subunits grows as LPO increases, accompanying the normal neuronal activities. Spike discharges can erase the memory – and these very spikes are driving sleep! This newly discovered LPO sensor mechanism at the core of sleep control suggests that sleep may be protecting, and restoring, neuronal membranes against oxidative damage. Indeed, the level of PUFAs in sleep-deprived brain is decreased, and slowing down the LPO clean-up process increases the demand for sleep[(6)]. Could similar mechanisms also be operational in the pain pathways (*ch.4.4*), perhaps explaining the analgesic, healing properties of sleep?

If even normal sleep seems to be somewhat controlled by the LPO status, then what about sleep disorders? Can LPO be elevated there? Indeed, it can. For example, malondialdehyde (MDA), one of the most popular LPO markers, is an indicator of obstructive pulmonary syndrome[(7)].

Circadian rhythms are directly related to sleep, as well as to a redox state of cells. Bipolar disorder is associated with both LPO sleep and circadian rhythm disturbance. MDA, as a proxy for LPO, was found to be elevated in the disorder. While circadian rhythms were not associated with LPO in the healthy controls, antioxidant enzymes such as SOD, peroxiredoxins, CAT and glutathione S-transferase (GST) have been shown to follow a circadian pattern of expression, lending further support to the LPO-disturbed sleep axis. Based on their observations, the authors firmly linked the LPO to bipolar disorder[(8)].

Other indications can be characterized by sleep pattern disturbances coinciding with elevated LPO, including various types of psychosis such as Schizophrenia, anxiety, depression, (chronic) stress, etc[(9)]. One obvious way to try and reduce LPO would be to employ antioxidants, right? Sadly, similar to their failure in pain management (*ch.4.4*), antioxidants do not seem to be terribly effective in improving sleep patterns or other symptoms. But this could be an antioxidant problem (see *ch.2.8*). Another option might be to simply replenish the damaged

---

* Making the machinery involved look very similar to the ALDH arsenal of anti-RCS weapons (see *ch.2.6.3*).

PUFAs with high quality fresh PUFAs. This strategy seems to work for AREDS supplements (*ch.4.1*). Preliminary data suggests that fish oil (*ch.3.2*) supplements may indeed be helpful in restoring healthy sleep patterns[10]. A similar conclusion comes from the study that looked at circulating levels of PUFA in people across a wide range of ages (35-96 years old). Participants with higher concentrations of very long-chain n-3 PUFAs were less likely to have long sleep duration[11]. Availability of neuronal membrane – building blocks should make it easier, and faster, for cells to use the Lands cycle to refresh their LPO-afflicted membranes (see *ch.5.8*). When left to their own devices, brain PUFA levels will tend to creep down with age[12], perhaps indicating a chronic systemic increase in brain LPO, which would also correlate with more disturbed sleep patterns in elderly people. It all seems to add up*...

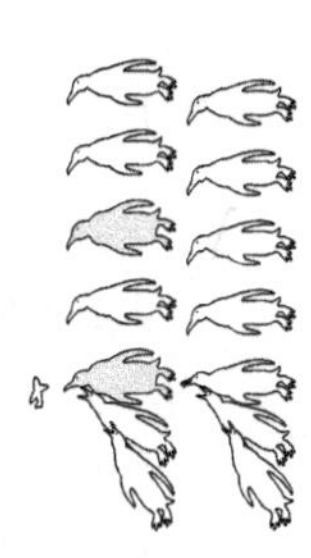

Inflammation is another sleep-affecting factor, and it may have two sides to it. On one hand, inflammation may send people into a secluded, semi-conscious, drowsy state of contact avoidance, which likely evolved to give one a chance to heal undisturbed, while minimizing the risk of passing an infection on to his mates. This state would encourage sleep. On the other hand, inflammation and immune dysfunction can disrupt sleep. Disturbed sleep is associated with elevated serum levels of pro-inflammatory cytokines interleukin-1 (IL-1), tumour necrosis factor α (TNF-α), IL-6 and IL-17, as well as oxidative stress markers MDA and 8-isoprostane, and an altered activity of immune cells such as macrophages, which all depend on LPO and enzymatic ARA processing (see *ch.4.9*). All these could lead to neuroinflammation, which would sustain the vicious inflammatory circle by activating microglia and astrocytes, thus leading to even more inflammation. The loop will circle around immune activation, neuroinflammation and LPO all affecting each other, disrupting sleep[13]. It sounds like inflammation

---

* Neurological and psychiatric diseases (see *ch.4.2*) are often accompanied by disturbed sleep patterns. As we have seen, these diseases, as well as some sleep disturbance-specific conditions such as narcolepsy, have LPO as a common denominator. Disturbed sleep and neurological diseases can thus be the two sides of the same LPO coin.

can both promote and discourage sleep – perhaps acute inflammation encourages sleep while chronic inflammation upsets it? The jury is still out to reconcile the divergence. But when the animals are deliberately deprived of sleep by keeping them awake 96% of the time, after 4 days of sleep denial their ARA-derived pro-inflammatory eicosanoids go through the roof. A particularly relevant prostaglandin, PGD2, crosses the blood brain barrier and induces systemic cytokine storm-like syndrome and neutrophil recruitment. Reducing the level of PGD2 in sleep-deprived animals (genetically or with drugs) led to a substantial reduction of inflammation[14].

A surprizing branching of the sleep-deprived brain damage topic relates to studies where people have failed to find serious oxidative stress related damage in the brain, and so looked elsewhere. Sure enough, sleep deprivation - related signs of inflammation-induced oxidative stress and injury, including lethal damage, were found in gut, lung and liver – rather than in the brain[15]! After all, perhaps all cells incorporating PUFAs into their membranes need a bit of spare time to remove the LPO-damaged molecules.

As the relationship between sleep and health continues to evolve, the glymphatic-system-of-waste-disposal disagreements may have no bearing on lipid peroxidation, as the role of LPO in sleep seems to have been carved in stone.

To end with an exhortation, centenarians in blue zones often have siesta naps and clock 7-9 hours of sleep each night. The main take-home message: do sleep more than 7 hours every day! We all need our 40 winks every night, - if only to neutralize the 40 most pernicious LPO products…

4.6. A puffa of puffed-up, oxidized PUFAs, and cardiovascular disease

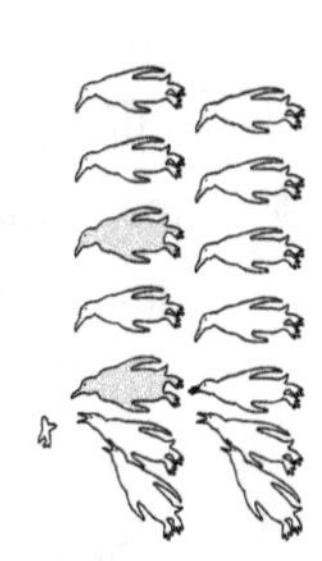

"The problem with heart disease is that the firstsymptom is often fatal."
— Michael Phelps

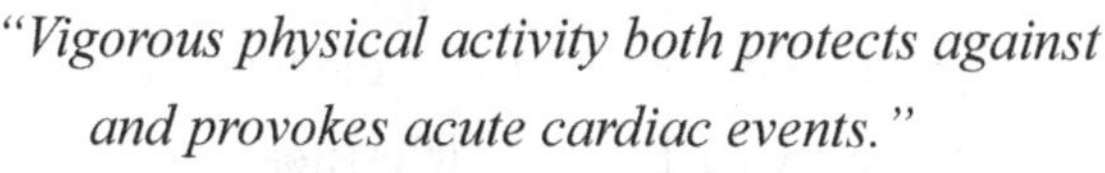

"Vigorous physical activity both protects against
and provokes acute cardiac events."

— Paul D. Thompson

Cardiovascular disease (CVD) is the number one killer responsible for about one-third of all global deaths, and will probably stay that way for a while, despite the recent success of lipid-lowering drugs. Currently in the US, 85% of individuals over age 50 have atherosclerosis. Autopsies of soldiers fallen in 1950-1953, average age 22, revealed that 77% had incipient CVD. Hundreds of thousands of research papers have been published, looking at every imaginable facet of the disease mechanism. And it all seems to converge on lipids: CVD develops when the fat and cholesterol transport systems start misbehaving themselves*.

To understand these body-wide lipid transport systems, think mayonnaise. Nature has sorted out carrying insoluble lipids in blood by emulsifying the fats. Hydrophobic TGs and cholesterol esters ("the olive oil") are wrapped around by amphiphilic PLs, free cholesterol and various apolipoproteins ("the yolk, vinegar and mustard"). Small

---

* As a lyrical aside, animals that retain the capacity to make ascorbic acid rarely develop CVD. Humans and apes cannot make their own vit C and so do have atherosclerosis. Vit C is required for building blood vessels, so Linus Pauling suggested that CVD is the result of protective systems going into overdrive trying to compensate for low ascorbate, a co-factor in collagen and elastin cross-linking. Pauling was famously taking 17 g of ascorbic acid per day, lived to 93, and did not die of CVD, so might have had a point. But watch out, as Vit C doses exceeding 2 g may give other problems.

intestine and liver are "the blender". Density defines destiny, as the fate of the resulting lipoprotein particles is linked to their density (a nod to classical separation techniques). In this most vital emulsion of our lives, chylomicrons are the least dense, followed by Very Low-Density Lipoprotein (VLDL), Low Density Lipoprotein (LDL) and the densest of them all, High-Density Lipoprotein (HDL). 50% of LDL is cholesterol, while VLDL and HDL have about 20%. This means that LDL can share its cholesterol with the surroundings, while HDL tends to mop it up and send it back to liver. Liver relies on this equilibrium to circulate cholesterol through the bloodstream (it holds on to 20% of the total body stock), at about 1 gram per day. Cholesterol-carrying LDL particles float around for days, while chylomicrons and VLDL, the lipid fuel carriers, transferring hundreds of grams of TGs around the body, are short-lived (an hour for chylomicrons, following food intake). Importantly, all types of particles are loaded with PUFAs. For LDL and HDL, size matters. Genetic variations in cholesterol ester transfer protein (CETP), responsible for filling the particles up with cholesterol, are key. Centenarians often have a CETP variant that makes larger sized particles[(1)]. The reverse is also true, as smaller sized LDL particles are predictive of atherosclerotic diseases[(2)]. This makes sense: the smaller the cholesterol-loaded LDL particles are, the faster they can latch on to, or slither through blood vessel epithelial lining, mingling with the inner elastic muscle layer and the structure-supporting cell layer outside. And the easier it would be for macrophages and other cells to gobble them up. As ever, though, things are not that black and white. CETP inhibitors failed in the clinical trials, so perhaps there is a less straightforward connection between the CVD risk and the LDL size.

The first person to confirm the role of cholesterol in CVD was Nikolai Anichkov (1885-1964) from Saint Petersburgh, who followed up on the 1908 Alexander Ignatowski rabbit experiments by proving, in 1913, that cholesterol was the compound causing atherosclerotic changes in the vascular wall[(3)]. For his experiments, Anichkov was dissolving cholesterol in "vegetable" oils, a point we will consider separately. But cholesterol is an essential membrane component, so how could it be bad?

The "response to retention" theory of atherosclerosis, which evolved from the original "response to injury" concept, posits that the first pathological step is the

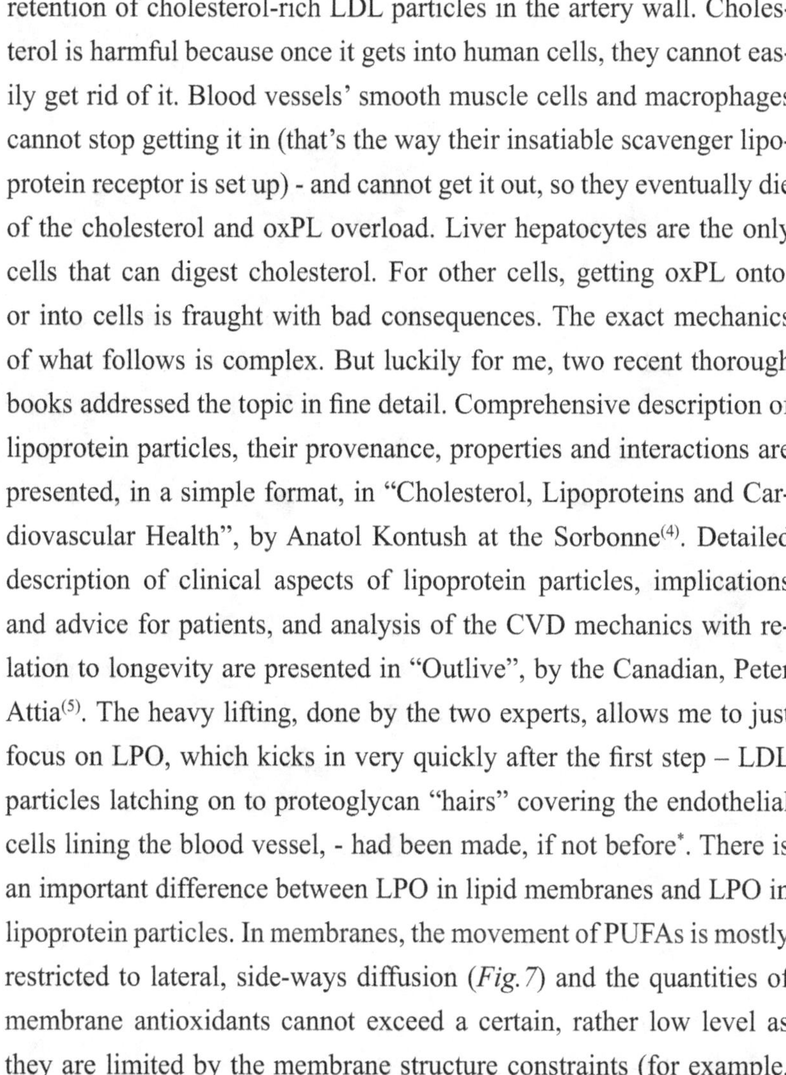

retention of cholesterol-rich LDL particles in the artery wall. Cholesterol is harmful because once it gets into human cells, they cannot easily get rid of it. Blood vessels' smooth muscle cells and macrophages cannot stop getting it in (that's the way their insatiable scavenger lipoprotein receptor is set up) - and cannot get it out, so they eventually die of the cholesterol and oxPL overload. Liver hepatocytes are the only cells that can digest cholesterol. For other cells, getting oxPL onto, or into cells is fraught with bad consequences. The exact mechanics of what follows is complex. But luckily for me, two recent thorough books addressed the topic in fine detail. Comprehensive description of lipoprotein particles, their provenance, properties and interactions are presented, in a simple format, in "Cholesterol, Lipoproteins and Cardiovascular Health", by Anatol Kontush at the Sorbonne[(4)]. Detailed description of clinical aspects of lipoprotein particles, implications and advice for patients, and analysis of the CVD mechanics with relation to longevity are presented in "Outlive", by the Canadian, Peter Attia[(5)]. The heavy lifting, done by the two experts, allows me to just focus on LPO, which kicks in very quickly after the first step – LDL particles latching on to proteoglycan "hairs" covering the endothelial cells lining the blood vessel, - had been made, if not before*. There is an important difference between LPO in lipid membranes and LPO in lipoprotein particles. In membranes, the movement of PUFAs is mostly restricted to lateral, side-ways diffusion (*Fig.7*) and the quantities of membrane antioxidants cannot exceed a certain, rather low level as they are limited by the membrane structure constraints (for example, only about one vit E molecule per 2000 PUFAs, see *ch.2.8*). ROS species are often generated in the membranes, because that's where molecular oxygen, ROS-generating proteins, metals** and redox molecules

---

* Apolipoprotein in HDL particles is the non-adhesive APOA type, which does not latch on to the blood vessel cells, coming in or out as it pleases. APOB, the main protein in LDL, is much stickier. It has plenty of opportunity for its mischief, for the vascular endothelium consists of over 10^{13} cells (and weighs 1 kg).

** Analogous to the Asley Bush metals in AD conjecture, in 1996 Jereside (*ch.2.6.1*). In non-bilayer assemblies such as lipoproteins, lipid droplets, adipose tissue or bottles of commercial fish oil (*ch.3.2*) PUFA-containing TGs and PLs are like a liquid in a barrel, with any ratios of antioxidants to PUFAs theoretically possible. To illustrate, a 25 nm (1/3000 of a hair width) LDL particle contains[6], depending on diet, time of day, lifestyle, genetics, chance, and the Zodiac signs, in molecules: LIN – 1100; ARA- 150; DHA-30 (out of the total of about 2600 FA esters, so about 50% (1280) are PUFA). There are also 3 to 14 molecules of vit E per particle or 1 vit E molecule per 90-430 PUFAs, and even less CoQ.

Our blood is very high in antioxidants, preserving PUFA-rich, LPO-sensitive lipoprotein particles. The caveat here is an about-face that one of the major LPO-terminating antioxidants, vit E, may do in the LDL environment, turning itself into a chain-propagating pro-oxidant (*ch.2.8*), because PUFAs in LDL are not structured as membrane bilayers. As LDL particles are long-living, there is plenty of time for the LPO to develop. Other factors play a role, too. Feeding fish oil to rats depleted their antioxidants and increased the level of LPO products[7]. As discussed in *ch.2.7*, DHA in membranes may not oxidize fast, but LDL environment is a different matter. Besides, pre-oxidized PUFAs may have been elevated in the fish oil used for feeding (*ch.3.2*), making the results of this experiment very similar to the rabbit study, where rabbits consuming pre-oxidized, rancid fats develop atherosclerosis faster[8]. While compared to rabbits humans may have a more robust barrier function in their gut uptake system, capable of denying the entry to the LPO products (perhaps explaining[9] the short lifespan of the gut epithelial cells, constantly bombarded and poisoned by the LPO products from food), the rabbit data confirms the role of toxic LPO products in atherosclerosis.

Rodents process their lipoproteins differently and are insensitive to high cholesterol so can only model a limited number of atherosclerosis features. Rabbits are closer to humans in this regard, but there are still major differences. For example, they can absorb food cholesterol directly into their blood vessels. Humans are, really, the only good model of human CVD. Urine isoprostane biomarkers,

---

rome Sullivan proposed the "Iron Hypothesis", suggesting that elevated levels of stored iron are associated with CVD. This explained the lower incidence of heart disease in pre-menopaused women.

the non-enzymatic products of n-6 LPO, are elevated two-fold in coronary heart disease patients compared to healthy controls[(10)], but is this the cause of CVD? It likely is, for ALDH2*2 phenotype, which has a reduced capacity to detoxify LPO products (*ch.2.6.3*), has been associated with higher risks of atherosclerosis[(11)].

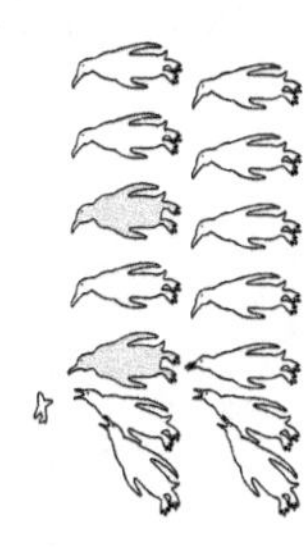

Box 7. Our body is pre-programmed to know that LPO is bad. Leave a "vegetable" sunflower oil bottle open in the sun for a couple of days, and apart from stench, you would notice an increase in volume. Multiple oxygen molecules react with PUFAs, increasing the mass. Around 1800, Nicolas-Theodore de Saussure used a primitive manometer to show that over months, the volume of walnut oil on exposure to air would take up 145 volumes of air. The increase in volume is one of the main factors driving the foam cell formation because, as mentioned before, thick means sick. But what about the "stench" part? Unpleasant smell perception is the way the body has of telling us to avoid things, which are indeed toxic when ingested, as described above. But when the undesired oxidized molecules form inside of the body, there is a defensive network at the ready, waiting to apprehend and neutralize the toxins. A broad range of different oxidation specific epitopes[(12)] (parts of the target molecules, or antigens, to which an antibody attaches itself) act as damage-associated molecular patterns (DAMP) and are scavenged by different pattern recognition receptors. These neutralize not just the atherosclerosis-associated oxPL species but also numerous other objects damaged by LPO-derived fragments. Proteins modified with LPO-derived lipid carbonyls, such as HNE and HHE, and lipid whiskers[(13)] (see *ch.2.9*) all belong to this category. Numerous scavenger receptors, such as CD36, recognize the oxidation patterns and latch on. While mice mostly use the hard-programmed innate immune system to arrest and remove the various types and forms of the LPO products, in humans both the innate and adaptive systems are employed. Warning us not to touch the rancid stuff, and then deploying such a large toolkit trying to remove various pro-inflammatory, pro-apoptotic, pro-senescent, and just outright toxic LPO products, nature is obviously programmed to recognize the danger of LPO. In addition, it constantly runs the PL reformatting using the Lands cycle (which in the eye takes the shape of the 10-day disc recycling, see *ch.4.1*), and generally spares no expense trying to weed out the toxic lipid derivatives. Sadly, it is an uphill battle.

These titbits, and multiple other evidence, firmly implicate the chain reaction in the CVD pathology (*Box 7*). But before we zoom in on the LPO chemistry, let me introduce the grisliest lipoprotein family member. Meet Lp(a) (reads "lipoprotein little "a""), a stickier version of LDL, heavily involved in atherogenesis and thrombosis. This fella goes around picking up oxidized phospholipids. When a PL molecule has its PUFA group damaged by the LPO, a bit of that fatty acid may fall off in a form of reactive carbonyl, such as HNE or HHE (*Fig.9*). The remaining truncated oxygen-containing stubs, still attached to their PL heads, are very pro-inflammatory. And as always, LPO begets more LPO. By accumulating those species, Lp(a) becomes the major bait for inflammatory cells, luring them to blood vessel walls. It also encourages smooth muscle cells to proliferate. The lifespan of Lp(a) is comparable to the lifespan of LDL, giving it enough time to wreak havoc. The less Lp(a) we have (rodents and rabbits do not have it at all), the better off we are. So hungry and efficient is Lp(a) in gobbling up the oxPL species, that measurements of Lp(a) to assess the degree of LPO or inflammation, is essentially an equivalent to measuring the oxPL levels. Liver is working around the clock trying to patrol the blood flow, attempting to clear oxLDL from circulation. Sadly, the bad guys prevail.

Once LDL and Lp(a) particles, with both pre-oxidized and intact PUFAs, have penetrated the vessel wall, they get stuck, - and modified. The immune system recognizes the intruders, and sends its foot soldiers, the monocytes, to clean it up. On entering the vessel wall, monocytes differentiate into macrophages. Macrophages engulf the LDL enemy by phagocytosis, employing multiple receptors: one for normal LDL, and several scavenger receptors for variously LPO-afflicted particles (OxLDL). For example, the CD36 receptor can recognise oxPLs bearing LPO-derived carbonyl groups. Once inside, the particles coalesce into lipid droplets, which macrophages try to destroy by hitting them with lipoxygenases or various ROS, such as H_2O_2, superoxide and bleach (hypochlorite). Accordingly, elevated levels of hypochlorite-producing myeloperoxidase and an inflammation marker CRP (*ch.2.9*) are associated with increased risk of CVD. However, this strategy backfires with the PUFA-rich LDL and Lp(a) particles, because looking from the LPO perspective, the chain reaction could not have dreamed of a better mistreatment! The ensuing LPO lights up and zips through the bulk of PUFAs. Multiple oxygen molecules form covalent bonds with lipids, increasing its mass and volume.

The macrophages cannot digest cholesterol, and its excretion, for a mop-up by HDL, is less efficient than deposition. Nor can they do anything further with the oxidized PUFAs. Increased volume of oxidized LDL material bloats macrophages up, turning them into foam cells. This happens inside of the vessel walls, resulting in constriction, - making the vessel narrower. Macrophages do a lot of pro-inflammatory signalling, and then die, with debris often consumed by new macrophages. Vessel cells can also take up LDL particles. Vessel cells death, and/or attempts to patch up the lesions in the vessel walls, may result in deposition of calcium salts, - exacerbating the damage. Even though some of that calcium can be removed by a vit D- vit K tandem, the deposition is still more efficient that the mop-up. Iron and copper have been detected in the lesions and can be released when various cells die, further promoting the oxidation of PUFAs. Indeed, iron removal by chelation decreased LPO and delayed atherosclerosis in rabbits[(14)]. Too bad tocopherol can be a pro-oxidant in the LDL environment*.

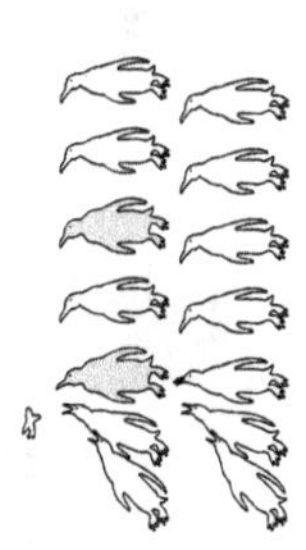

All these processes produce numerous detectable LPO products, such as urine isoprostanes mentioned above. The whole gamut of LPO fragments can be found in lesions themselves, including PUFA hydroperoxides and hydroxides, isoprostanes, various reactive carbonyls such as HNE, HHE, MDA and acrolein. The carbonyls can often be detected as conjugates with proteins[(16)], and as mutagenic DNA bases (*Fig.19*). APOB of LDL particles readily reacts with carbonyls, and the modified apolipoprotein has a different receptor activity, encouraging macrophages to gobble LDL particles up uncontrollably. Conjugates of carbonyls with various proteins are highly proinflammatory and damaging to endothelial cells, compromising their barrier function, and

---

* Genetic abnormalities in haemoglobin structure, such as sickle-cell anaemia and beta-thalassemia, may result in elevated blood iron levels. Abundance of PUFA-rich lipoprotein particles and iron overload make the LPO processes described above even faster, contributing to CVD complications. Beta-thalassemia patients had substantially increased levels of oxidized and truncated LIN and ARA species as well as reduced vit E levels compared to healthy controls, and the use of iron-chelating agents was found to correlate with decreased LPO levels, further implicating LPO in the CVD pathology[(15)].

leading to adhesion and activation of neutrophils.

Irrespective of whether the LPO events are the primary driver of atherosclerosis, or a major secondary pathway, the signs of LPO in CVD are ubiquitous. But the role of PUFA oxidation does not end here. Damaged vessel walls, stressed macrophages, LDL and Lp(a) particles with oxidized PUFAs and plentiful LPO products, sloshing about free or as conjugates, all combine forces to send a powerful pro-inflammatory signal, calling for help, including more macrophages. Then the next big trouble comes. A huge inflammatory conflagration ensues, with big chunks of blood vessels and the surroundings bursting into flames. Enzymatic PUFA oxidation plays a very important upstream role in inflammation (see *ch.4.9*). The above processes contribute to growth of plaques, constricting the blood flow. Lesions can rupture, sending debris into the bloodstream. The clots block the blood flow. This thrombosis result is heart attacks, strokes, pulmonary embolism, and various other problems.

So vast is the amount of data produced by the atherosclerosis research, that it is easy to not see the forest for the trees, as the above brief non-scientific description of but a small number of relevant mechanisms attests. So how to reconcile inflammation, PUFA oxidation and endothelial problems? Let's use a non-biological resource: linguistics. A colloquial name for atherosclerosis, "hardening of the arteries", brings oil paintings to mind (*ch.1.2*). Incessant oxidative damage, fat deposits, cross-linking* and immune erosion turn our wonderfully elastic and flexible circulatory system into linoleum...

---

* Cross-linking of proteins, particularly of the lifespan-lasting, non-renewable structural ones like collagens and elastins that make up key parts of our bodies, - blood vessels, cartilage, bones, skin, tendons and ligaments, - is one of the hallmarks of aging. Various processes, like lysyl oxidase – driven formation of lysine dimers and Maillard chemistry, are involved in the cross-linking, and LPO – derived RCS like MDA and HNE are one of the major driving forces. The process makes these elastic structures more rigid as we age, the reason, according to Steven Austad, why we do not see many 80-year-old gymnasts. In conditions such as diabetes, the collagen stiffening is accelerated due to elevated levels of LPO and MDA.

4.6.1. "Vegetable" oil vengeance

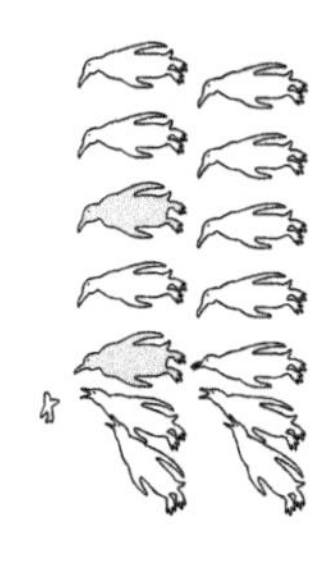

"It's a bit of a myth that too much cholesterol causes heart attacks."
— Steven Gundry

"A feeling of emulsion swept over me"
— SJ Perelman

Enrico Fermi and Andrei Sakharov were both known for accepting a physical value, or a number, only upon performing their own quick *back-of-the-envelope* calculation on scraps of paper to see if it made sense, for their own piece of mind so to speak. I am sure many physicists are like that. A friend of mine is like that too, but he is a biologist. He therefore cannot get away with ballpark estimates and has to resort to some basic experiments instead. He once put himself on a high cholesterol diet, by breakfasting, lunching and dining on eggs, bacon and cheese, while lamenting the fact that he could not get any cholesterol supplements from his local pharmacy. A month later, he switched to a cholesterol-free vegetarian diet for a month. As the two blood cholesterol readings were essentially identical at 6 mmol/L, he grew a bit suspicious of the test reliability. Help came unexpectedly. For personal reasons, he had a stressful month, immediately following his dietary self-testing. He lost appetite, sleep, and weight. Then things came back to normal. Blood test, done in the same clinic, revealed lowering of his cholesterol to 5 mmol/L, validating the first two data points. Whether reducing cholesterol this way is a healthy thing is a different matter.

Combined efforts of Conan Doyle, Agatha Christie and George Simenon might be required to untangle the plots permeating the world of "vegetable" oils and its influence over the current dietary guidance. Some of that intrigue is mentioned in (*ch.2.5.2*). The accepted dietary advice is based on the following premises:

1. Cholesterol is bad and causes CDV.

2. Saturated fats exacerbate this and should be eliminated.

3. "Vegetable" oils reduce cholesterol and can thus prevent CVD, so are highly recommended.

Interestingly, all three statements are misleading, and dousing foods and animal feed with toxic LIN-containing seed oils, aggressively lobbied by large commercial interest groups, is to a large degree responsible for a plethora of modern diseases such as obesity, diabetes, metabolic and inflammatory diseases. These and many other consequences of the population-wide addiction to "vegetable" oils reduce lifespans. Not only does seed oil consumption exacerbate the CVD incidence and outcome (while reducing cholesterol – yet another argument in favour of PUFAs and LPO in the CVD pathology). It ruins the omega balance(17). A rare example of a fat-based FDA-approved drug is Vascepa, an ethyl ester of EPA, which targets some CVD-associated features such as elevated TGs. Tentative mechanisms of action proposed all seem to converge on one feature of supplemental EPA. It restores the omega-balance(18), and the benefits follow quickly. More specifically, it may be competing against ARA for COX and LOX enzymes, resulting in decreased output of ARA-based pro-inflammatory eicosanoids (see *ch.4.9*).

Seed oil – supplied LIN also causes abnormal levels of lipids in blood (dyslipidemia, a major risk factor for CVD), and badly distorts the mechanics of ARA signalling, resulting in chronic inflammation. To conclude,

1. Cholesterol is a vital building block, not an unwanted by-product. Keeping cholesterol low, or reducing the LDL to below 6, by drugs or diet, doubles the risk of dying over the next 8 years(19).

2. There is no evidence that saturated fat is associated with mortality from cardiovascular disease(20). In fact, the reverse may be true(21).

3. Vegetable oils lower cholesterol by disrupting the proper distribution of fats and cholesterol, and assist in building inflammatory atherosclerotic deposits in our arteries, thus accelerating both the plaque formation and systemic and local inflammation(18,22,23).

So, maintain your omega-balance, and watch out for LPO. Which is easier said than done, - and heaven forfend you try to do it by increasing your vit E intake…

4.7. Skin in the game

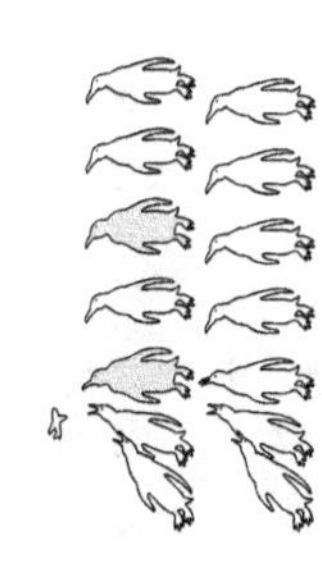

«Здоровье краше всех румян.»
— Александр Пушкин

"All the carnal beauty of my wife is but skin-deep."
— Thomas Overbury

Icebergs and glaciers do not lend themselves well to photo sessions. Pictures fail to capture what is clear only when directly looking at the real thing – the floes emanate this unearthly, alien-blue light from within. How could that be conveyed? Perhaps that is what the Snow Queen's skin was like, mesmerizing poor Kai. The situation is even more complicated for warm-blooded humans. The soft rosy glow coming off the healthy radiant youthful flesh is now accompanied by a myriad of balmy hues and cues, so fully appreciating it would require all five of our senses, if not more. I initially considered having a go at describing the peachy texture here but had to give up. According to Willem de Kooning, "Flesh is the reason oil paint was invented". By and large, the world's art and poetry, when depicting beauty, actually portray human skin, for beauty is indeed only skin deep. So much was accomplished over centuries, on paper and canvas, by such great artists, in so many languages and styles, that my paltry contribution, in crooked Pidgin English, would not make a dent. Besides, there may simply be no need to portray healthy young and blooming skin, nor withered aged skin, for, equipped with our senses, to bastardize Justice Stewart's comment, "We know it when we see it" (*Fig.3*).

Let's zoom in, again only skin-deep, on some bio-technical specifics. Any organ of our body deserves to be called "unique", for they all are. With any body part malfunctioning, life may turn into misery. Still, in the inventory of tissues or organs, skin stands out. Surrounded by hundreds of millions of cubic kilometres of oxygen, and half the time exposed to light, this largest organ in the body forms a shield which

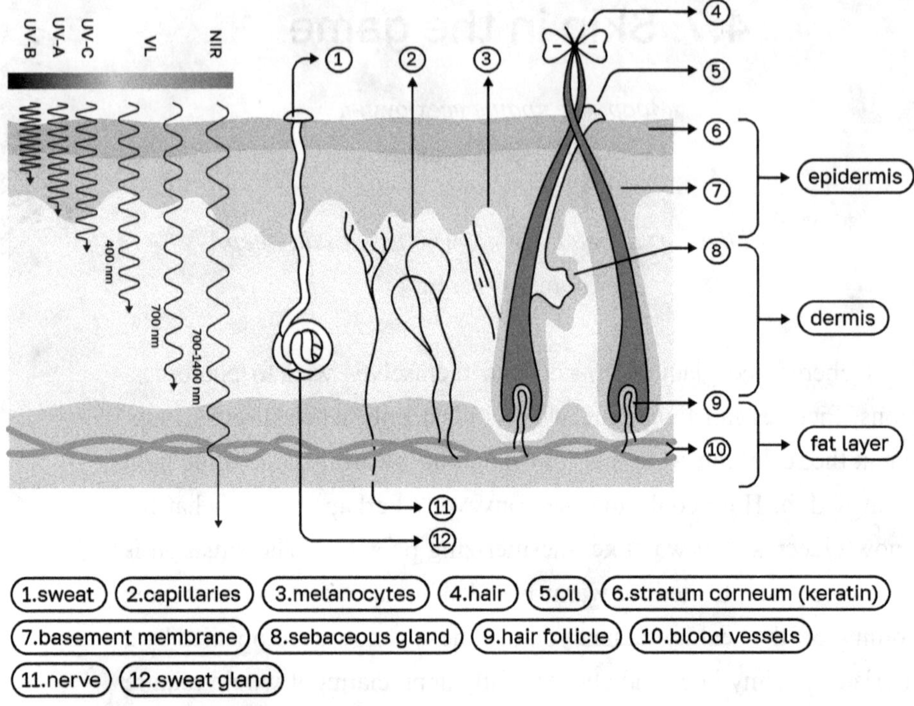

Figure 24. Skin, our largest organ, is rich in PUFAs, often exposed to light, and always to oxygen, so that LPO is all but inevitable. Different wavelengths reach different depths, but it is fair to say that there is no safe wavelength.

protects, senses and regulates. And many of those functions involve PUFAs*.

Two major layers of mammalian skin are the outermost epidermis (from Greek, *epi*, above) and the underlying dermis. The external part of the epidermis – stratum corneum, - consists of lipids and keratin, a protein also found in hair, nails and various body surfaces (*Fig.24*). The most predominant lipid in the striatum corneum is ceramide (50%, *see Fig.13*), followed by cholesterol derivatives (35-40%). LPO-vulnerable PUFAs, in the form of free acids, make up about 10%[(1)].

* Limited space does not permit to look in detail at a distant relative of skin, the mucous tissue. In terms of surface area, the mucous membranes area (400 sq m) far surpasses the skin (2 sq m^2). This slippery hydrogel contains carbohydrates, proteins and lipids which vary in composition depending on which organ or tissue it lines, trapping irritants and invaders. Mouth, sinuses, eyes, stomach, lungs and respiratory tract, intestines and reproductive system are all covered by mucous membranes. LIN and ARA are the key PUFAs there, making the mucal composition similar to that of skin.

Corneocytes, the flattened, stratum corneum-forming cells, filled with keratin and lacking nuclei and organelles, are dead. Embedded in a lipid matrix, these cells are shed every 2 weeks and are replaced by new ones from the lower epidermis. Epidermis has a barrier function that keeps the body hydrated. It contains skin colour-determining melanin, small nerve fibers, but no blood vessels. Below the epidermis, there is dermis. Cells at the interface of the epidermis and the dermis are keratinocytes. They actively divide and move up into epidermis. Dermis contains connective tissue, blood vessels, sweat (eccrine or apocrine) and oil (sebaceous) glands, hair follicles and other structures. It sits on an adipose fat layer called hypodermis.

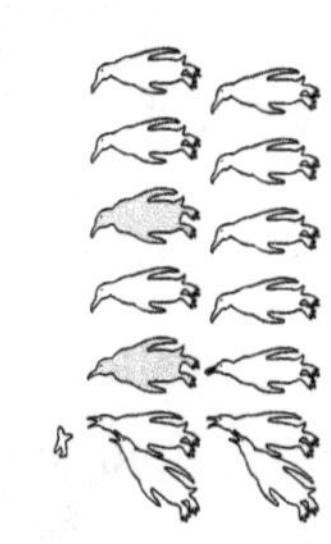

PUFAs make skin rather vulnerable to the onslaught of oxygen. Across the epidermis, the predominant fatty acid species is LIN (22%, the most abundant fatty acid in skin), with OLE the second most abundant (at 15%), followed by ARA (6%) and DGLA (1%). This corresponds to the skin PUFA composition of about 75% LIN and 21% ARA. The presence of tangible quantities of DHGL, an enzymatic conversion product of LIN which can then be further converted into ARA, suggests that the conversion of linoleic into ARA in the epidermis occurs locally.

4.7.1. Barrier function and rusty shield

"Skin care must be good enough to eat!"
— Joanna Runciman

LIN is important for skin's barrier function. Animals deficient in LIN "dry out" to death, despite spending most of their time drinking, as their skin turns into a "strainer". The problem is easily fixed, and animals demonstrate a rapid return of the barrier function when given LIN in their diet, or just by topical delivery of LIN to the skin[(2)]. Eating sunflower oil, or smearing it on the skin, is all it takes. No wonder LIN is the major component of most skin creams and lotions. But how does it work?

Skin cells are like "bricks" sealed together by a matrix "mortar". Corneocytes are enveloped into glutamic acid-rich protein (mostly involucrin) layer, called "cornified cell envelope". Glutamic acid forms ester bonds with ceramides, lipid units containing very long chain (28-36 carbons) fatty acid (VLFA) residues (*Fig.13* and *Fig.22*), which, unusually, have hydroxyl groups at the terminal (opposite to the carboxy) ends of their acyl chains. The mechanism is not fully understood, but LIN is instrumental for this esterification, which works through a unique mechanism mediated by lipoxygenases(3). The process starts when LIN forms an ester bond with the free terminal hydroxyl group of a ceramide-bound VLFA residue. LIN is then oxidized enzymatically, by one, or sometimes by two, different LOX enzymes, 12R-LOX and eLOX3. The preferred substrate for both is free ARA, but in this case, they oxidize LIN while it is still attached to a ceramide. The resulting hydroperoxy (from 12R-LOX), and epoxy (from a follow-up action by eLOX3) LIN derivatives are then disconnected from the ceramide by an unknown extracellular lipase. The reason for this enzymatic oxidation is unclear, but somehow it helps the terminal hydroxyl group of VLFA hydroxyl ceramide derivative to form a new ester bond with Glu, creating the protective intercellular lipid filling. Freed oxidized LIN derivatives, after additional transformations, become reactive towards amino groups of the cell envelope, forming Schiff bases and other adducts with the envelope proteins. A fault at any of the multiple steps involved would compromise the barrier function. Some of the possible mutations in the genes involved are lethal, while a few result in various types of ichthyoses. Moreover, as we already saw, peroxy-PUFAs (LOOH) on contact with transition metals can initiate LPO, provided there is more PUFAs close by. The structure of this extracellular lipid-rich matrix "mortar" allows for free fatty acids, as well as cholesterol, to intermingle with it. In the epidermis, LIN, ARA and DGLA make up about 25% of all fatty acids. In pathological conditions such as atopic dermatitis, free PUFAs are elevated, implicating LPO in the pathology.

There are essentially no phospholipids around corneocytes, but plenty in the cells below. LIN is not just there to help build the ceramide matrix. Skin cells rather recklessly have plenty of LIN, as well as some ARA in their PLs. Skin brings several incompatible things close to each other: PUFAs, transition metals and oxygen make awkward bedfellows. Even in the dark, they struggle to mind their own business. But turning the lights on, UV or blue, really portends trouble. Penetrating right into dermis, this light activates Type I and Type II photooxidation processes which can accelerate the LPO in cell membranes as well as PUFA oxidation in the matrix, as will be discussed in the next section. Even OLE may succumb to these oxidation conditions[4].

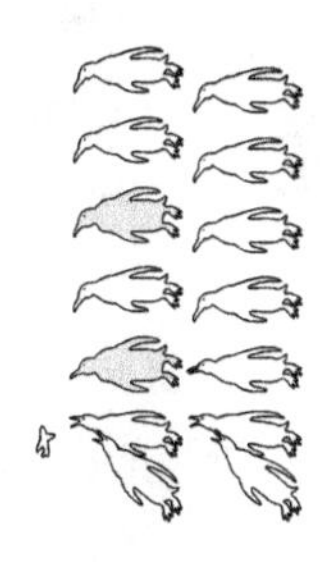

While the inner organs are exposed to oxygen levels which are orders of magnitude below the atmospheric (*Back-of-the-envelope-1*), our skin is facing a constant 21% oxygen onslaught, day in day out, dressed or naked. And with so much molecular oxygen sloshing about, its insidious relatives ozone and singlet oxygen will sure turn up as well, if in smaller numbers, ready for mischief. The wicked family of party animals runs its subversive action 24/7, always on a lookout for target molecules to oxidize. It has been known since the 1850s that human epidermis takes up atmospheric oxygen. While the contribution to the total respiratory oxygen intake is negligible, it was recently estimated that under certain conditions up to 0.4 mm of the upper skin layers are almost exclusively supplied by external oxygen, with only a minor input from the oxygen transport in the blood[5]. On average, 1-2% of inhaled oxygen is converted into ROS, mostly due to mitochondrial activity, but in the skin, up to 5% of all oxygen turns into ROS[6].

It is rarely recognized[7] that there are about as many sweat glands in the skin as there are nephrons (two million), the structural and functional units of human kidneys. The latter is considered the main filtering and waste removal system of the body, but the skin is not too far behind. And unlike our kidneys, safely tucked behind various protective moats, mounds and baileys, the skin is right at the forefront. When the sweat flows profusely, apocrine glands, the main type of sweat gland, can pump out up to 4 litres of sweat per hour. Sweat contains, among other things, lipids, PUFA oxidation products and our good old friends transition metals. Keratinocytes and skin fibroblasts are certainly "aware" of the looming danger, and stock up on antioxidant defences. Millimolar levels of glutathione, the familiar team of enzyme antioxidants (SODs, GPx, catalase, glyoxalase, peroxiredoxins) and small molecule antioxidants (vit E, ascorbate, melanin, carotenoids), and other repair systems are all kept on a high alert by the skin cells. Deficiency in CuZnSOD leads to accelerated skin aging in mice. As a result, keratinocytes are more resilient than most animal cells to H_2O_2 or organic peroxides. But sadly, the input from the rogue actors can overwhelm even the reinforced defences.

Many tissues have trouble disposing of the garbage material, which accumulates over time, forming scummy lipofuscin, or "aging pigment" granules inside the cells. This garbage can end up on the outside when the cells die. Various chemical reactions are involved, but LPO-derived reactive carbonyls (*Fig.9*) are one of the important contributors, cross-linking various types of compounds through a 1,4-Michael addition mechanism with suitable functional groups (amines, thiols, imidazole rings of histidine, etc). Many biomolecules get fused together, including proteins, nucleic acids, lipids and carbohydrates. The adducts have "unusual" bonds which the cellular garbage incineration factories, - lysosomes and peroxisomes, - have trouble digesting. Particularly if the materials are formed outside of cells, or end up in the extracellular space when their host cells die and lyse. A gradual build-up of this "wear and tear" pigmented junk affects many tissues, including skin, liver, kidney, heart, retina, and neurons, with post-mitotic cells particularly vulnerable. Lipids and LPO products constitute up to 50% of lipofuscin, which also contains cross-linked proteins and transition metals. This means that lipofuscin is capable of catalyzing even more damage in a vicious cycle. Once formed, these ever-accumulating products are hardly degradable, and can trigger senescence. The common denominator is always the abundance of LPO products, but the exact composition of such lipofuscin deposits depends on its tissue of origin.

Lipofuscin in muscle cells differs from the "age spots" in the skin, which contain, in addition to the protein-LPO mesh, melanin derivatives and blood-derived compounds. Both muscle and skin pigment deposits would be different from eye-related lipofuscin, which contains unique ingredients derived from vitamin A, such as A2E (*ch.4.1*). Skin "wears its age on itself", by putting its age spot blemishes on display. To generalize, Lewy body diseases, neurological conditions involving abnormal plaque depositions, including Alzheimer's (A-beta and Tau), Parkinson's (aSyn), lysosomal storage diseases (lipofuscinosis) and many more, would have their own unique types of lipofuscin-like deposits. Invariably, though, LPO products are the common denominator. Gradual accumulation* of waste products with age is so pervasive and features in so many pathologies, including aging, that it is sometimes called a "garbage catastrophe"[(8)]. Accumulation of such indigestible molecular garbage is not unique to humans. For instance, measurement of lipofuscin deposits in eyestalks of crustaceans is used as an index of their age.

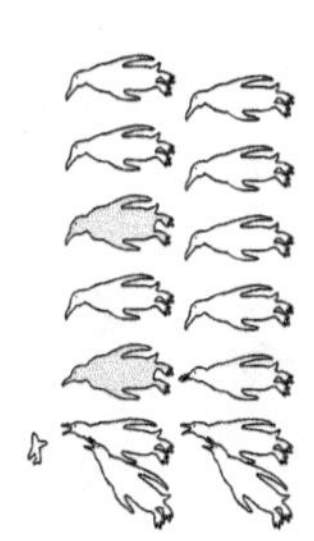

An important part of lipid-related skin physiology involves sebaceous glands (*Fig.24*). On average, adult skin has about 5 million hair follicles, and each follicle has a sebaceous gland outlet. Other types of sebaceous glands are not connected to a hair follicle. For example, Meibomian glands are involved in tear formation (*ch.4.1*). Found on all areas of the skin except palms and soles, sebaceous glands secrete skin- and hair-lubricating sebum (from Latin, tallow). Their action kicks in around puberty, and when overactive, may result in blocked sebum ducts, leading to acne. Functions of sebum include emulsification of sweat so that it is not quickly lost in drops, which delays dehydration. In rainy conditions more lipids are exuded, helping repel rain droplets. Sebum's main components are triglycerides, wax, squalene and vit E. Vit E and squalene absorb some UV light and even ozone, which destroys them. However, sebum is continuously replaced. Sebum is much leaner in PUFAs, presumably because it is more directly exposed to the elements than the skin. PUFA content in sebum is substantially different form the epidermis, with only 1% LIN, and other PUFAs combined making up another 1%. PUFAs are actively degraded by the sebaceous

---

* Mitochondria are a major source of ROS but vital for tissue performance. In some cases, such as in muscle, cells invest heavily into maintaining functional mitochondria by cleaning up oxidative modifications (see *ch.4.3*). Yet elsewhere in muscle cells accumulation of lipofuscin is a robust marker of aging[(9)].

cells. To keep sebum fluid, there is a very high level of shorter fatty acids with 12, 14, 16 and 17 carbon chains. Monounsaturated fatty acids (MUFA) also make up a large fraction of sebum, with OLE at 15%, and palmitoleic acid (16:1, n-7) together with a uniquely human monounsaturated sapienic acid (16:1, n-10) making up 22%[10], although even MUFAs can get oxidized if the conditions are right. Free fatty acids (16%) render sebum antibacterial properties as skin acidity (pH 4.5-6.2) is a barrier to microbes that might otherwise penetrate through the skin.

4.7.2. Skin smell

"Your skin smells like the light."
— Jason Isaacs

"The best thing to do with a bad smell is to get rid of it."
— Carol Kendall

Consider a popular idiom, "to smell blood". Predators are attracted to the odour of blood, while it is often aversive to the prey species. But blood itself has no smell, so what is going on? Well, you are actually smelling the Fenton reaction! When exposed to oxygen in air, iron in hemoglobin generates ROS, which oxidize PUFAs in red blood cells into lipid peroxides (*Fig.9*) that decompose into volatile secondary PUFA oxidation products, which give off the characteristic smell, even when present at disappearingly low levels. Alternatively, LPO can be initiated by other mechanisms, including enzymatic (lipoxygenases), polluting environment, light of various wavelengths, or combination of the above), and the peroxides formed would then be decomposed by iron. A related "metallic" smell emanates from metal objects - but tellingly, not before we touch them. A strong whiff of the "musty" odour would waft around if a finger were soaked in a solution of iron (II). As with blood, the good old Fenton reaction generates species that oxidize, or further decompose, our skin PUFAs, with several highly volatile molecules, including fungal-metallic-smelling 1-octen-2-one. The smell is the same if, instead of touching metals, some blood is smeared over the skin, on account of trace iron in sweat. And it works the same with copper, and other transition metals[11].

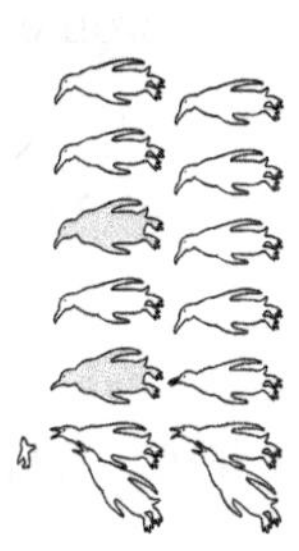

Moving on to less macabre subjects. The attachment begins so pleasantly! Mothers universally agree that the most delicious smell in the world is that of their babies' heads. Reminiscent of warm, freshly baked buns and blossoming meadows, it generates the surge of social bonding- and love-evoking hormone oxytocin in mothers' brains. Somewhat surprisingly, it can also trigger mothers' aggression, but inhibiting it in men, likely decreasing the chance of the outside hostility, while boosting the defence(12).

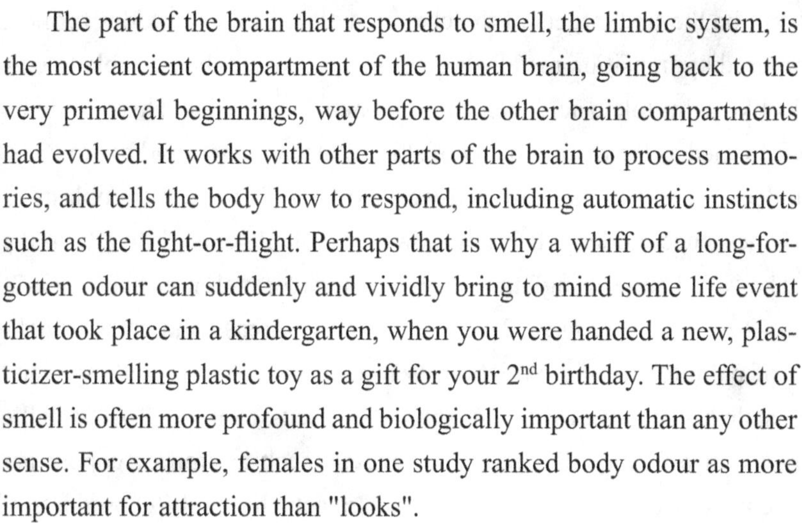

The part of the brain that responds to smell, the limbic system, is the most ancient compartment of the human brain, going back to the very primeval beginnings, way before the other brain compartments had evolved. It works with other parts of the brain to process memories, and tells the body how to respond, including automatic instincts such as the fight-or-flight. Perhaps that is why a whiff of a long-forgotten odour can suddenly and vividly bring to mind some life event that took place in a kindergarten, when you were handed a new, plasticizer-smelling plastic toy as a gift for your 2nd birthday. The effect of smell is often more profound and biologically important than any other sense. For example, females in one study ranked body odour as more important for attraction than "looks".

Recognizing "bad" smell is an evolutionary adaptation, a warning that whatever gives it off should not be eaten, or even touched. This deeply ingrained system strongly recommends that we steer clear of rancid foodstuffs. Those who did not have their system issue strong enough recommendations, or ignored the warnings, did not live to produce descendants who could continue disregarding the advice. This way, heeding the advice is now hardwired. And that's good, because many stinky foodstuffs, including LPO products, are highly toxic when ingested. So, when skin lipids start going rancid, we are rather sensitive at detecting that from quite a distance. In addition to oxygen, light and transition metals, skin microbiome may also contribute to fatty acid catabolism, generating multiple LPO products, only some of which would be volatile enough to produce odour. For the volatiles profile, the exact position of a skin patch also matters. Experiments with dogs show that the profile varies between different skin areas, such that dogs trained to recognize a set of compounds for one place might not recognize compounds wafting off a neighboring part. The human studies of the volatiles from the back and the forearm revealed two different profiles, with some overlap[(13)].

The key culprits for the "aged smell" are LPO products. What matters here is not which LPO compounds are the most abundant, but which are the most volatile, - or smelly. LIN and ARA, the key PUFAs in skin, are the main LPO "fuel", and as such could still be the first targets of the chain reaction. Lipid peroxides formed, once decomposed by transition metals or other means, would then be able to oxidize MUFAs like palmitoleic acid, generating nonenal and its ilk. There could be more 4-HNE formed as the main product of skin LPO than other lipid byproducts. 4-HNE is indeed present in the skin, and was shown to force skin fibroblasts to switch into senescence mode[(14)]. It is not, however, volatile enough to add to the aged odor. The "aged smell gang" molecules are hexadecanal, nonanal, octen-2-one and trans-2-nonenal[(15)] (*Fig.9*). 2-nonenal, a musty, greasy and grassy-smelling product of palmitoleic acid oxidation, is formed increasingly with age, and also renders stale beer its wet cardboard smell. As discussed in *ch.2.4*, "metallic smell" conveys an olfactory illusion. When metals such as iron or copper are brought into contact with skin, the metallic smell is actually the LPO products of PUFA and MUFA oxidation.

Aged smell is not rated as unpleasant or even particularly intense, it is just uniquely recognizable[*]. Nonenal levels go up from about 40 years of age, yet the distinct scent is associated with the 75+ age group. Light alone cannot explain the "aged" smell, for even though kids and young people spending days in the sun do develop some LPO-associated odour, it differs from the "age smell" in both composition and intensity. Sun-bathing does not substantially exacerbate the "aged" smell in elderly people, even though it generates a very faint temporary "burnt" odour in kids. It therefore must involve processes taking place in the epidermis and the dermis over many years. Aged mitochondria and accumulated lipofuscin deposits, awash with high oxygen and transition metals, would be the two sus-

---

* Green leaf damage smell and chopped cucumber aroma are also associated with LPO.

pects capable of constantly leaking LPO-triggering ROS species.

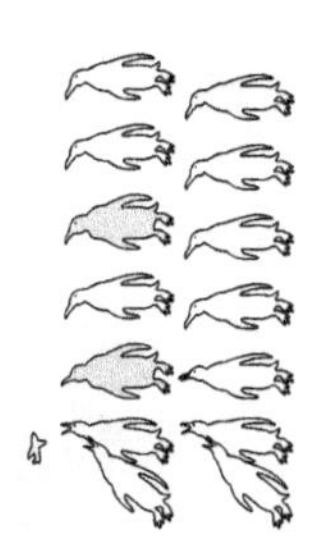

Japan is the biggest market for cosmetic products which assist the aged population in tackling this issue. Even the term "aging odour" was coined in Japan (kareishū, 加齢臭). Several factors may be behind this, including the Shinto-based drive for all things pure. As humans are seen as fundamentally pure by Shintoism, age-related deviations from this tenet, such as aging odour, need to be fixed. On average, Japanese people live long lives, so the market size for such products is quite big. As the majority of the "Aged" smell gang are aldehydes and ketones, which engage in well-known chemical reactions, many commercial products utilize various chemical tricks to arrest and immobilize the perpetrators. Since the 1990s, spray and lotion products have been manufactured which claim to suppress the "aging" odour.

Skin aging, unlike aging of other tissues, depends more on the extrinsic than intrinsic processes. Relatively to other tissues, extrinsic factors (such as the environment and UV radiation) play a proportionally larger role in skin aging. Oxidative degeneration, the prime cause of skin aging, easily overwhelms various adaptive anti-aging efforts, while senescence plays a less prominent role[(16)]. But unlike the damage that had been done and would require complicated, if at all possible, cleanup, the aged skin smell is the result of an ongoing, not a completed, process. As such, it would be safe to assume that should a method be worked out to stop the LPO smouldering (see *ch.6*), the malodour problem could be mitigated.

4.7.3. Sun, skin and PUFA

"A day without sunshine is like, you know, night."
— Steve Martin

"There's no such thing as a healthy tan."
— Walayat Hussain

As discussed above, the trifecta of PUFA, metals and oxygen make very uncomfortable bedfellows, even in the dark. Now imagine what would happen if for a good measure some light were shone upon that bed.

Skin-penetrating UV and visible light (*Fig.24*) may be strong enough to "radicalize" water molecules (*ch.2.3*) and UV can break up H_2O_2 into hydroxyl radicals. However, light can also do a lot of damage in a non-direct, mediated way, by energizing the "third party" molecules named photosensitizers. There are many natural photosensitizers already present in the skin, such as flavoproteins and heme containing proteins. Aged spots may provide an ever-increasing supply of other nasty things. External sources of photosensitizers include cosmetics and various foods, such as riboflavin (vit B2), St. John's Wort, psoralens from celery and parsnips, some antibiotics, and others[17]. Two types of processes can be mediated by photosensitizers once they get energized into excited (*) state:

Light + photosensitizer = photosensitizer (*)

Type I is a direct damage to a biomolecule by the excited photosensitizer, leading to a radical formation:

Photosensitizer (*) + PUFA = PUFA-radical,
+ molecular oxygen = LPO

Type II requires an intermediary energy carrier, for example an oxygen molecule:

Photosensitizer (*) + O_2 = $^1O_2$ (singlet oxygen),
+ substrate = oxygenated product.

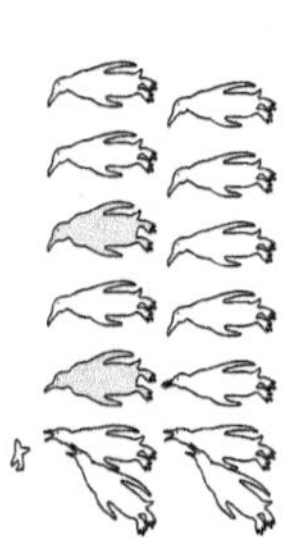

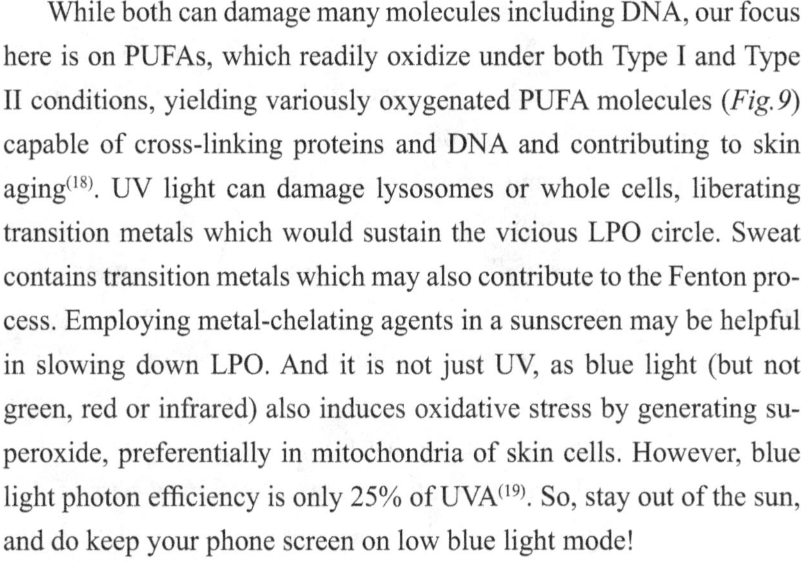

While both can damage many molecules including DNA, our focus here is on PUFAs, which readily oxidize under both Type I and Type II conditions, yielding variously oxygenated PUFA molecules (*Fig.9*) capable of cross-linking proteins and DNA and contributing to skin aging[18]. UV light can damage lysosomes or whole cells, liberating transition metals which would sustain the vicious LPO circle. Sweat contains transition metals which may also contribute to the Fenton process. Employing metal-chelating agents in a sunscreen may be helpful in slowing down LPO. And it is not just UV, as blue light (but not green, red or infrared) also induces oxidative stress by generating superoxide, preferentially in mitochondria of skin cells. However, blue light photon efficiency is only 25% of UVA[19]. So, stay out of the sun, and do keep your phone screen on low blue light mode!

Light exposure is not all bad news, though, as the devil is in the wavelength. Light does have a positive effect as well, beyond the famous vitamin D production. While we will talk about the inflammatory aspects of PUFAs in *ch.4.9*, red light has anti-inflammatory and pain-relieving effects on the skin. This about-face by the irradiation on skin well-being was investigated by Niels Ryberg Finsen, who became interested in the topic through frequent sun-bathing in attempts to mitigate the symptoms of his innate lysosomal storage disease. Awarded the 1903 Nobel prize for his work on using the light skin therapy, he was reluctant to accept it, citing his inability to understand the mechanism of action. The positive effect seems to come from the increased blood flow to tissues, which may boost the antioxidant supply helping clear inflammation and heal wounds, as well as stimulate mitochondrial cytochromes and, perhaps, activate hormesis (*Box 4)*. The best time for sunbathing seems to be at sunset and sunrise.

Having lost much of their body hair, our ancestors developed pigmented skin as protection against the sun. Migration into Europe from equatorial Africa made that not just obsolete but an obstacle to vitamin D synthesis, so pigmentation was abandoned. Yet pigmented skin is not a bulletproof, as people with dark skin do get skin cancer occasionally. While Queensland, Australia (37 cases per 100,000 in 2022), is the world capital of skin cancer, with New Zealand (32) in the second place, those countries in 3rd through 8th places are less obvious: Denmark (30), Netherlands (27), Norway (26), Sweden (23), Switzerland (22) and Germany (21), while the world's average is 3.4. But there is no sun in Norway, period. Unless you count aurora borealis. What's going on?

In the early noughties I lived in San Diego, working at a company where about 50% of the workforce was of Teutonic origin. And many of them operated on a rather peculiar work schedule. They would arrive between 5:00 to 5:30 am in pitch darkness (makes sense: less traffic), then work with the trademark assiduity and industriousness for eight hours, as required. Did they continue living on CEST perhaps? Oh no, they would do that to maximize their beach time! After lunch, they would march off straight to the ocean. Parchment-ness of skin on well-toned, swarthy bodies would tick like a Geiger counter, clocking in the UV A and UV B photon hits. But at least these guys stayed indoors for the first half of the day. Elsewhere, Mediterranean summer is the best time, and place, for Viking-spotting. The northerners are out on the beach en masse, from Algarve to Turkish Riviera, stockpiling vitamin D for the looooong polar night, while totally overestimating the durability of their often-tattooed skin PUFAs and DNA. They all use sunscreen* of course. But…

"I just don't think there's a sunscreen that gives you enough protection."

— Brooke Burke

---

* Some historical speculation on sunscreens. Our planet's magnetic field is fickle and flippant, as it flips regularly. What if that happens whilst you are on a transatlantic flight? Anyhow, the Laschamp event, circa 42,000 years ago, saw the magnetic field change its polarity, - getting really weak in the process. As the result, solar irradiation reaching the surface increased dramatically, killing things. They speculate that this may have been the straw that broke the Neanderthal's back. And what about our homo sapiens ancestors? Well, they may have used clothes – and the ochra sunscreen!

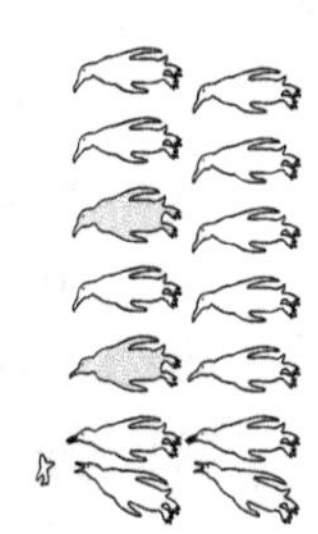

Besides, skin protectors could do more harm than good. One of the recent big reveals was how some sunscreens, while preventing sunburn, would contribute to sunlight-related cancers. A case in point was a sunscreen ingredient Padimate-O, an industrial chemical employed to generate ROS when illuminated. Lipid peroxidation was not on biochemists' radars back in 1990s as people preferred DNA readouts as a measure of "light toxicity", but all the ingredients required to start LPO were there all along[20].

Having a cigarette while sitting on a gunpowder barrel would have looked risky even before smoking was banned. It simply sounds dangerous, even when no fire or heat is brought in direct contact with the explosive. It just does not sound right, so why take a chance? Gunpowder combustion is a chain reaction, and we treat it with caution. Yet this is not the case with sunscreens. Titanium dioxide is hailed as an inert mineral, used in cosmetics industry as a thickening, opacifying and sunscreen ingredient. The industry claims it shields skin from UVB, and even from some UVA radiation, and that the material is structured in such a way as to prevent its d-element properties from doing mischief. But how improbable would the radical formation be? Sounds like a tall tale.

Indeed, what is described as "inert" in cosmetics earned a whole different set of epithets in a seemingly unconnected area of "smart", "self-cleaning" surfaces. How do you clean windows on the 120th floor of a skyscraper in smoggy downtown without risking lives? Just cover them with TiO_2! This time, rather than inertly shielding and sun-screening an object, TiO_2 is expected to actively get activated by sunlight, then passing it on to the plentiful H_2O and O_2 in the neighborhood to generate radicals (titanium is a transition metal, *ch.2.4*), which duly destroy the grime, grease and goo, leaving behind clean and shiny glass. Today, it is a big industry[21].

But wait a minute: does it behave so differently in cosmetics and materials? Well, maybe it doesn't. Maybe it does exactly what a transition metal would do, in the vicinity, or on, or in the skin? So, plenty of

Fenton- and sunlight – initiated ROS formation to be expected for this "inert" sunscreen ingredient. Slapped right on the skin, - which is full of PUFAs. Skin is also often treated with creams and lotions which almost always have LIN as the main ingredient (it is also increasingly becoming the main ingredient of what we eat, and hence of ourselves). PUFA-containing skin cream is slathered on PUFA-rich skin, and before it wears off, a layer of TiO_2-containing "protection" is applied. And then the whole thing is exposed to sunlight, hence LPO galore*.

But it can get even worse.

4.7.4. Tattoos

"Don't think it, ink it."
— Mark Victor Hansen

"Tattoos are for life, not just for a season, so if you're planning on getting inked you might want to be wary of jumping on any trend bandwagons you may live to regret"
— Sophie Gallagher

The number of people who have hired someone to deposit unknown chemicals under their skin steadily grows, with 32% of Americans and 26% of the Brits sporting at least one tattoo. The palette of offered colours is also expanding. The growth is highest among young adults, meaning the chemicals will be lurking in their skin for a long time. Typically, people get inked to make sure their tattoos are visible to others, so naturally this often overlaps with beach time and sun

---

* Yet the list of d-elements shows the way. Zinc is inert in Fenton (*ch.2.4*), so to save your PUFAs and your DNA, zinc oxide, ZnO, may be a safer alternative to dangerous sunscreen ingredients. The silver bullet may well be made of zinc. There are products on the market already. Zinc is more accessible than titanium, the latter until recently coming to the Western market mostly from the utilized Soviet nuclear subs, - hardly a renewable source. And if zinc oxide sounds a bit like "common people", the groovy types should not despair. Stay ahead of the pack by applying a really snazzy concoction, made of flashy gold nanoparticles and some oil, - no, no snake oil, *snail oil*, blinding the hell out of the fellow beachgoers. You think I am kidding? Check it out[(22)].

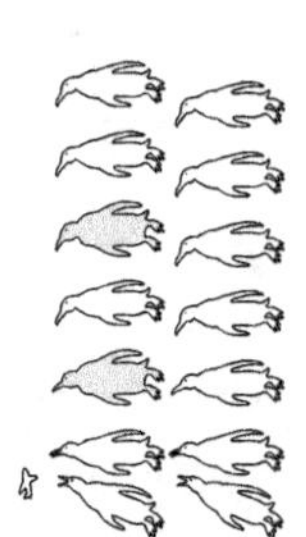

exposure. From the LPO perspective, several things should concern us here. Many of the dyes are photosensitizers (some photo unstable), which can initiate LPO and other damage (see *ch.4.7.3*). Carbazole, azo dyes and their decomposition products, for example, are toxic or/ and carcinogenic (tattoo carcinogenesis). Photosensitizers and metals in tattoo pigments are lying in wait, and when exposed to sunlight, yellow and red tattoos sometimes cause phototoxic skin reaction that is likely to involve both type I and type II (oxygen-dependent and oxygen-independent) mechanisms. Insidious, light-independent Fenton activity of dyes containing transition metals, discussed in *ch.2.4*, is an orthogonal source of damage, as many inks contain dangerous metals*: cadmium sulphide (yellow), cadmium selenide (brick red), mercury (II) sulphide (burnt orange of Longhorn, Austin), chromium (III) oxide (green, chromium acid turns into it), cobalt aluminate (blue spinel), and iron, nickel and lead compounds are all used in the art(23).

Laser-based removal of tattoos may also be fraught with toxicity and LPO, but this time the damage is body-wide, because the removal process just re-distributes ink and pigment particles, scattering them around the body into lymph nodes and God knows where else, when billion-watt laser pulses (different wavelengths required for different ink colours) rupture the dermal cells. TiO_2-containing inks cannot be destroyed by lasers, while iron oxide based ones permanently turn black upon laser irradiation(24). As only the success stories are typically advertised, the "laser tattoo removal" may give false confidence in the reversibility of tattooing. Laser-botched tattoos, and accompanying changes in skin pigmentation are rarely mentioned.

Whilst we are on the body art topic, another pertinent area we should look at with piercing sadness is the choice of metals for the art of piercing. The popularity of skin-penetrating metal objects grows, and the metals of choice include gold, titanium, niobium and surgical

---

* Overlapping categories of somewhat less bright cosmetic tattoo pigments, and vibrant tattoo inks are sometimes used indiscriminately. Both categories may contain dangerous chemicals.

grade steel. All these characters feature prominently in *ch.2.4*, on account of their membership in the transition metals club. Some more active in Fenton than other, they may still, over time, put their ability to exist in different oxidation states (there are ten for platinum!) to a bad use, generating ROS*.

Skin, this thin, delicate, tender shield is all we have to separate us from oxygen, which, - remember? - had oxidized the entire planet. That's its full-time job. As if this were not enough, it also has a half-time job: shielding us from light. And to do this, somewhat illogically, it is filled to the brim with highly oxidizable PUFAs, not unlike a flimsy party balloon, loaded with hydrogen, exposed to sunlight and oxygen on the outside. Surely any help would be welcome if we wanted to keep our skin younger for longer… Read on.

---

* Implants may pose even a bigger problem. Figures vary, but around 800,000 knee replacements and 600,000 hip replacements are performed in the USA annually, and the numbers are expected to rise as the population ages. Patients with implants often have elevated levels of transition metals including chromium, cobalt, molybdenum, niobium and titanium in their blood and CSF, so the brain gets them, too. This coincides with neurological effects post-surgery. Wear and tear degrades implants over time, releasing metal particles and ions into tissues. This is known as metallosis. Apart from neurological problems, inflammatory, pain-related and muscle weakness effects are also known, all stemming from the elevated LPO levels. On the tissue level, the damaged organs include heart, kidneys, liver, endocrine system and nervous system, as well as vision problems and hearing loss. Alloys used in dentistry may also include chromium, cobalt, gold, nickel, palladium, platinum and titanium.

Removing metals by chelation is very inefficient as chelating agents have their own toxicity profile, and often metals trapped by chelates are more active in Fenton than the free ions (see *ch.2.4*).

4.8. Emperor's new "Ptoses"

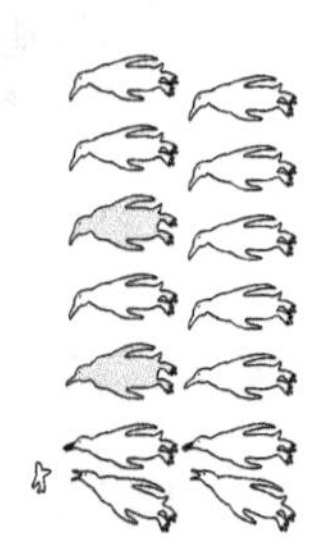

"Rebranding is not just about changing your logo; it's about reinventing your story and resonating with a new audience."
— generative AI

"Progress is man's ability to complicate simplicity."
— Thor Heyerdahl

Test yourself with a cryptic crossword, designed by an anonymous science-funding body:

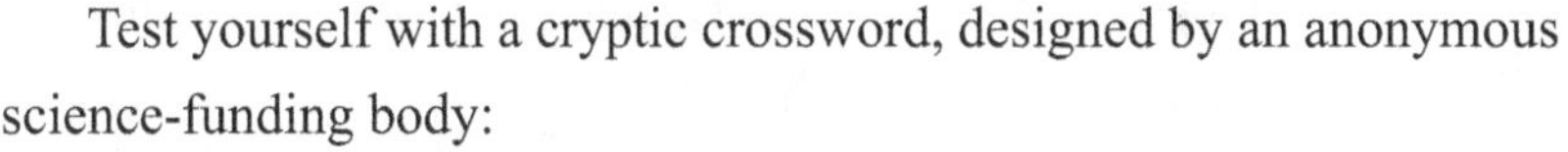

Q: *A novel cutting-edge multi-$ Bn technology. Clues: 14 letters, 1st, 3rd and 9th are "N". 4th, 10th, 12th are "O". The second letter is "A", and the fifth letter is "T".*

A: *Colloidal chemistry.*

Imagine dipping a Persian longhair cat, or a Chow Chow, in water. Or just shaving them (do not try it at home). The result, while much smaller in size and less fluffy, would still contain the essence of your pet. There are some informative photos on the web attesting to this. So, at the risk of ruffling some fur or feathers, how about we took a current fluffy faddy tabby, which grows fluffier by the day, fluffed up by the tornadoes of PR and tsunamis of funding, and shaved it with Occam's razor to check on what might be lurking beneath?

To help weather the tornados, let's hold on to some solid historical foundations. Once upon a time, in the late-70s and mid-80s, there were many colloidal chemistry Departments. Every self-respecting University had one. Tenured professors lectured, published, and applied for grants, duly received. PhD candidates were interviewed. Conferences were organized. Scientific progress was being made, and life was good. But then, skies started turning worryingly dull, portending change. Cynical types might call it perestroika (from Russian, *«дебилдинг»*). Funding dwindled, and so did the tenures. Alarming comments began to pile up, along the lines of "The field of colloidal chemistry had been worked out every which way over the last hundred years… Nothing left to discover…". But it only takes one person to make history, and let that individual remain anonymous. Having just had another grant application, - to study the chemistry of colloidal gold, - rejected, that person had his golden "Eureka!" moment dawned on him: "These are not colloids, oh no. These are, er, um, - oh yeah, - "n-a-n-o-particles!" This flash threw the funding gates wide open, and the newly born baby Nanotechnology emerged through them to rule the world, with nanobots driven by nano-sensor-controlled nano-motors powered by nano-energy flowing through nanotubes and nanowires, all made of nanomaterials. This nano-rebranding mega-insight, likely inspired by Feynman's "There is plenty of room (for funding) at the bottom" financially turned out to be a runaway, **giga-success**. It has rejuvenated the field so well that these days, only the old, dying out breed like the author remember the word "colloid".

A quick litmus test to check if the above holds water. In 1926, Theodor H.E. Svedberg won the Nobel Prize for his studies in colloidal chemistry, and for inventing the colloids-separating ultracentrifuge. 97 years later, in 2023, Yekimov, Brus and Bawendi won the Nobel Prize for the discovery and synthesis of quantum dots, a type of nano particles. A 97-year-long lull for colloidal chemistry, and then - renewed interest, a re-born field, and fascinating new discoveries.

History may be repeating itself in the PUFAs domain. It has been known since antiquity that oils and fats store poorly. Antioxidants such as ground up leaves of rosemary and olive trees (hence the green colour of the high-end olive oil brands) were employed to prevent rancidity of olive oils. Systematic studies of lipid chemistry were started in the 1800s by the French chemist Michel Eugene Chevreul. He must have been the first person to appreciate the importance of lipids in aging, - having begun to study this topic at the age 100 (two years before his death, at 102, in 1889). LPO studies also began in the 1880s (see *ch.4.6*). With the discovery of the first organic radical in 1900 by the Russian – born Moses Gomberg, radicals entered the field of chemistry for good. The Germans Fritz Haber and Richard Wilstätter postulated the existence of hydroxyl radicals in the 1920s. Everything was ready for the final push, and the British Rubber Producers' Research Association scientists J.L. Bolland (1949) and L. Bateman (1954) connected the dots. The concept of the chain reaction of PUFA autoxidation (LPO) fully hatched[(1)].

The stage was set to look at the relevance of these findings to biological systems. Henry Fenton showed the importance of transition metals in radical generation back in 1876. It became clear early on that such metals (*ch.2.4*) play a very important role in LPO as without them, LOOH derivatives were shown to be rather non-reactive. The arsenal of potential LPO initiators was expanded with the discovery of LOX in the 1960s and COX in the 1970s. MDA had been known for a while, but with the 1960s discovery, by an Austrian biochemist, Hermann Esterbauer, of 4-HNE, the field of toxic downstream LPO products really took off. There are now several societies and scientific clubs dedicated to this group of compounds, highlighting their importance. A massive amount of work aiming to stop the chain reaction followed - endogenous (*ch.2.8*) (made by the body, like GSH) and exogenous (supplied with food, like vit E) antioxidants, chelating agents to remove the toxic transition metals from circulation, etc., etc., were looked at, with 70 000 papers on vit E alone since 1922.

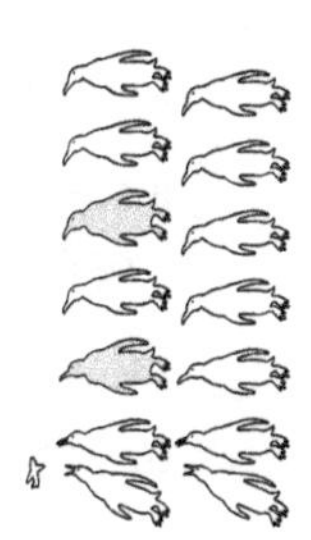

But even the very bestest of these efforts often had trouble getting published in top journals. Esterbauer's key 1991 review on 4-HNE[2], cited 9000 times, ended up in Free Radical Biology and Medicine, a journal which back then had an impact factor of around 1 (currently 8.1, to a substantial degree due to the Esterbauer papers; the impact factor of Nature is currently 50). And the funding rule of thumb is brutally simple: work published in second-tier journals receives second-tier funding…

The LPO field was plagued with these problems, partly because historically the work was mostly published in obscure journals related to areas of food, oils and agriculture, the topics at the bottom of the life sciences hierarchy. The unfairness was blindingly obvious to the best minds in the LPO community, who stood gawking with envy at the ivory Nanotech towers and other successful darlings of the science funding agencies (*ch.1.1*). Something radical (no pun intended) needed to be done to break through the glass ceiling… but what? How to thrust the topic into the mainstream, reaching wider audiences? How could one get into the top tier Cell and Nature journals, propelling the field all the way up and reaching much wider audiences - and funding? These journals want to see a sexy twist, and that's what was missing in the LPO area. Which one of the three LPO-related manuscripts would you expect to be accepted by one of the most famous scientific journals, Nature: (a) "On some rarely addressed nuances of vicarious olefin oxidation", (b) "Novel universal and powerful mechanism of cell death modality", (c) "Routes to an uncommon optical isomer of 4-Hydroxynonenal"?

A poster child of biology, apoptosis (from Greek, *apo* "away from", *ptosis* "falling", like leaves from a tree) is a programme that, once "clicked on", executes cell death, with obvious importance to cancer and aging. By 2022, 560,000 papers on cell death and 515,000 on apoptosis were published. Imagine how much funding went into that. Klondike! El Dorado! But LPO, no matter how you initiate it, can also injure or kill cells, because there are always metals and oxygen present in us aerobes, and the defences are never adequate… Could this fact perhaps be highlighted, made more salient, for the general public?

The first shot on goal was by two Salk professors, Pamela Maher and David Schubert. If only I knew, chatting with Dave about the oxygen-triggered cell death mechanisms, drink in hand, at an aging conference in the Bahamas in 2019, that he had just one year left… He would tragically pass away in 2020. Around 1990, Pamela and Dave postulated "a unique oxidative stress-induced programmed cell death pathway"[3], in which cysteine uptake was compromised, leading to a depletion of the LPO-inhibiting glutathione machinery. Just the right set of conditions in which LPO would thrive, damaging or killing cells. But this description is rather dull, so they sexed it up a bit, calling it Oxytosis (*tosis*, from Greek, indicates condition or state). Condition of (too much) oxygen. Who knows why the concept did not take root? Could it be that the name was off-putting, with "oxy" having bad undertones - of damage and rust? Or perhaps Oxy-tosis seemed an oxy-moron: oxygen gives life – and oxygen takes life. "Tosis", too, might have been perceived as the wrong term – compared, say, to a much more appropriate "ptosis". Hm… Maybe for the killer concept to take off, the word would need more killing power, more metallic strength, more style? As streamlined as a full metal jacket?

4.8.1. Strike whilst the iron is hot

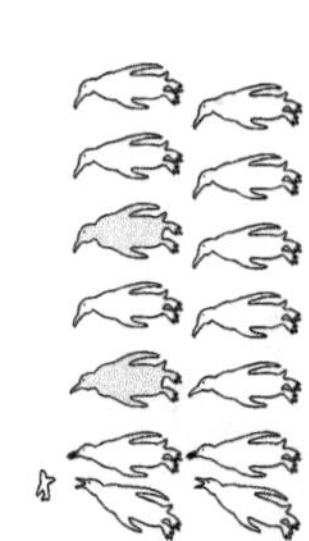

"It's the LPO, stupid!"
— after James Carville

"What's in a name? That which we call a rose,
by any other word would smell as sweet."
— William Shakespeare

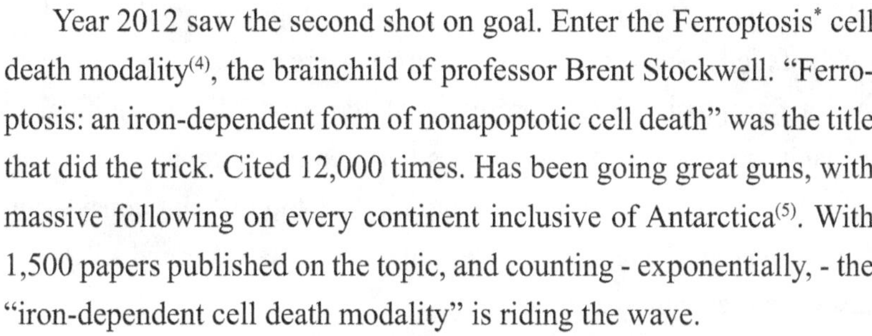

Year 2012 saw the second shot on goal. Enter the Ferroptosis* cell death modality[4], the brainchild of professor Brent Stockwell. "Ferroptosis: an iron-dependent form of nonapoptotic cell death" was the title that did the trick. Cited 12,000 times. Has been going great guns, with massive following on every continent inclusive of Antarctica[5]. With 1,500 papers published on the topic, and counting - exponentially, - the "iron-dependent cell death modality" is riding the wave.

And this is all great news for the LPO field, it really is. The work is still pushing the borders, churning out plenty of useful data. Yes, the claims pertaining to ferroptotic mechanisms may sound a tad pretentious. The compounds claimed to activate or inhibit this "novel type of cell death modality" often fall into one of several categories: either redox-active (LPO-suppressing antioxidants), or inhibitors of natural defences (blocking the glutathione machinery by interfering with cysteine import or through other tricks), or transition metal-arresting chelating agents. But so, what? If this is the price to pay for getting the work noticed, or funded, - to then venture into the areas that matter for us all, - how to stop the LPO which drives numerous age-related diseases, - then so be it. Having jumped on the bandwagon, various groups keep discovering wonderful things that would have been unlikely to attract attention, nor funding, in the pre-ferroptosis era. The nexus between Acetyl-CoA subtype (ACSL4) responsible for PL syn-

---

* Spelled F-e-r-r-o-ptosis, pronounced: F-a-n-t-a-s-y-ptosis, j-u-s-t -- k-i-d-d-i-n-g

thesis, GPX4 (an LPO product – removal tool) and LPO, as well as the role of the LOX enzymes in initiating LPO, reported by Brent Stockwell's group at Columbia[(6)], the interplay of enzymatic and non-enzymatic mechanisms on LPO, studied by Derek Pratt in Ottawa[(7)], and the work of Valerian Kagan in Pittsburgh on the role of LOX oxygenation and oxidised ARA derivatives in triggering pro-apoptotic and pro-ferroptotic events[(8)] are examples of such findings.

And, considering how much of a battlefield modern science is, what with dwindling resources and increasing competition, there is still room for gallantly gracious acts. Having underestimated the power of "ptosis" over "tosis", and the importance of referencing metallic strength on their attempts to popularize LPO, the authors of the failed-to-take-off oxytosis pathway were no doubt pleased when one of the major ferroptosis players, professor Marcus Conrad, magnanimously extended his hand to reconcile the differences between the ferroptosis and oxytosis schools[(9)]. The reconciliation process shouldn't be too difficult considering both "cell death modalities" are one and the same, our good old friend LPO. But chivalry is always a great and welcome thing.

While copying may not be. If a stockbroker hits on a winning formula and makes a few billion dollars, there is no use in trying to borrow his secret equation: with a limited reward pool, it only works because there is a single user – if everyone else were to use the trick, no one would gain as the pool would be shared equally… Some people claim, based on this, and on issues such as unpredictably complex human behaviour and lack of control experiments, that economics is not a true science… But I digress. A we-too, in science, is tricky to pull off, and accordingly, it seems unlikely that another "novel concept of non-apoptotic cell death modality" – cuproptosis[(10)] would emulate the success of ferroptosis. It is highly original and all, no question. Shouldn't a patent be pending? But then again c'mon, guys, there are 38 transition metals in the Mendeleyev table. And many of them can be found in vivo, hitting cells with Transitionmetalptosis…

Rather than forcing scientists to improvise in their attempts to come up with sexy F- or C-words, maybe the top Journals need to become less sexy-ist – so that people would not need to invent gargantuan titles nor embellish their areas of expertise by painting them in a more attractive way by a bit of **Rembrandting**, - focusing instead on the gist of things. What could be simpler?

4.9. Oxidized PUFAs and inflammation

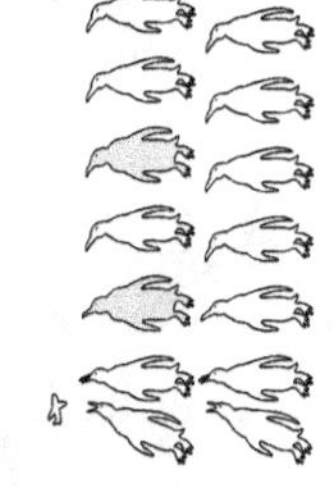

"Inflammation is in the background of every single major illness we know now."
— Julie Daniluk

"Inflammaging refers to the chronic, low-grade inflammation that characterizes aging."
— Mark Hyman

Acquired immunity relies on antibodies to neutralize specific invaders. While it takes time to learn to recognize them through natural exposure or vaccination, these memories are stored for life so that a swift action can be taken upon a second encounter.

But the first line of defence is innate immunity. It is non-specific and does not distinguish much between different types of threats. When the body detects damage or infection, innate immune cells trigger inflammation, by releasing chemicals that call on white blood cells. Those cells, known as phagocytes, engulf and neutralize many an intruder, be it a big cell, a little molecule, or anything in between. And PUFAs, particularly ARA, are key.

There are hundreds of inflammatory diseases, most of them recognizable by the suffix "itis". A protective response aimed at neutralizing the triggers, inflammation serves to make the body aware that something is wrong and needs attention, yet it can often be more insidious than just the famous Hippocrates' "redness-swelling-heat-pain" dictum. Multiple lipid and non-lipid mediators have to be highly synchronized, and impaired coordination portends trouble. Swelling of lung airways (bronchitis) may be suffocating. Crohn's disease, or colitis, prevents gastrointestinal tract from absorbing nutrients and water. Arthritis produces severe joint pain. Neuroinflammation is lurking behind the majority of neurological or psychiatric and other mental disorders(1).

Often, the events start with a knock on a cellular door (*Fig.25*). The caller could be a calcium ion, an LPO product, glutamate, LPS, an eicosanoid, a cytokine, acetylcholine, or many other things, - a rather eclectic group of visitors*. Let's skip the fine molecular details that involve receptors, channels and kinases. A lipase – say, cPLA2[2], - peers through the peephole, and acknowledges the caller by latching on to the ER membrane. Membrane PLs always have some ARA residues attached, and one of those gets sucked into the lipase, which disconnects it from the PL core and releases the ARA into cytosol. The next step, depending on the cell type, and on who exactly knocked on the door, involves a COX[3], or LOX[4] enzyme which adroitly picks up that ARA from cytosol**. The trapped ARA is then converted, by our bad old friend the hydrogen abstraction process, into one of many dozens of eicosanoids – COX forms prostaglandins and thromboxanes, while LOX yields leukotrienes and lipoxins, - a powerful signalling crew which spreads from the mother cell and interacts through various receptors like GPCR, regulating vascular, renal, gastrointestinal and reproductive systems (*Fig.26*). Eicosanoids call for thrombosis, inflammation, bronchoconstriction and all sorts of other processes elsewhere in the body. To be fair, some messengers, like lipoxins, DGLA-derived 1-series prostanoids like PGE1, as well as some DHA-derived docosanoids, can resolve inflammation***. But the great majority are pro-inflammatory[5]. Other PUFAs that can serve as the input feedstock for COX and LOX include DHGL and EPA (PUFA). Elevated formation of eicosanoids, known as an eicosanoid storm[6], exacerbates the inflammatory response by promoting the excessive release of cytokines (interleukins, interferons, etc) and other

---

* Physical actions that precede or trigger chemical signalling include exposure to light or irradiation, temperature variations and tissue injury. A recent addition to the list is pressure, - cells appear to be "measuring" spatial confinement imposed on them using the stiffest organelle, the nucleus. And the ensuing action seems to be activation of cPLA and release of ARA. Figuring out the mechanistic details of the process is the current cutting edge in immunology.

** A third major group of enzymes to convert ARA and other PUFAs into eicosanoids is cytochromes, but as they do not abstract bis-allylic hydrogens, they are not discussed here. An alternative mechanism has been proposed, whereby the hydrolysis of ARA by cPLA2 only happens when a COX or a LOX dock with the ARA residue-holding lipase, bypassing the free-ARA-in-cytosol stage. The jury is still out on this one.

*** Vagus nerve can exercise "spooky action at a distance", transducing inflammatory or anti-inflammatory signals half the way across the body.

pathways. And cytokines can knock on the cell door, calling for more eicosanoids. These events often proceed in cascades, such that the output of one event is the input of the next, amplifying the downstream effect with positive feedback.

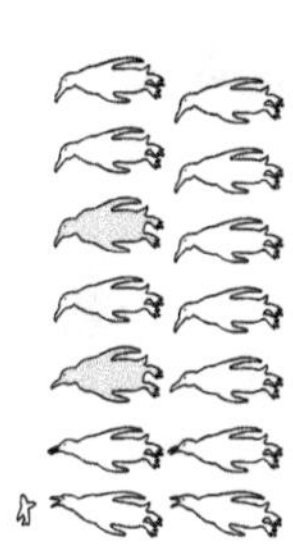

A bit of inflammation is not bad, in fact it is desired, as infections, cancer and other challenges need to be fought. But there is a goldilocks zone, and the problems arise once inflammation leaves the sweet spot and starts wielding a sledgehammer, going into an overdrive, like in inflammatory storms. Or if it never subsides, smouldering body-wide as in inflammaging, as our homeostasis becomes increasingly misaligned, stressing and eroding cells and tissues. How can it be kept in check*? The best way would be to stop it way upstream. And that's how nature does it. The most powerful anti-inflammatory weapon the body has is cortisol, which does exactly that: nips the process in the bud, by inhibiting (through an annexin-1 intermediary) phospholipases like cPLA2.

That's exactly what a stress hormone should do, - one cannot afford a luxury of snuggling up in a cosy cave to sleep through a bout of flu if faced with a clear and present danger – a war, a predator, a flood, when a fight-or-flight response is needed, prioritizing an immediate survival over anything else. It is well known that in the Soviet army during the Great Patriotic war, hardly anyone contracted the flu, – despite of sitting in swamps, or snow, in wet boots, for days and sleepless nights on end. A recent addition to this list suggests that even a small dose of fear could have a therapeutic effect, reducing inflammation(7). Mighty powerful is Mr Cortisol, but therein lies a problem. Stress is a multi-component thing, and cortisol, single-handedly, pulls on many strings. If supplied as a drug (typically as prednisone or a more potent dexamethasone) to treat inflammatory conditions, it would mitigate

---

* Complicated inflammatory signalling networks make mitigating inflammation challenging. If an old person twists an ankle and it swells, this acute inflammatory event will resolve in a week or two, yet their inflammaging status will remain. Why one resolves but not the other? There are many questions still remaining.

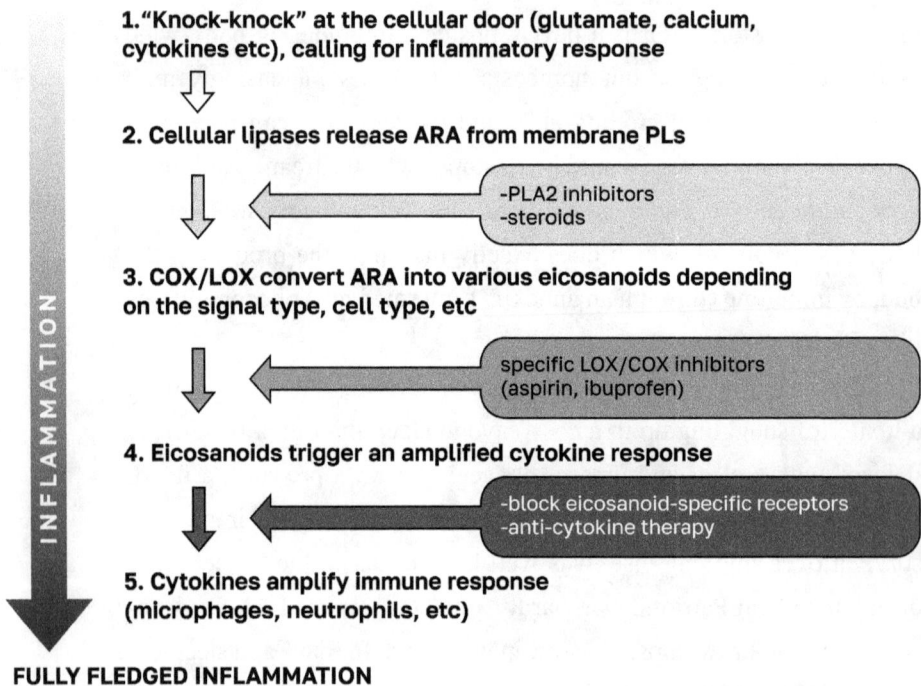

1."Knock-knock" at the cellular door (glutamate, calcium, cytokines etc), calling for inflammatory response

2. Cellular lipases release ARA from membrane PLs

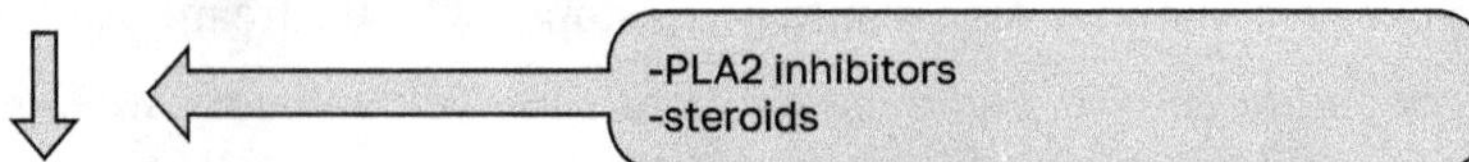

3. COX/LOX convert ARA into various eicosanoids depending on the signal type, cell type, etc

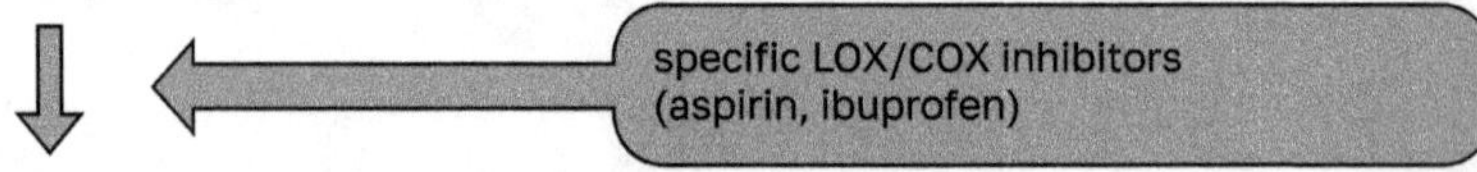

4. Eicosanoids trigger an amplified cytokine response

5. Cytokines amplify immune response (microphages, neutrophils, etc)

INFLAMMATION

FULLY FLEDGED INFLAMMATION

Figure 25. How inflammation starts, and why is it difficult to keep it in check. **1.** Cell receives an inflammatory message. **2.** In response, a lipase releases an ARA molecule from a membrane PL. Lipase inhibitors can dangerously decrease levels of lysolipids, compromising lipid turnover, signalling and the Lands cycle, while systemic inhibitors (cortisol) create too many problems elsewhere. **3.** Depending on the knock-knock signal, cell type, etc, COX or LOX convert that ARA into particular eicosanoid. Inhibition of specific COX or LOX risks upsetting a fine balance, - between thromboxanes and prostacyclins, prostaglandins and leukotrienes (aspirin-sensitive asthma), etc. **4.** Eicosanoids trigger various responses, but there are simply too many eicosanoid receptors to inhibit, and too many cytokines to intercept, often in inaccessible locations. **5.** Number of players multiply, so each next step is more difficult to stop than the previous one, and to nip it in the bud, the inhibition should happen way upstream. But where? See ch.6.

them all right by inhibiting cPLA2, while also causing many systemic problems, pertaining to blood sugar level, heart rate, blood pressure, immune, metabolic and chronic issues and sleep-wake cycle. That's why it is used as the last resort remedy, prescribed for short periods of time only*. Cholesterol-lowering drugs often downregulate the master steroid synthesis pathway, resulting in decreased levels of cortisol, which is made from cholesterol. As a result, multiple inflammatory conditions such as lupus, psoriasis and arthritis can no longer be controlled by anit-inflammatory action of cortisol and so get elevated(8). So, raising cholesterol could be a good natural way of keeping one's inflammation in check(9). Try as they might(10), as of February 2025, pharmas big and small failed to get a non-steroidal inhibitor of cPLA2 approved. Whatever the reason, - target complexity, location, side-effects, or various compensatory mechanisms, - the only cPLA2-targeting drugs available are the toxic steroids.

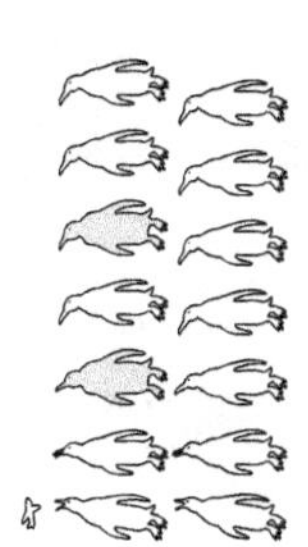

The next set of drug targets downstream of cPLA2 would be COX and LOX enzymes (*Fig.25* and *Fig.26*). There are plenty of compounds to pick from, such as paracetamol (acetaminophen) and a famous group of Non-Steroidal Anti-Inflammatory Drugs (NSAID), which include COX inhibitors aspirin and ibuprofen. Alas, this approach is not without its own problems. Aspirin permanently deactivates COX enzymes, and so is "expensive" for cells to deal with, unlike other NSAIDs like ibuprofen, which act reversibly. But a bigger problem is that COX and LOX inhibitors differ in their relative efficiency. 5-LOX, which gives rise to asthma-triggering leukotrienes, can be inhibited by zileuton and other compounds(11). 12-LOX drives platelet formation and thrombosis, but there are not many known inhibitors (curcumin from turmeric being one)(12). 15-LOX family is involved in various physiological and

---

* Cholesterol-lowering drugs often downregulate the master steroid synthesis pathway, resulting in decreased levels of cortisol, which is made from cholesterol. As a result, multiple inflammatory conditions such as lupus, psoriasis and arthritis can no longer be controlled by anit-inflammatory action of cortisol and so get elevated(8). So, raising cholesterol could be a good natural way of keeping one's inflammation in check(9).

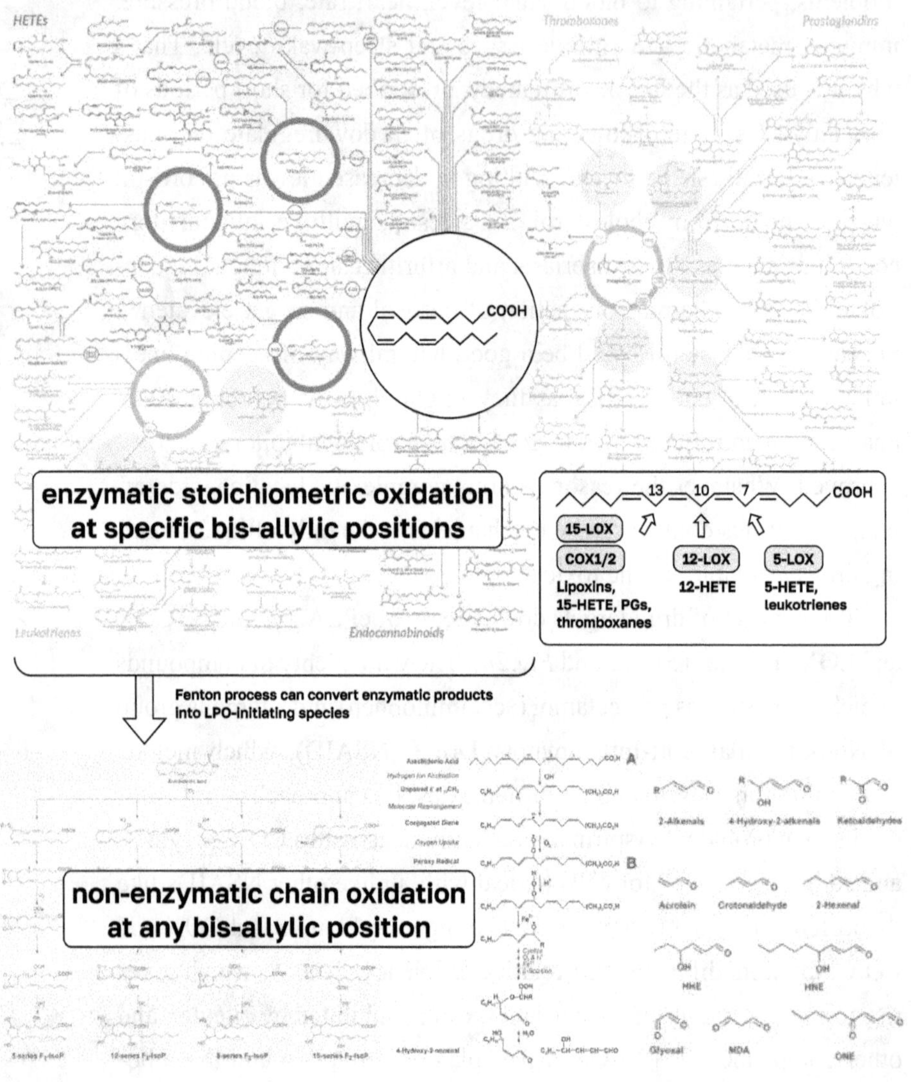

Figure 26. Enzymatic oxidation products of ARA are too numerous to be described individually. They are mostly pro-inflammatory, with notable exceptions such as lipoxin B4. The sheer number of these pro-inflammatory eicosanoids rules out the approaches targeting individual compounds, or even individual COX or LOX enzymes. There is an additional sinister side in many eicosanoids. Both LOX-produced LOOH derivatives such as various HpETEs and COX-produced endoperoxides such as PG-G2 and PG-H2 on contact with transition metals can decompose into LPO-initiating radicals. Various isoPs can be pro-inflammatory. The two groups, - the enzymatic and nonenzymatic derivatives, - have overlapping properties (see main text), and COX and LOX can get the LPO started, in this roundabout way.

pathophysiological processes such as cancer, yet there are currently no approved drugs acting on it[13]. Some inhibitors are active against several LOX enzymes, but to a different degree. The inhibitor tool bag is thus pretty mixed, - so trying to inhibit one, or several, COX and LOX enzymes, or attempting to down-regulate all COX and LOX enzymes at once like a cortisol molecule would do may throw the fine eicosanoid balance out of kilter. And this could be bad. In aspirin-sensitive asthma[14], aspirin knocks out COX enzymes, reducing inflammation by PG (good) and vasodilation by a prostacyclin (bad) yet leaving more ARA* substrate available for 5-LOX to be converted into pro-asthma leukotrienes (very bad), with the net effect of exacerbated asthma.

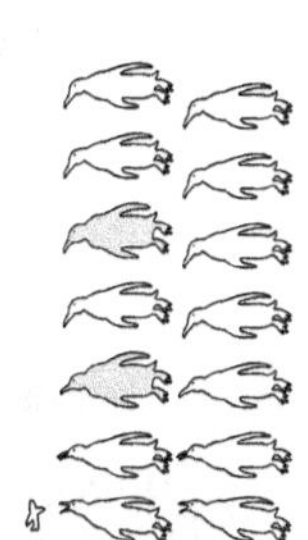

Leukotrienes are extremely powerful, operating at nanomolar concentrations. As bronchoconstrictors, they are at least 1000-times more effective than histamine. Various pathologies can be worsened by NSAIDs, - metabolic, gastrointestinal, skin, liver and kidney, - through similar mechanisms, stemming from changes to finely tuned eicosanoid ratios. Yet there are many conditions where NSAIDs can be helpful: pain relief, reduction of inflammation, fever, blood clots… As discussed in *ch.2.5*, dietary n-3 PUFA deficiency (and, as a consequence, higher relative levels of n-6) causes cognitive defects. These can be reversed not only by n-3 supplementation, but also by COX inhibitors, suggesting that pro-inflammatory PG play some role[15]. Is an NSAID a friend or an enemy then? A frenemy, as it all depends on the context, - and the fine balance of various eicosanoids.

Moving further downstream in our search of inflammation checkpoints suitable for inhibition, numerous eicosanoid-specific receptors exist in membranes of various cells. Receptor-specific inhibitors could be used to down-regulate inflammation, but since the eicosanoid genie is out of the bottle and is at large in the body, such a downstream control would be much less efficient than attempts to inhibit COX and LOX. Likewise, intercepting cytokines with antibodies or blocking

---

* Other PUFAs give rise to similar compounds, including DGLA and EPA. DHA-derived docosanoids are discussed below.

their receptors would be technically possible, but less efficient than blocking the pathways leading to them way upstream.

We have already discussed a concept of chirality (*ch.2.9*). In solution, molecules can approach reactive sites of each other from more angles than is possible in catalytic pockets-constrained enzymatic reactions. There are two hydrogen atoms at each bis-allylic site of a PUFA, either of which could be abstracted by an ROS species non-enzymatically. In contrast, enzymatic oxidation results in the abstraction of one specific hydrogen, with defined stereochemistry. Accordingly, there is substantial overlap between isoprostanes and eicosanoids, with one important difference. Eicosanoids are chiral (or optically pure – as they have only one of the two possible optical isomers), while isoprostanes are a racemic mixture (containing both isomers). This is akin to having one glove versus a pair. As eicosanoid receptors are chiral, 100% of a particular eicosanoid would interact with a specific receptor, while only 50% of a structurally identical, bar the stereochemistry, isoprostane would (e.g. PGD2 versus D2-isoprostane). Analytical techniques can distinguish these optical isomers (*ch.2.9*), allowing one to assign the origin of a particular compound to either enzymatic processing, or LPO.

Enzymatic and non-enzymatic PUFA oxidation are inter-connected at multiple levels. As previously mentioned, LOX can initiate LPO by producing LOOH which can induce the chain reaction in the presence of metals. Eicosanoids and isoprostanes can multiply by inducing their own positive feedback loop of ever-increasing inflammation: they trigger the formation of pro-inflammatory cytokines like TNF-alpha and IL-6, which can initiate LPO, generating more isoprostanes, - and so ad infinitum*. This catch-22 often plays out in human diseases and aging. Depression, anxiety and PTSD are associated with both LPO and inflammation (see *ch.4.2*)[(16)]. Injecting mice with LPS results in elevated cytokines – and depression[(17)]. Yet many cancer immunotherapy treatments aim to crank up inflammation as a way of eliminating the rogue malignant cells[(18)]. So bad is the ensuing depression that prophylactic antidepressants are often prescribed as part of the treatment regimen.

---

* An additional nexus point between inflammation and LPO is the RCS gang. Inflammation-derived ROS ignite LPO. MDA, 4-HNE and other LPO by-products can activate immune cells, stimulating immune cells into more cytokine production.

Naked mole rats and bats live long lives (see *ch.5*) yet get almost no cancer. Elephants have evolved a similar adaptation (by amassing 40 copies of protective, apoptosis- and DNA–repairing, anticancer gene p53, versus a meagre single copy in humans* and bats, who have an extended suite of DNA repair adaptations, helping to cope with increased demand for fixing the damage resulting from their fast metabolism[19]. Bats can batten down their inflammatory hatches to weather down the infection storms with softer inflammatory responses. This dialling down helps them get through without risking any tissue damage through an inflammatory overdrive. Because often it is this overdrive, and not the initiating infection, that kills us. Infections often kill by over-activating our inflammatory responses: up to 90% of COVID severe cases and fatalities came from lung inflammation (through sky-high prostaglandins) and up to 10% - from thrombosis (through exorbitant thromboxanes)**.

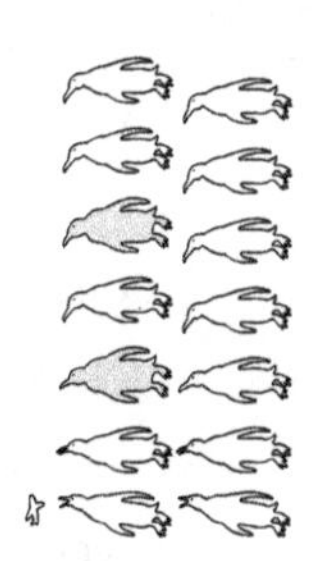

4.9.1. Quagmaresins?

"The Emperor's clothes were not invisible;
they were simply non-existent."
— H.C.Andersen

"It's very difficult to find a black cat in a dark room,
especially if there is no black cat."
— Anonymous

Life coexisted with LPO long enough to learn how toxic the oxi-

---

* But whales, another species that rarely get cancer, have just one p53 copy, just like humans. Perhaps their huge intake of n-3 PUFAs is partially responsible?

** Still, bats do employ the eicosanoid pathways, and in some infections, such as the deadly White-Nose Syndrome, their pro-inflammatory eicosanoid output is elevated.

dation products are. Animals evolved to recognize the telltale signs and respond, trying to mitigate the problem. We saw multiple recognition mechanisms for oxidized PUFAs in *ch.2.9* and *ch.4.6.* So carefully is the PUFA oxidation status monitored, that essentially any oxidatively (enzymatically or non-enzymatically) modified PUFA, with oxygen atoms attached to pretty much any carbon on the backbone, would be flagged up as a distress signal, triggering a response. The conveyed message is simple: "SOS. Come and help to stop the LPO". Suppose there is an ongoing LPO event, which causes inflammation and leaks oxPUFAs left and right*. As one of those gets detected, the repair systems come and fix the problem. LPO is stopped, inflammation resolved. Yes, the call for help was conveyed by an oxPUFA. But can this compound be said to have resolved the inflammation event? Multiple other types of oxPUFA, or other compounds, could have played this role… It is conceivable that there might be compound-specific receptors for specific oxPUFAs. But it is difficult to see why that should be the case, considering that an LPO event generates multiple, if not all possible, permutations of oxPUFAs. It might make more sense to recognize them as a class, like CD36 and CRP do, rather than individually. It was initially believed that inflammatory molecules, having completed their repair work, would just passively disperse away – bringing the inflammation event to an end. This changed in 1984 when Charles Serhan, working in the lab of the Nobel prize-winning discoverer of prostaglandins and other eicosanoids Bengt Samuelsson, reported that lipoxins, made from ARA by COX and LOX enzymes, could actively stop (resolve) an inflammation event[(20)]. An analogy to illustrate the difference was of a crowd dispersing itself, versus riot police actively breaking the mob up. A lot of effort has gone into this new area of Specialized pro-resolving mediators (SPM) since, with dozens more COX- and LOX-processed lipid derivatives described, making up four SPM classes: lipoxins (from ARA), resolvins (E-series from EPA, D-series from DHA), maresins (from DHA) and protectins (from DHA)[(21)]. Subclasses include DHA-derived neuroprotectins, discovered by Nick Bazan, with a role to play in neurological conditions such as AD[(22)]. More than 2,600 papers were published on resolvins alone. A common denominator of SPMs has always been the strikingly low concentration re-

* COX and LOX enzymes, at least to some extent, can operate on pre-oxidized PUFAs, generating a smorgasbord of products, some present at a near-single molecule level.

quired to resolve inflammatory events, orders of magnitude lower than the typical levels of pro-inflammatory mediators*, making the reliable detection of SPMs an analytical challenge (see *ch.2.9*). The SPM field, however, is not without its controversies if not to say inflammatory storms, with no resolution in sight. The SPM field, however, is not without its controversies if not to say inflammatory storms, with no resolution in sight. The SPM-related claims have been recently challenged on several grounds, including (a) biosynthetic pathways, and the questioned capacity of leucocytes to make SPMs; (b) gaps in our understanding of SPM-detecting GPCR receptors; (c) a disconnect between the availability of SPM starting materials, the stage of inflammation, and the SPM levels detected, and (d), the biggest of all, - the irreproducibility of the reported SPM detection protocols. Basically, in several instances the detected analytical peaks claimed to signal the presence of SPM molecules were deemed to be noise. This quagmire may be challenging the role of SPMs as mediators of the resolution of inflammation**[23].

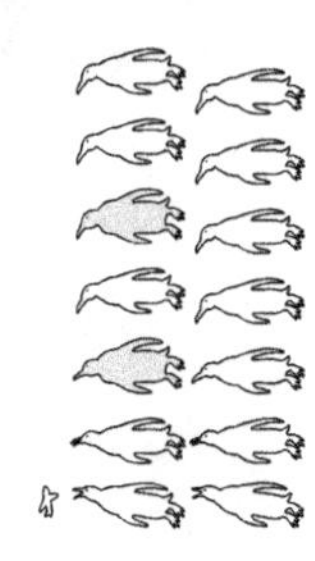

---

* I will quibble with an occasionally mentioned figure of one SPM molecule per 10 L being active, - as humans only have five litres of blood, one molecule should thus help two people, evoking succussion. However, autocrine mechanism, of a signalling molecule acting on the same cell that would produce it, or a close neighbour, would make the concentration per litre value irrelevant. Still, a more accepted active range is in single nanomoles, or single picograms per ml, an equivalent of 10^7 molecules per 10 L of blood.
Very low active concentrations pose another challenge. Eicosanoid levels measured in blood upstream and downstream of a localised inflammatory event may differ substantially. What then would be the right location to look for them?

** These niggling issues are no obstacle for commerce. Non-enzymatic OxPUFAs (hydroxy-DHA, hydroxy-EPA), squeezed from anchovies, claimed to be (precursors of) SPM, are sold ("SPM Active"), on the strength of two case studies (Metagenics website) as chronic pain mitigators, even though non-randomised trials did not find reduction in inflammation metrics (CRP/ESR)[24].

5. PUFAs and LPO are primary contributors to aging

"You wouldn't mind being at once introduced to the Aged, would you?"
— Charles Dickens

"My Spirit shall not strive with man forever, for he is indeed flesh; yet his days shall be one hundred and twenty years."
— Genesis 6:3

It is obvious that aging is a progressive failure of the maintenance of tissue homeostasis, hence the impaired function. A gradual decline, which does not kill by itself, but opens the door, - the older the wider, - for other nefarious actors to sneak in. "You know one when you see one", - yet beyond this quip, aging is not that easy to define, let alone quantify. Consider a 35-year-old competitive triathlete. He is based in Arizona, consumes energy bars and drinks, and loves barbeque. He played American football in college, and his brain is reaping the harvest sown by the long-forgotten concussions. TBI (see *ch.4.2*) is slowly developing, so that his cortex is at 62 years of age, but which direction the pathology would take, - PD, AD, other, - is still unclear. The process is accelerated by his consuming lots of fresh strawberries, which he does not bother to wash, - a mistake, on account of neurotoxic pesticides in their skins (see footnote*, p.90, *ch.2.6.4*). His cardiovascular setup is in the spectacular form of a super-fit 20-year-old. His blue eyes, aged from sun exposure and inchoate AMD, are at 55 years of age. Where his skin is regularly exposed to the merciless sunlight it appears like the skin of a 60-year-old, but the patch with the coloured tattoo, which constantly leaks ROS species (see *ch.4.7*), looks even older. The non-exposed skin matches his chronological age. His left knee is OK, at about 40 years old, while the right knee is overworn to about 65 years of age and may soon need a replacement. His metabolic condition, on account of energy bars, junk food and what not, is pre-diabetic, at 65 years of age, while his penchant for burnt-ends (*ch.3.3*) is gradually inflaming his intestines, accelerated by the plentiful dietary "vegetable" oils, and his distaste of n-3-PUFA-rich seafood. There is some bronchial irritation from chlorinated water in swimming pools, pushing him to a pre-asthmatic state. What biological age is he? It is just a number, - but which? 35? As old as his most worn-out organ? An average across all parts?*

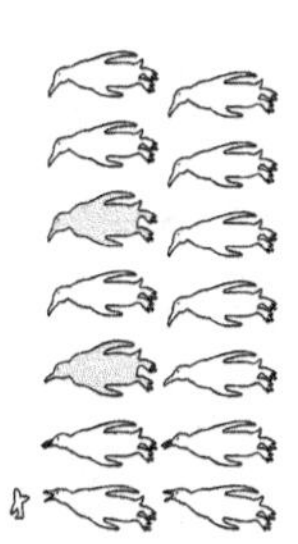

There is a booming field of age biomarkers intended to help answer such questions. However, it's early days, and the field still has a long way to go. A recent addition to the toolkit, called epigenetic clock, looks particularly promising. But even that technique is still a work in progress, as DNA methylation values for the same person may vary depending on whether the sample was taken before, or after, a nap, or a cup of tea…

While eagerly awaiting for a test to be properly optimized, let's step back and look at aging from a PUFA angle. Maximum lifespan of individuals within the same species, barring accidents and bad luck, varies by around 50%. Employing the most powerful tools we have, such as genetic modifications, drugs or caloric restriction can extend some species' lifespans by around twofold. This holds for various phyla, as exemplified by the popular nematode, fly and mouse anti-aging-workhorses. However, variation between species, even within the same family, can reach orders of magnitude. It spans single days to 15 years in various nematodes, one day (mayfly) to 50 years (queen thermite) in insects (although metamorphosis can make the comparisons with other phyla challenging), and one year (shrew) to 200 + years (bowhead whales) in mammals, a difference of at least two orders of magnitude. The vertebrate record holder seems to be a Greenland shark, at 270 + years. The outdated way to rationalize these differences would rely on the rate-of-living hypothesis, which posits that smaller organisms have higher basic metabolic rate (BMR, dubbed "fire of life") and so "live faster", on account of their larger surface to mass area requiring increased heat exchange. Numerous examples, such as birds versus terrestrial animals, break this pattern. But even within

* The "average" may seem the least suitable here. See footnote (*) on p.52.

mammals, maximum lifespan is not as strongly correlated with body mass as with mammalian basic metabolic rate. African elephants (6,000 kg) and gorillas (200 kg) have different BMRs, but a very similar maximum lifespan (around 70 years). In this sea of variables, one parameter seems to hold: irrespective of the size of the mammalian species, a gram of its tissue would expend about the same amount of energy before it dies at maximum lifespan[1].

Higher energy use per gram of tissue in smaller animals compared to larger ones would create more ROS and LPO faster, accelerating the aging process, although genes can mitigate this to an extent: the relative rate of mitochondrial ROS and LPO-derived RCS production in birds are substantially lower than in rats[2]. Higher output of ROS should seemingly be accelerating aging in physically active individuals compared to sedentary life style. But compensatory mechanisms would see to it that more mitos would be made in response to regular training, and so the rate of respiration per unit volume would not increase[3]. In addition, hormetic effects (*Box 4*) of moderate regular exercise or active lifestyle would help decrease the rate of systemic aging as assessed by markers such as mtDNA damage[4]. This reconciles the finite number of heartbeats per human lifespan (estimated at 1-3 x 10^9 beats) with the benefits of active lifestyle: increased numbers of mitos help reduce the pulse rate when training. The resting heart rate in physically active individuals is reduced, too (compare cyclist Miguel Indurain's 28 beats per minute, to 60 for a regular person), keeping the total number low. Given their increased lifespan, one would expect larger, longer living animals to have higher endogenous antioxidant levels. Yet, the reverse is the case[5].

Now to PUFAs. Grafting an LPO branch of the oxidative stress theory onto the trunk of the rate-of-living hypothesis yields bitter fruit. While many people tended to this cultivar, the chief grafters who greatly increased the harvest are Tony Hulbert and Paul Else from Australia, and the Spaniards Reinald Pamplona and Gustavo Barja. The key insight for reconciliation of long lifespans, BMR, and low antioxidants first turned up in an obscure paper from the 1970s: from mice to whales, the more metabolically active an animal, the less DHA in its heart[6]. And DHA, as discussed before, is the easiest PUFA to get oxidized. This was later extended to other tissues (brain being an exception, see *ch.2.5*). In mammals, every doubling of body mass would decrease the DHA content by 12-24%, in line with the corresponding 19% decrease in BMR[1]. In fact, the PUFA composition predetermines the BMR – this correlation is named the membrane-pacemaker theory of metabolism[7]. One reason mice live to three years and humans to one hundred might have to do with the fact that mice indiscriminately place DHA into all lipid membranes including in mitos, while humans don't[8]. And the mito-produced RCS can reach far beyond their place of birth[9]. In agreement with this, in long-living mammals the lower levels of membrane PUFAs correlated with lower levels, and lower sensitivity, to LPO[10]. To reduce LPO, long-living birds evolved to deal with PUFAs in an unusual way. As n-6 ARA is less oxidizable than n-3 DHA, albatross, a long-living fish-eater, has 30% more n-6 than n-3 in its skeletal membranes. In contrast, its adipose tissue TGs had five times more n-3 PUFA than n-6 PUFA, in line with its dietary PUFA intake[10], suggesting the control of membrane composition is important in relation to a species longevity.

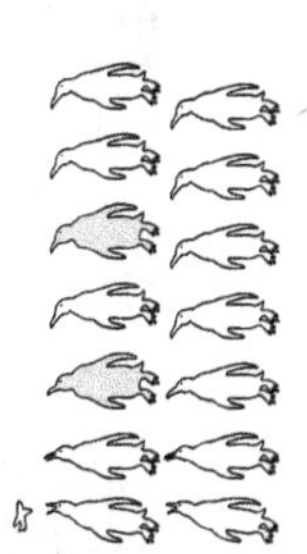

Animals that hibernate enjoy longer lifespans. As hibernation can induce LPO[11], such animals are known to change their PUFA profiles in favour of lower PI PUFA species prior to hibernation[12].

There is a correlation between maximum lifespan and PUFA composition for different species. But could PUFAs play a role in longevity variations between strains of the same species? Wild mice have extended average and maximum lifespan compared to genetically heterogeneous laboratory strains. The level of long PUFAs and LPO in skeletal muscle and liver in long-lived strains was found to be significantly lower compared with laboratory mice when kept under identical conditions*. As all mice were fed the same diet, the difference in membrane PUFA composition must have been down to genes, not diet[14]. Jumping the gun: despair not. Despite the fact that the genes are involved, and a very poor level of understanding of how the PUFAs get selected and

---

* This also holds true for the ugly darling of longevity research, the naked mole rat, which for its size has a very low load of DHA, and PI[13]. Centenarians show reduced LPO, yet higher levels of salubrious DHA[1] may be correlated with neurological and age-related conditions.

transported to a destination tissue (*ch.2.5*), there could still be the way to manipulate the process (see *ch.6*).

Numerous studies report on age-associated increase in membrane PUFA levels, as well as elevated PI and LPO in elderly, with higher levels associated with various background diseases[15]. LPO leads to body-wide increases in rigidity of lipid bilayers. Across different species, longevity is associated with reduced susceptibility of lipid membranes to LPO.

PUFAs vary in a systematic manner with body size in birds and mammals: the larger the animal, the lower the membrane peroxidation index[16]. With a caveat: this holds for all tissues but brain, which, large or small, requires a very high level[17] of DHA and ARA. Membrane peroxidation index (MPI) reflects PUFA composition. DHA is approximately 6 times more susceptible to LPO than LIN, and 320 times more oxidizable than OLE (*ch.2.5*). Another important membrane metric is membrane unsaturation index (MUI) (*Back-of-the-envelope-11*). By adjusting the PUFA composition in non-neuronal membranes, nature managed to keep them unsaturated enough, without increasing their MPI.

Low MUI, associated with longer lived animals, translates into lower levels of RCS, and so less damage to DNA, proteins and PLs such as ethanolamine-PL. RCS are relentless accomplices, and drivers, of the aging process. Increasing the expression of 4-HNE-neutralizing glutathione transferases improved both the stress resistance, and the longevity of C. elegans[18]. However, the supply of extra antioxidants, enzymes and small molecules alike may be more important for short-lived than for long-lived species.

PUFA composition in membranes can explain (a) the shorter lifespan of small mammals; (b) longevity of naked mole rat; (c) longevity of wild mice in captivity; (d) longer lifespan of birds compared to mammals; (e) other things not covered here, such as longevity of bee queens versus workers, and calorie restriction (CR); (f) exceptional lifespan of our own species.

Summarizing the importance of PUFAs and LPO as common denominators in longevity and aging should therefore be straightforward. Did you hear the one (hundred-word story) about the army? Corporal Douglas sawed off a wooden board in the platoon latrine. Sergeant Fitzgerald went in, and fell through. The remaining seventy words were blurted by the sergeant while clambering out… The role of LPO in aging can be described equally succinctly, in 100 words is: *It is the LPO, stupid!* Times twenty.

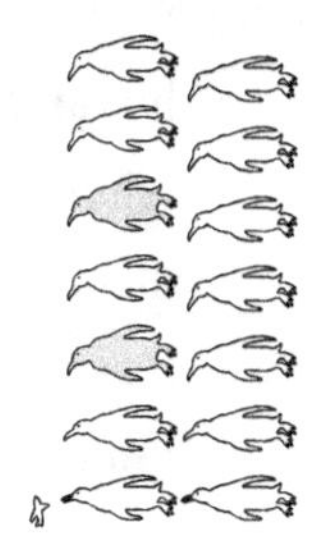

Land mammals are often compared to birds when looking at PUFAs, aspects of bioenergetics, and longevity. Yet a more fitting comparison might be between rodents and their much closer relatives, bats. Moving from a 2D (ground) to 3D (sky) habitat encourages evolution to kick in, often with amazing results. Bats are small yet can live more than 40 years, outperforming the naked mole rats hands down, while living way more actively. Flying is demanding, and while bats' resting heart rate could be 400 beats per min versus the mole rats' 250, when airborne (all night long, or during long migrations), this can go up to 1,100. Higher levels of PUFAs in bats help them survive hibernation, otherwise the jury is still out on the details of their PUFA profiles[(1)]. A particularly striking feature of bats – an ability to dial down their inflammatory status, - is considered in *ch.4.9*.

5.1. LPO, the quintessence of senescence

"The dead man walking."
— A guard announcement, walking a condemned prisoner to execution

"What is the worst of woes that wait on age?"
— Byron

In 1961, Leonard Hayflick discovered that, contrary to the prevailing thinking at the time, human cells have only a limited capacity, - a few dozen times*, - to divide. The eponymous Hayflick limit concept

*Hayflick limit and replicative lifespan, increase under senescence-delaying, low oxygen conditions. This is likely due to reduced oxidative stress and LPO, which slows telomere shortening and reduces cellular damage.

was not immediately accepted[19], but turned out to be a forerunner for a very hot area of modern biology, known as senescence (Latin *senescere*, to grow old), when the cells stop dividing, but stay alive. Abrupt transition into senescence is distinctly different from the aged cells: the latter, by progressively accumulating stochastic damage (mutations, lipofuscin, damaged proteins and organelles etc) gradually lose their function. As Hayflick suspected, senescence turned out to be very important in aging and cancer. Counting cell divisions is labour-intense, so the field had to wait till 1995 when the major senescence biomarker, beta-galactosidase, was discovered[20]. Easy to use, this colour test, positive for cells that reach Hayflick limit, helped launch the field, though one has to be careful with the cell data (*Box 8*).

Box 8. DO CELL CULTURE STUDIES OF LPO CORRELATE WITH IN VIVO SITUATION?

"You see, but you do not observe."

— Arthur Conan Doyle

Hayflick's, and other cell culture results could have been affected by the methodology, as cell culture data is not directly extrapolatable into the in vivo realm. Interpreting the cell data related to antioxidants can be challenging. Suppose measurements show an elevated level of GSH. Does that mean the cells are under high stress, and so more GSH is being made to compensate? Or the stress is low, and so there is a lot of excessive GSH leftovers sloshing about? To make sense of this, one has to look at multiple parameters. Besides, several underappreciated pitfalls should be mentioned. Cell culture studies are often carried out at 21% ambient oxygen. Even when not studying redox processes, this is fraught with problems. 300 million years ago, during Carboniferous period, the oxygen level climbed to 35%, - and the forest fires were raging non-stop. 21% is an unnaturally high oxygen level, so cells might revert to ROS-dependent signal transduction pathways normally not operational in vivo[21]. Also, antioxidants in the media are typically very different from those in cells' natural environment. Moreover, antioxidants can react with other media components, and oxygen, to produce ROS. Many published results have been later re-interpreted and ascribed to these artifacts[22]. An LPO-specific pitfall stems from the fact that PUFA composition in cell medium is often dramatically different from the natural habitat. The standard medium has 10% foetal bovine serum, which only provides 1% of PUFAs and 0.3% of LIN compared to what is available in the body. Cells cannot make n-3 and n-6 PUFAs so resort to making MUFAs and Mead acid to offset the deficit, throwing the membrane PUFA composition out of kilter. So how could one investigate LPO in cell cultures, when cells are hit with orders of magnitude more oxygen, yet have no PUFAs in their membranes[23]? Many studies fall into this trap. The rarely used fix: reduce O_2 levels and normalize PUFA composition. Meanwhile, liposomes, the bare-bones skeletons of cell membranes, can be more informative than cell cultures when studying aspects of LPO.

The helmswoman of the field for more than three decades, up to her untimely death in 2024, was Judy Campisi of the Buck Institute. She and James Kirkland at Mayo Clinic advanced the field greatly. Tens of thousands of papers tell a fascinating story of the senescence phenomenon, - the twilight of cellular life.

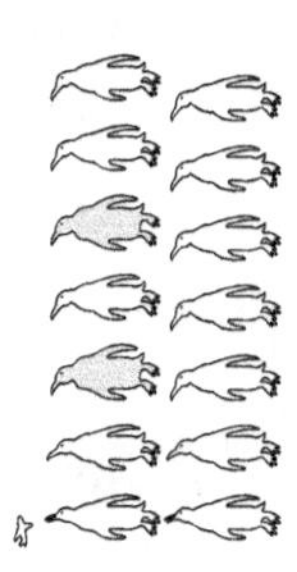

Since 1961, this highly regulated transformation was variously ascribed to a certain number of cell-division-associated scars on the plasma membrane preventing further divisions; to stopping pro-cancerous DNA mutations from proliferation, once a threshold level had been reached in the cell genome; and to telomere shortening, among other things. Cellular senescence is a programmed change in a cell state, irreversible proliferative arrest and shift towards a pro-inflammatory phenotype. Both ionising radiation and ROS drive cells into senescence. It can also be induced by a mito dysfunction, and as a result mitos often switch from normal respiration to glycolysis. Senescent cells have more mitos, and they are larger in size. Yet their membrane potential is lower, as is the mitophagy (mito recycling). However, their ROS production and LPO are increased(24).

Lest I digress too far from the main track: LPO plays the major role in the senescence pathways. Oxidized PUFAs and LPO products are elevated in senescent cells, and senescence is strongly associated with LPO and in particular with its most toxic offspring, the RCS(25); apprehending those blocks cell senescence(26). Senescent cells are not just sitting there in the dormant state, consuming resources while shirking their duties. They are prolific at contaminating the environment with a hotchpotch of LPO products, expectorating them left and right(27). On account of their pro-inflammatory phenotype, they fill the surroundings with nasty signalling molecules which inflame healthy neighbours. Ionising radiation, one of the factors causing senescence, is associated with the bystander effect (*ch.2.3*), when the damaged cell releases molecules that diffuse away and damage healthy cells nearby. Senescent cells, too, can irreversibly convert their healthy neighbours into the dysfunctional state of senescence, creating a cytokine – and chemokine – producing, chronic pro-inflammatory environment. Judy Campisi had long suspected that this senescence-associated secretory phenotype (SASP), and the bystander effect, might both depend on the oxidized PUFA products to carry out the message[(28)]. Both non-enzymatic (LPO) and enzymatic PUFA oxidation into eicosanoids by COX and LOX enzymes are involved[(29)], possibly playing a role in cytokine storms and inflammaging (see *ch.4.9*). In one of her last papers, Judy described a peculiar subtype of prostaglandin J2, which, on account of an alpha, beta-unsaturated carbonyl moiety, is as reactive (read: toxic) as 4-HNE, and promotes senescence and SASP activation[(30)].

Yet not all senescent cells are bad. Compounds produced by some of them (senescent secretome) help with embryonic development and wound healing, among other things. Just like inflammation, some is desired while too much is bad. Moderation! Senescent cells accumulate exponentially after the age of 60, and up to 10% of cells in tissues can be senescent in the elderly. Antioxidants, not surprisingly, cannot turn the tide, if not to say a deluge, of senescence[(31)]. An exciting new direction, spearheaded by Kirkland, is focussing on selectively eliminating senescent cells, using various approaches and compounds known as senolytics. First-in-human clinical studies in idiopathic pulmonary fibrosis patients employed a cocktail of senolytics including dasatinib, a cell killer used as an anticancer drug, and a plant flavonol, quercetin. The latter may be both an antioxidant and a hormetic agent, and can down-regulate LPO markers in mice[(32)]. The human trial showed promise, but it was an open label study such that both the doctors and the patients knew they were taking the cocktail. Larger, randomized trials are needed[(33)].

5.2. Aging: a collection of imperfections

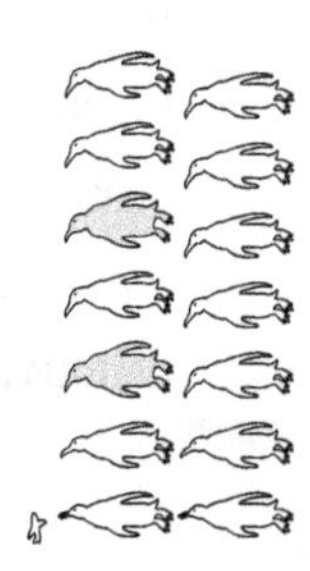

"Moanday, Tearsday, Wailsday, Thumpsday, Frightday, Shatterday, till the bitter end."
— James Joyce

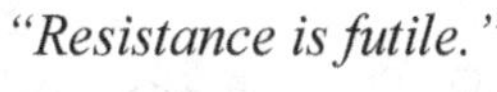

"Resistance is futile."
— The Borg, Star Trek

The natural aging process gradually erodes the efficiency of physiological functions. Such a progressive decline suggests that the factors that cause it are present throughout the lifespan, starting at a young age, or earlier. Following up on the Denham Harman mid-70s experiments, the husband-and-wife team of Leonid and Natalia Gavrilov put forward the High Initial Damage Load hypothesis, positing that 20% of damage is done by birth, - the price to pay for a highly error–prone embryonic development process[(34)]. While the environment and lifestyle habits have a very large impact on aging, they are not the decisive factors. Over the human history, there must have been cases of the most optimal combinations of the environmental factors. Say, a coddled king, being fed on black caviar, broccoli, pomegranates, figs, cherries and pecorino cheese, with a moderate alcohol intake, enjoying eight hours of sleep, great hygiene and dental care, active lifestyle with no exposure to in-laws or direct sunlight, having sex at least 5 times per week and keeping his brain active by dabbling in math and poetry. He is as dead now as a hedonistic epicure, cigar in mouth and drink in hand, betting big at the casino's green table in the middle of the night or what not, with cavalier attitude towards his diet, detesting anything his mum told him could have been good for his health. In fact, thanks to hormesis, the hedonist might have outlived the king, as the example of Jeanne Calment, the longevity record holder and a life-long smoker, attests. Yet the data from the largest ever longitudinal longevity study, involving 100 billion human subjects over the last 200,000–300,000 years, is implacable. Luck, genetics, and environmental variations combined are used up by around 122 years of age. Since death is not directly programmed by genes (imagine if it were, and then the master gene would mutate by chance, or we could deactivate it specifically…), and the influence of environment and chance maxes

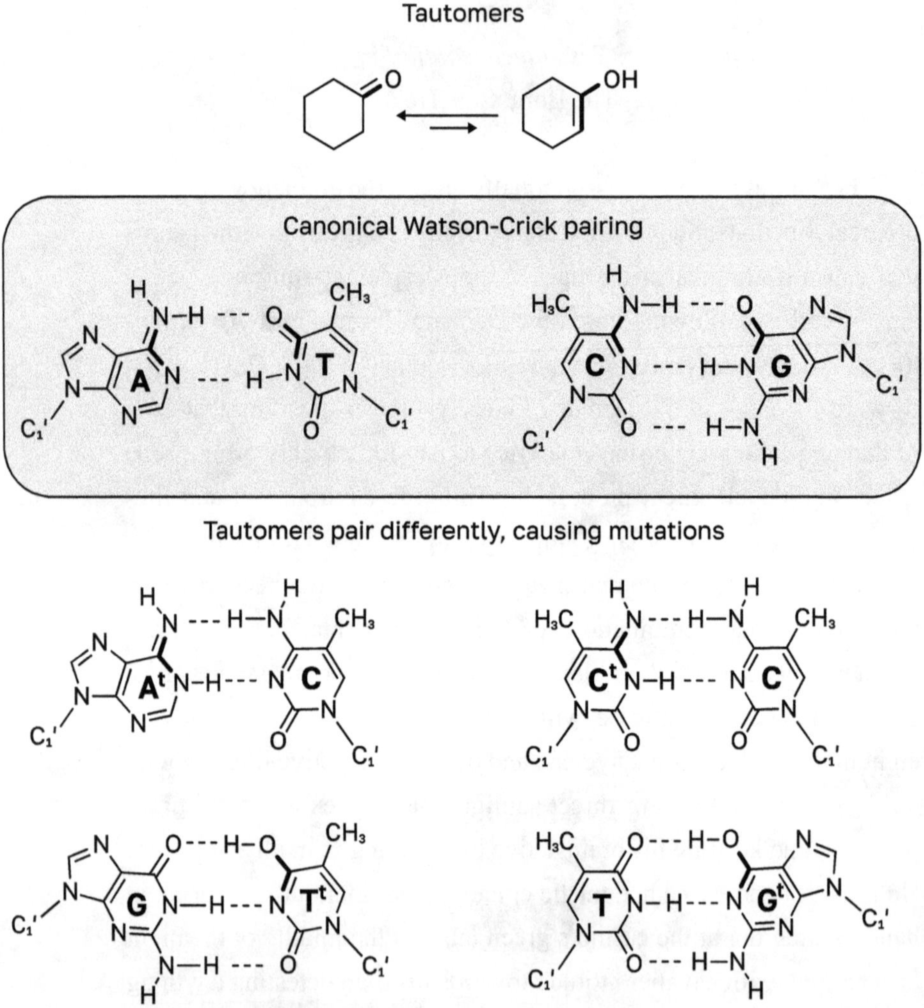

Tautomers

Canonical Watson-Crick pairing

Tautomers pair differently, causing mutations

Figure 27. How tautomers cause mutations. Canonical A,T,G and C structures are 10,000 times more favourable then their tautomers At, Tt, Gt and Ct, but there will always be some less favourable in the genome's 3 Bn bases. They pair differently to Watson-Crick, causing mutations.

out, it looks like the inexorable accumulation of garbage and random errors (*Box 9*) is behind the lifespan limit.

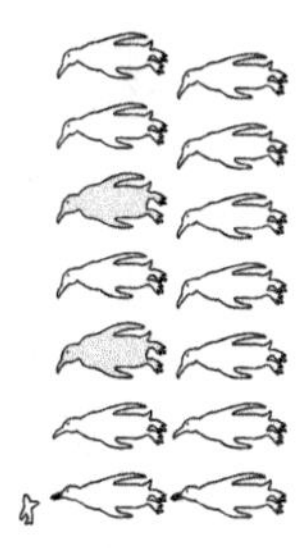

Box 9. DNA STRUCTURE AND RANDOM MUTATIONS

"History is a sequence of random events."
— Neil Armstrong

One of the most "lawful" types of random damage is imbedded into the mechanism of DNA copying. The famous Watson and Crick paper on DNA structure has the following acknowledgement: "We are much indebted to Dr. Jerry Donohue for constant advice and criticism, especially on interatomic distances"[(35)]. Donohue explained to Watson what the correct structures of the four DNA bases, A, T, G and C were, - and that turned out to be the last bit of the double helix puzzle to fall into place. The key term here is tautomer (*Fig.27*). An assembly of atoms, while remaining in their original places, can slightly reshuffle their single and double bond "connectome" into a similar looking, but different, structure. Some forms may be much preferred to other forms, as the unequal equilibrium arrows show, as the keto form is more stable than the enol structure. This equilibrium means that in a very large number of molecules, the great majority would have the predominant structure, yet a small percentage would fleetingly exist in a different form, randomly distributed, as defined by the equilibrium constant. In DNA bases, the typical ratio of canonical to tautomer forms may be as high as 10,000 to one. But given the number of bases in our genome, at any given time there will be 300,000 (!) "mutant" forms, which would pair in a non-Watson-Crick way, giving rise to mutations. Most of them will occur in the parts of the genome which do not matter much, but some… 35 trillion cells (and their mitos) in the body would be making DNA or RNA copies daily, diverging more and more from each other, generating the mosaics of slightly different cells through this process of random accumulation of errors, as envisaged by Peter Medawar. Not all errors would be detrimental, but some… To draw an analogy with road accidents, in the UK, for about 40 million licensed vehicles of all types, there were about 333,000 traffic accidents in 2022, or 900 per day, 5 of them fatal. Roughly one accident per day per 40,000 cars. Some no doubt were down to reckless driving. But for the remaining majority, it was just random. No guilt, no sense, no reason.

As long as cells continue dividing, mutations will keep popping up. Basically, every human being should die of cancer, but not everyone will live long enough. Environment, such as radiation and chemicals, may accelerate the process, but they are not absolutely necessary, because the mutation-generating mechanism is implanted into the very fabric of evolution. It drives evolution, but it drives cancer, too*. As is shown on *Fig.19*, LPO helps generate more DNA mutations. And once cancer takes root, LPO turns into one of its weapons.

The minimum viable product concept, prioritizing quick software release and doing things on the cheap versus excessive refinement, is not a Silicon Valley invention. It is 3.5 billion years older, as evolution only develops things to the state of being "good enough". Thus, even though there is no "aging gene" that turns on to turn us off, the imperfections are essentially encoded into enzymes, by never optimizing them to 100%. And this is bad for longevity. LPO can add to the mutagenic process. As explained in *ch.4.3*, RCS can react with DNA bases forming DNA base adducts that pair in a non-canonical way, leading to mutations. However, this process is mostly relegated to mitos, as their DNA is much closer to the LPO hotbed, and less protected compared to its nuclear counterpart.

LPO can also worsen enzymatic faults (*Box 10*). As shown on numerous examples in other Chapters (*ch.4.2, 4.3*), LPO-derived RCS avidly react with whatever comes their way, changing the properties of those proteins, or reducing the enzyme efficiency. And often making these molecules difficult to digest or remove by the cleanup systems.

Still, various imperfections affecting enzymatic reactions and DNA tautomerism are "linear". Going beyond stoichiometry would place the rogue actors head and shoulders above the other bad guys. And that's what the chain reaction of LPO does, for it is not linear but exponential. Imagine a battalion of troops spraying the enemy lines with fire from their automatic machine guns. Formidable though this might be, this linear, stoichiometric assault is no match if the enemy were to respond with a nuclear weapon. And nuclear weapon does not just symbolize the chain reaction here. It is the chain reaction incarnate…

---

* As noticed by Albert Lansing in the 1940s, children of older parents tend to be less fit and have shorter lifespans. But experiments in animal models show that selectively breeding the progeny, born at the young age, of the longest-lived parents for several generations does increase the strain's lifespan(36). The increase is due to improved defence and cleanup systems, - basically, marginal improvements. Perhaps there are innate limits to evolution, which can optimize things, but not come up with revolutionary new solutions, such as the enzymatic control of LPO.

Box 10. BILLIONS OF IMPERFECTIONS

"Small leaks sink great ships."
— Benjamin Franklin

"But there is nothing in biology yet found that indicates the inevitability of death."
— Richard P. Feynman

Enzymatic processes are pretty efficient, but not infallible. Even at a 99.99% precision level (and many enzymes fall below that), *Back-of-the-envelope-2* gives about 10^{15} errors per cell per day. Most of the errors are just aborted attempts to complete the process. But if 0.01% of errors lead to the wrong product, that gives 100 billion mutant molecules, per cell, per day... Cells, of course, are not watching this idly. According to Vadim Gladyshev at Harvard, really bad damage is tracked down and cleared – by expelling, or digestion. Damage that is less critical is left alone, as the price of perfection is very high and is only expended maintaining vital structures like egg cells. When cells divide, this imperfect material gets divided between the mother and daughter cells[(37)]. The process is not linear, as even in yeast, as shown by Thomas Nyström at Gothenburg, the mother cell gets to keep more of the garbage, giving the daughter a chance to start new life with a fresh slate. An unexpected chaperone in this process of filtering is a Sir2 protein, of the resveratrol fame[(38)]. That the cancer cells are so keen to divide is partly due to the incessant need to dilute the garbage.

5.3. Iron age, copper age: does Iron curtain, by teaming up with LPO, block the path to longevity?

"The world is an ever-living fire."
— Heraclitus

"Der Tod kommt stets zu früh."
— Christoph Pöppe

There is a sweet spot for the optimal ROS and antioxidant levels, both exceeding or falling short of which would be detrimental. Metals have their goldilocks zone too, - think anaemia and hemochromatosis. In humans, both too high and too low haemoglobin seem to predispose to AD[39]. This fine balance relates to all biomolecules. Both too little, and too much antioxidants can be detrimental, to give peroxiredoxin and GSH machinery as examples[40].

Similar to oxygen, iron is both essential, and harmful, for life (*ch.2.4*). Iron accumulates with age, and excess iron is linked, through the Fenton reaction, to multiple pathologies, such as hemochromatosis (iron in liver, heart and pancreas), NBIA, neuronal conditions related to ataxia, dystonia, PD, neuropsychiatric abnormalities, optic atrophy, retinal degeneration, - basically, to lifespan shortening*.

High levels of iron in the elderly could lead to release of labile extra iron in cases of damage, elevating LPO in older patients and exacerbating brain injury[41]. Ashley Bush of the Florey Institute, one of the most cited Australian neurologists, links ROS-generating action of iron or copper and various proteins with a large number of neurological, retinal and mitochondrial diseases, including AD, ALS,

---

* In general, accelerated aging models do not seem to be terribly informative. It is relatively easy to extend the shortened lifespans, but we do not know if it is aging, - or are they simply sick? Amputate a leg, and a mouse will die of hunger; reattach – and now it lives longer than the lame one. Do legs increase lifespan then? Will a five-legged animal live 25% longer? Aubrey de Grey argues that failed accelerated aging models could be revealing. A knockout of an important gene fails to reduce a lifespan - now that would be an informative result indeed.

PD, prion diseases and cataracts[42]. Pre-menopausal women are less affected by neurological and heart (*ch. 4.6*) conditions compared to men as bleeding may keep the metal levels low[43]. Post-menopausal women are more affected by AD, but as always, the reason may not be clear-cut. Is it to do with them getting abruptly hit with iron post-menopause, while men might have been adjusting over years? Or with the fact that women tend to live longer? And is the latter due to women having less unhealthy habits than men, - or due to higher levels of antioxidants, lower rates of ROS production, or different hormones*?

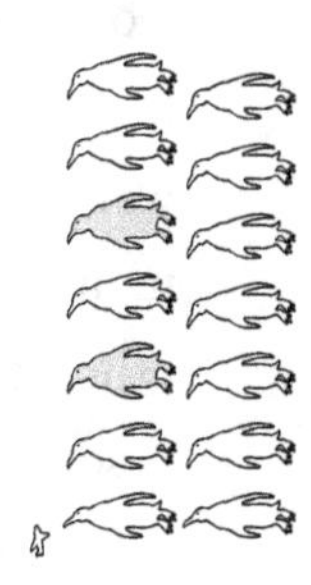

With so many variables, we should be excused for feeling rather defeated in our attempt to fully understand the mechanism… Still, there seems to be a correlation between cancer and high iron levels, but only if blood PUFAs are elevated simultaneously[45], implicating the LPO in cancer aetiology. Challenges measuring total Fe aside (high level of ferritin is OK, as that Fe is safe), this returns us nicely back to the LPO topic.

5.4. Damnaging flow of LPO: PUFA oxidation increases with age

"Small errors over time build into a mountain of trouble."
— Unknown

«В непрочный сплав меня спаяли дни. Едва застыв, он начал расползаться.»
— В.С. Высоцкий

Everywhere in the body, - in lipid bilayers of cells and organelles, in lipid droplets and lipoprotein particles, - wherever there are PUFAs,

---

* As mentioned in 2.5, women are more efficient than men at converting ALA to DHA. Measurements in trans people suggest that this is not a genetic but rather a hormonal (estrogen) effect[44].

- one can see the meandering trajectories of LPO chains, ceaselessly crawling or darting, day in, day out, through the membranes, leaving destruction in their wake. Oxidizing PUFAs, churning out toxic agents, spilling pro-inflammatory molecules. Ghoulishly, when we die, our DNA machinery and enzymes quickly grind to a halt. But not LPO, - for as long as there still are PUFAs and oxygen to sustain the fire (*ch.1.3*).

Thousands of research papers describe age-associated increases in PUFA oxidation. And the number of studies investigating LPO-involving pathologies is tenfold that, with most if not all of those diseases being related to aging. On the sub-cellular level, LPO hits all places where PUFAs congregate, such as in mitos, where there is an age-dependent increase in CL oxidation[(46)], accompanied by a decline in cytochrome oxidase activity and mito function. "The more LPO the older the age" is, invariably, the common finding of the in vivo studies, from fruit flies to humans[(47)]. Tellingly, in long-lived dwarf mice the levels of ROS and RCS are much lower to begin with, and take longer to build up[(48)]. And the cause-and-effect relationship is straightforward: LPO is not a side-effect of aging. Rather, LPO drives aging and age-related pathologies, and propagates the changes in the elderly people[(49)]. Time and again, correlations are reported between the levels of toxic RCS and frailty[*(50)].

As discussed in other chapters, cells do eventually succumb to LPO, but not without a fight. A large effort is expended to mop up the ROS with antioxidants. However, their action is stoichiometric (linear), so LPO, being an exponential event, manages to evade the antioxidants. The next line of defence is to try and disarm various lipid peroxides (LOOH) and RCS species. A network of antioxidant and detox systems includes the GSH and GPX machinery, several classes of cytosolic and mitochondrial ALDH enzymes to convert toxic aldehydes into

---

* There are tantalizing exceptions, the nature of which remains to be elucidated. Naked mole rats have higher oxidative damage levels, and lower antioxidant levels[(51)]; as mice age, their LPO levels rise until they reach the naked mole rat levels[(52)]! Perhaps in naked mole rats hormesis and repairs play a bigger role, and/or an extra effort is expended to keep the LPO in key locations (invisible to systemic LPO measurement) in check. Or, for them, this level is the "norm" and higher level are required to kill them: consistent with this, they do get lipofuscin as they age. The composition of their LPO products could be less toxic as they have less DHA. Or maybe they simply have less oxygen to keep LPO going, as their environment is pretty hypoxic.

harmless acids, glyoxalases, and different aldo-keto-reductases which reduce the carbonyls into non-reactive alcohols[53]. This effort, however, is expensive to sustain, and cells are miserly with resources, - but done half-heartedly, the defences are eventually out-dodged by the LPO. Helped by the vicious circle of damage catalysing more damage, and so ad infinitum, - or more precisely, till the bitter end. Active and less active cells alike – muscles[54], retinas, neurons, aging skin[55] and all the rest, - have a lot of ongoing LPO, and suffer badly as a result.

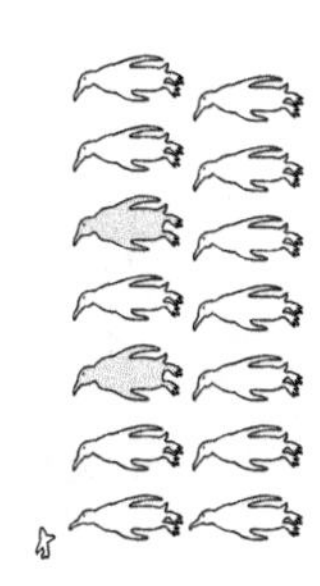

Even though the area of age-related biomarkers is still in its infancy, some LPO-related telltale signs of aging are relatively easy to measure (see *ch.2.9*). Older rats exhale more pentane and ethane, and this technique should be easier to apply to cooperative humans than to wriggling rodents. A lot of hope is currently pinned on isoprostanes (isoPs), the stable LPO products. Many analytical methods have been developed and optimized by a large international effort, spearheaded by Ginger Milne of Vanderbilt University who runs the best isoP measurement facility in the world, and heads the lab where isoPs were discovered more than 30 years ago by the late Jackson Roberts II, and Jason Morrow.

A sub-group known as F2-iso-Ps is elevated in plasma, kidney and liver of older rats. In urine of elderly humans, various forms of F2-iso-Ps can be elevated by more than two-fold, and the increase can be somewhat attenuated by caloricrestriction[56], or plant-rich diets.

5.5. Lipofuscin: the **agefying** age spots are both the consequence and the cause of LPO

"Lipofuscin, the wear and tear pigment."
— Anonymous

"When we are oxidized, we call it aging."
— Steven Austad

Age pigments are like routers in a large house, ensuring the wi-fi signal is always strong. If for whatever reason the LPO dwindles – they make sure it is rekindled, by virtue of their photosensitizing properties (particularly in the eye and skin) and high transition metal content. But growing steadily, and the older the faster, the lipofuscin deposits gradually clog up and suffocate cells, especially post-mitotic. In motor neurons of centenarians, it makes up to 70% of the volume[40] – a lot, considering those neurons can be a metre long*. Typically sized at 1-5 μm, the intracellular granules grow in size and number with age. They initially form in lysosomes, peroxisomes or other compartments, for instance through a failed utilisation of damaged mitos, or from the botched recycling of oxidized PUFA-rich outer segments in the RPE. Harmless whilst inside lysosomes, they spill out when the autophagy fails to fully digest the material, and spread wide, disturbing proteasome and just about any other cellular process, and self-catalysing the formation of more. Ceroid, a relative of lipofuscin, accumulates in various diseases such as neuronal ceroid lipofuscinosis (NCL). Retinal varieties include eye-specific lipofuscin (containing vit A) and drusen (no vit A). Extra-cellular ceroid from dead cells, a feature of atherosclerotic lesions, fills up macrophages (see *ch.4.6*). With the super-gluing units coming mostly from LPO and through Maillard chemistry, lipofuscin is a massive factor in aging[57].

---

* Lipofuscin is found in almost every atrophied cell of old age: in myocardial fibres, hepatocytes, Leydig cells in testis and in all types of neurons. In heart muscle, its accumulation is associated with muscle waste and is known as "brown atrophy".

Blood-brain barrier (BBB) forms the protective shield around the brain, only letting in a select group of molecules. As it ages, lipofuscin cross-linking makes more rigid and leaky, so that unwanted things such as albumin start seeping through, causing brain inflammation and various neuronal pathologies*.

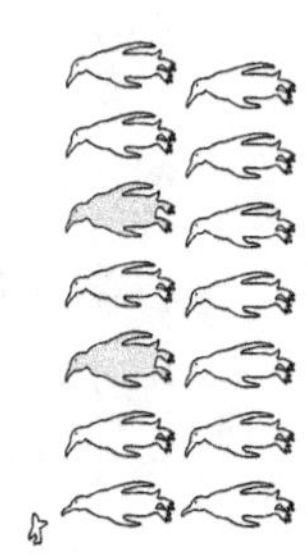

5.6. Hitting the goldilocks zone of LPO balance

"Man is born free, but everywhere he is in chains."
— Jean-Jacques Rousseau

"We forge the chains we wear in life."
— Charles Dickens

There are some paradoxical observations, where deficiency in antioxidant defences is associated with increased lifespan, with examples from popular fly, worm and rodent workhorse biological models. This was observed for both natural antioxidants such as peroxiredoxins and GSH-related machinery, as well as for vit E, which in field vole reduced both LPO and lifespan(58). Seemingly a contradiction to the Harman theory, it may rely on a hormetic mechanism to crank up alternative cellular defences. Keeping ROS such as H_2O_2 below certain level seem to be detrimental to longevity(59). The opposite is also true, at the same time! Increasing the levels of mito catalase delays aging(60), but increasing it too much is detrimental(61). Depending on their genetic background, mice lacking mitochondrial MnSOD are either stillborn, or die within 10 days of birth. What does it portend for the therapeutic window then, is it half open or half closed? To reconcile, it seems that only mild antioxidant deficiency may have positive effect. High ROS levels, for extended periods of time, cause severe damage and

* BBB can also temporarily turn leaky under stress.

early death, whereas low levels of ROS provide hormetic signalling and are therefore useful. This is also overlapping with multiple redundancy circuits, which are "analogous not digital", meaning a degree of individual variability. Unlike "more pre-determined" enzymatic processes, oxidative damage including LPO is much more random. And this ties up nicely with the fact that very genetically uniform nematode worms, placed in a highly defined and unchanging environment, still show a very variable individual longevity. As all cells in nematodes are post-mitotic (do not divide), these observations are likely to miss the effects of ROS and LPO on cell proliferation. In humans, centenarians tend to have higher levels of antioxidants such as carotenoids in plasma, but this may have to do with diets richer in fruit and veg rather than effects of carotenoids themselves[(62)]. Moderate exercise, a life extension method, may also act through hormetic mechanisms initiated by elevated levels of exercise-linked ROS and LPO.

While ROS species, apart from doing damage, play a signalling role and are important in recruiting the immune system components, various LPO species are just bad, and it is best to keep their levels to a minimum. And being the chain reaction products, they multiply exponentially, as compared to all other types of damage which are stoichiometric. Yet the repairs are stoichiometric too, and so once the LPO erupts, the defences cannot cope. Tentacles of LPO reach out every which way, smothering the vital processes and spurring the main contributors to aging. Epigenetics (the DNA methylation part) and telomere shortening can both be affected by LPO products, through DNA base damage (the methylation-undergoing CpG elements of genes as well as telomeres are guanosine – rich, and G reacts readily and irreversibly with RCS, see *Fig.19*), and through the impairment of DNA – processing protein machinery. Other aspects of aging, such as inter – or intracellular garbage accumulation, inflammation, vascular problems, cellular homeostasis and cell death are influenced by LPO even more directly. And, as discussed, this damage accelerates the formation of even more damage.

The profound effects of LPO chain reaction on all aspects of aging cannot be underestimated*. It is a unifying factor, the common denominator of our decline.

---

* As discussed in several chapters, LPO often operates in tandem with enzymatic PUFA oxidation, particularly with regards ARA. This is of particular importance for aging[(63)]. Controlling both, preferably in one fell swoop, would be the Holy Grain of PUFA

We do not notice these chains until they become too strong to be broken. Is it totally hopeless? (Spoiler alert: see *ch.6*).

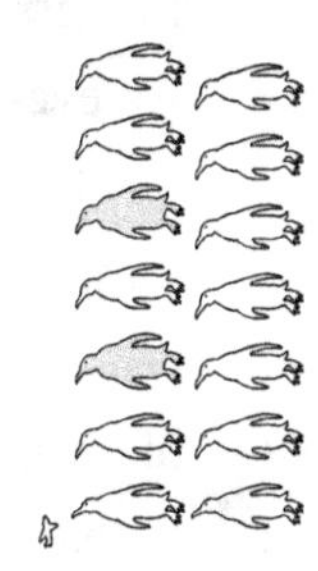

5.7. Long tentacles of LPOctopus stir up troble in every cell

"You can run, but you can't hide."
— Joe Louis

"No matter where you go, there you are."
— Confucius

Staring down the barrel of aging, we see the relentless onslaught of LPO, with the newly minted RCS rank and file marching forward day and night, as described in the previous chapters. We may not be aware, but our body is certainly mindful of the clear and present danger. Otherwise, why would it be expending 5% of its total energy output just for removing oxidized fatty acids from the brain alone[(1)]. Multiple systems are hard at work trying to contain and darn up the damaged patch. Various enzymatic and small molecule antioxidants can inhibit the LPO (*ch.2.8*). Some amino acids within membrane proteins, such as tyrosine and tryptophan, can also terminate the chain. Cysteine residues can accelerate propagation, while the side the methionines take, - pro- or anti-LPO, - is currently not entirely clear[(2)]. Glutathione machinery, dehydrogenases, reductases, heat shock proteins[(3)] and other systems are all there to preserve the membrane integrity and minimize the LPO product toxicity, but they easily get steamrolled over once LPO gets going. An orthogonal defence mechanism is based on Lands cycle (*Fig.11*), a very energy-expensive pathway which patrols lipid membranes 24-7, removing PUFAs from position 2 of PLs. An amazing thing about this cycle is its indiscriminate recycling of membrane

---

well-being, - and our resistance to aging.

PLs, irrespective of whether the PUFAs attached to PLs are intact or damaged[4]. An eye-specific form of the cycle is described in *ch.4.1*[*].

The lyso-lipid species formed then get re-acylated with PUFA residues at the same position, perhaps with the exact same fatty acids that had just been disconnected. This enzymatic step only uses undamaged PUFA. It sounds like a terrible waste, and it is, requiring large energy investment to keep the cycle rolling. Perhaps the cells opt for taking no prisoners, constantly sweeping the floor clean of debris in an earthquake-active location, and later deciding which items are FUBARred to be reused, and which can go back on their shelves. Even TGs in the adipose tissue undergo a similar recycling, again at the huge energy cost, but all in all minimizing the consequences of LPO-inflicted toxicity and trying to contain the chain reaction below a certain threshold state beyond which the damage is not allowed to progress, or if you prefer, degenerate. As the main fatty acid depository in the adipose tissue is TGs, this TG-FA re-forming cycle, consisting of lipolysis (disconnection of FAs from glycerol) and glyceroneogenesis (re-connection), is different from Lands cycle, which operates on PLs[5]. Organisms can also reduce LPO by dialling down their PUFA levels, as plants do under stress and some mammals do before hibernation. Cells can temporarily move their PUFAs to a safer storage location, such as lipid droplets, while weathering the LPO storm[6].

All these mechanisms strive and toil to slow down or halt LPO progress, and they may briefly succeed, for the time being. Riffs may be furrowing through the lipid membranes, resembling the deeply carved grooves of the San Andreas fault that scar the landscape a few miles south of Tijuana. But for now, rancidity and Armageddon have been prevented, LPO has been reined in, and all is temporarily quiet on the **Festern front**. Lysosomes and peroxisomes[**] get busy, digesting the oxidised PUFAs, and/or expelling their corpses to the outside for removal through the kidney. The apocalypse has been postponed once again, with prayer and profanity. But not for long, not for long.

The problem is that not all the garbage goo can be digested or removed. Inex-

---

* Perhaps this should not be surprising. With so many different PUFA oxidation products possible, it is unlikely that a specific phospholipase could have evolved for each possible type.

** And mitochondria, which burn fatty acids through the beta-oxidation cycle.

orable and irreversible build-up of waste, deposited as varieties of lipofuscin (or browning, using Steven Austad's term), can catalyse more LPO, making the attacks more and more vicious. We have looked at a selection of organs and diseases in previous chapters, but the problem is much wider. Metabolic[7] and pulmonary[8] conditions, hearing loss[9], baldness[10], liver[11] and kidney[12] pathologies, muscle tissue issues[13], compromised sperm function[14], red blood cell malfunction[15], etcetera - the **LPOctopus** reaches out far and wide, smothering everything with an iron tentacle (*Fig.4, Fig.20*). To cover these, and many other conditions would require a monograph several times the size of the one you are reading now. Yet distilling this ocean of data down to a punch line suggests that LPO is truly the common denominator of damage and the biggest problem still to be tackled, whichever way one looks at it.

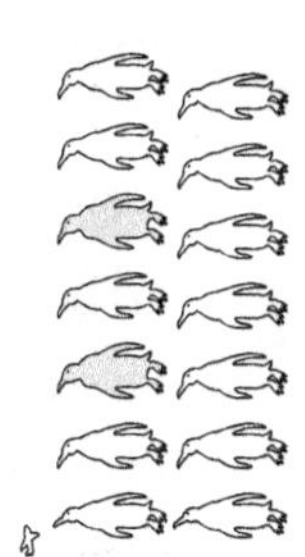

Or hears about it... If we were in a quiet lab, those of us with sharp enough hearing might even be able to percieve a commotion: the jiggling of lipids in the membrane, the popping of ROS, and the humming and soughing of the chain reaction, like rustling beetles in a matchbox. With a microscope that could livestream radicals in a ravaging LPO as blue dots, zooming in on a lipid membrane*, we would see this coal-black night sky with scattered, flickering ROS Death Stars. As the LPO speeds up and spreads wide, the number of stars skyrockets into myriads of bright blue dots, twinkling non-stop all over the visual field. Beautiful! And the constellation they seem to form is – wait a minute, it looks crab-shaped, - oh, no, can this be? – it is - Cancer**...

When buying a pair of jeans, one would expect them to wear out, but where exactly would they fray might be more difficult to predict. At the hems of the pant legs? Back pockets? Laps? How about the exact hole shapes, size and location? This would depend on your body shape, and the environment. Just like the LPO. Omnipresent, but what

---

* Which appears pitch-black due to light refraction, resembling a night sky with zero light pollution; there is even a popular membrane-studying technique known as BLM, "black lipid membrane"

** LPO is massively involved in cancer. This fascinating and complex topic is huge and merits a dedicated narration, so it is not covered in this book.

would give way first is very individual, and likely impossible to predict. Using this jeans analogy, classifying diseases by organ, or organelle, may be somewhat misleading. Things are additionally complicated by the fact that diseases do not arise fully fledged, like Botticelli's Venus emerging from the sea in her shell*. A better way might be to reach upstream, from where the trouble trickles down and spreads, and turn off the leaking tap. By making your jeans out of wear-and-tear-resistant tarpaulin perhaps.

The next chapter suggests a way to turn off the LPO faucet.

6. Heavy handling of lifespan-shortening problems

"I don't want to achieve immortality through my work. I want to achieve it by not dying."
— Woody Allen

"To live forever or die in the attempt."
— Joseph Heller

The previous chapters may have painted a bleak, pessimistic picture. We cannot live without PUFAs, oxygen, or metals. Yet the inexorable laws of chemistry, - laws as old as the Big Bang if not older, prescribe for oxygen to combine efforts with transition metals to destroy PUFAs. Damaging the membranes and generating a bunch of toxic RCS nasties along the way, through an LPO chain, - the only non-enzymatic chain reaction in biology. LPO is not controlled by enzymes, so evolution could not have worked out the way to keep it in check directly**. And

---

* Another complication is the misalignment between the number of symptoms (in their tens – blood pressure, body temperature etc), - and the number of known diseases (tens of thousands; there are 6,000 orphan diseases alone), making (early) diagnosis challenging.

** Various indirect ways, – COX and LOX initiating the LPO chain by generating LOOH; enzymatic control of antioxidant levels; regulation of metals and oxygen; and many others,– are sadly no match for the wayward and recalcitrant nature of LPO.

adding insult to injury, antioxidants are inefficient against this oxidation, and sometimes toxic. Nature is OK with such an arrangement provided we last long enough to leave offspring. Even if chance and luck would bestow a long life on someone, ultimately even that person would still succumb to oxygen, rust, grind to a halt and pass away, as attested by the examples of a handful of supercentenarians.

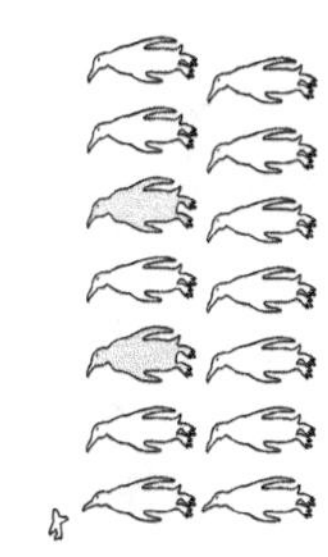

Could there be a way to mitigate LPO?

6.1. Good vibrations, good isotopes, good isotope effect

"I thought it might have a practical use in somethinglike neon signs."
— Harold Urey

"Not all chemicals are bad. Without chemicals such as hydrogen and oxygen, for example, there would be no way to make water, a vital ingredient in beer."
— Dave Barry

Biochemistry deals with interactions of valence electrons occupying the outermost layers of electron shells in atoms. Nuclear chemistry focuses on reactions that happen inside atoms. The latter are more violent as there is more energy "stored" in nuclei than in electrons* Valence electrons are easily accessible, always on tap. Unlike the nuclear particles, which, due to their huge energy content, are stored out of the harm's way, behind a difficult-to-open, massive armoured vault safe door of the atomic nucleus. That's why, where a biochemistry lab would be content with heat blocks and hot plates, a nuclear chemistry

* This is to do with the strong nuclear forces that hold protons and neutrons together being significantly more powerful than the electromagnetic forces that govern electrons.

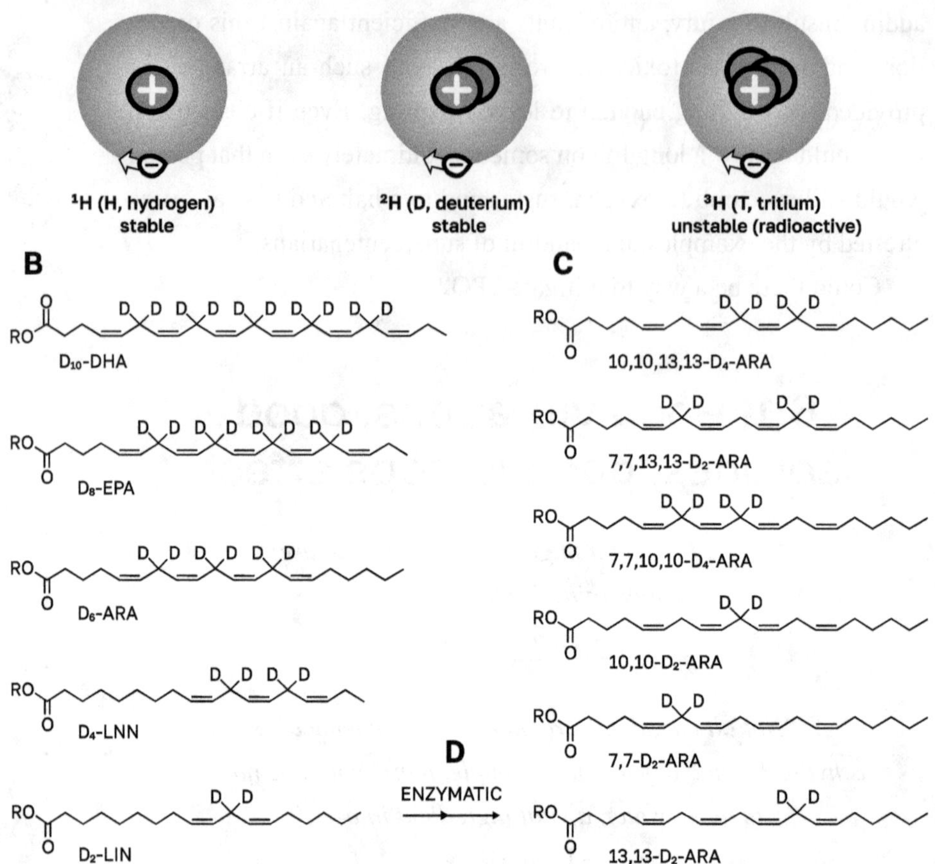

Figure 28. A, Hydrogen isotopes. Tritium would give the strongest IE, but cannot be employed as it is radioactive. **B,** some D-PUFAs, made at scale and tested in various animal models and in humans. **C,** a library partially deuterated ARAs, made to study site-specific enzymatic oxidation at different bis-allylic positions. **D,** enzymatic desaturation and elongation of D2LIN to D2ARA, observed in animal models and humans.

---

lab would require atom smashers and particle accelerators.

An interesting phenomenon emerges at the overlap of the two chemistries. Masses of protons and neutrons are roughly the same, at one atomic mass unit (1 amu). For a nucleus of a given chemical element, the relative numbers of positively charged protons, and gluing-them-together, non-charged neutrons can vary. Step too far from the optimal ratio though, and an atomic nucleus becomes unstable. Nuclei of uranium-238 (atomic mass 238), made up of 92 protons and 146 neutrons, are relatively long lived. But take away 3 neutrons, and the lighter version, ^{235}U (92 protons, 143 neutrons) becomes significantly less stable, - and a much sought-after fissile material for various energy-demanding applications, civil and beyond. The two varieties are called isotopes (from Greek, *isos* equal, *topos* place). But if we do not step too far from the optimal proton-neutron ratios, we can have several stable atoms of the same type, yet with different nuclear masses.

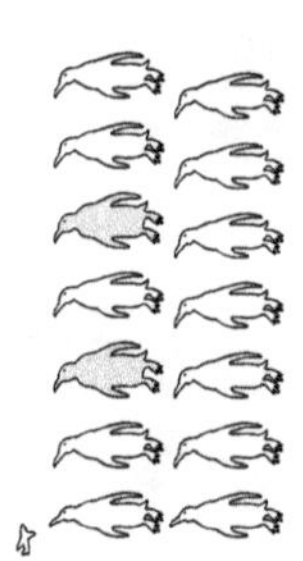

Biologically relevant examples include two stable carbon isotopes ^{12}C and ^{13}C, and two stable varieties of hydrogen: ^{1}H, with one proton; and the heavy version, with a proton and a neutron (^{2}H), also known as deuterium (D), discovered by Harold Urey in 1931*. Its natural abundance (0.0156 % in sea water) means that roughly every 6,500th atom of hydrogen is, in fact, deuterium.

An important contributing factor to the strength of a chemical bond is vibration. As an analogy, if a sphere attached to a spring oscillates, the rate of the oscillation depends on the mass of the sphere. The heavier the sphere, the slower the vibration, - and the stronger the bond. Ignoring the pitfalls of using soccer as an analogy to explain quantum mechanics, consider two identical footballs, one filled with water, and the other with air. They look identical at your feet, yet the effort required to kick them the same distance would be substantially different. Heavier isotopes of the same element are a different ballgame as they vibrate slower, forming stronger bonds. Of all stable isotopes, H and D have the biggest mass difference (1 versus 2, or 100%, see *Fig.28*). A pH measurement easily reveals the bond strength difference (*Back-of-the-envelope-12*). This good vibration of D is known in chemistry as the isotope effect (IE).

---

* Talking about stepping too far from the optimal proton-neutron ratios. A hydrogen atom may have not one but two neutrons, giving it a mass of three: ^{3}H. Known as tritium (T), this variety is unstable (radioactive), and so cannot be used for long term biological applications, unlike deuterium (see below).

6.2. D-PUFA: the α-male of the ω-fats

"There is pleasure of recognizing old things from a new viewpoint."
— Richard Feynman

"If your experiment needs statistics, you ought to have done a better experiment."
— Ernest Rutherford

As discussed in *ch.2.6*, the slowest, rate-limiting step of the LPO process is the abstraction of hydrogen from the CH_2 moieties located between two double bonds, known as bis-allylic methylene groups. In agreement with a chain being only as strong as its weakest link, if this step were made even slower, the overall rate of LPO process would decrease. Because C-D bonds are more resistant to breaking compared to C-H bonds, replacing the hydrogen atoms of these CH_2 groups with deuteriums (to form CD_2) would reduce the rate of abstraction, slowing down the LPO process, owing to the IE. And since deuterium has natural abundance, - there is more than a gramme in a human body (*Back-of-the-envelope-13*), in water and other molecules, - no enzymatic "rejection" would occur, as the body recognises deuterium as a normal part of life and does not favour H at the D's expense. What if we chemically plugged the oxidizable, bis-allylic sites of PUFAs with deuterium "fire retardants", converting them into D-PUFAs (*Fig.28*)? As essential nutrients, PUFAs have to be supplied with food, so there would be no danger of a body making its own H-PUFAs, diluting the supplemented D-PUFA pool. Might that slow down the LPO???

The very first test to answer this question was performed by a UCLA professor Catherine Clarke in 2010 in Experiment Number 1, employing her preferred workhorse: yeast. The yeast she used could not make their own PUFAs but would avidly take them up from the medium, incorporating them into all their membranes. What the yeast did not know was that they were engineered to lack one of the main LPO-stopping antioxidants, CoQ. They could live OK without it, - but stressing them, for example by giving them some normal (but alien to them) PUFAs, would be lethal. Oblivious to the mortal danger, the mutant yeast would happily incorporate the PUFAs, - and then the ensuing LPO, exacerbated by the lack of the key antioxidant, would swiftly kill the bugs. Feeding the yeast the D-PUFA version made them 150 times more resistant to LPO[1], making the Rutherford quote above, frowned upon lately by the statistics-loving biomedical community, fully justified for this particular experiment.

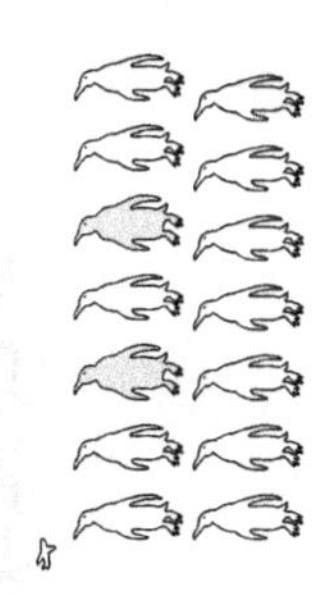

D-PUFAs protect against LPO in a way that is very different from how the antioxidants work. Imagine a settlement, surrounded by a palisade of wood logs dug close together into the ground. A group of peaceful settlers is huddling inside, trying to fend off the non-stop attacks of the outside crowd of mean unfriendlies. The settlers fire at the attackers, eliminating them one by one. But the unfriendlies can shoot the settlers dead, too. Worse, they are constantly trying to set the flammable palisade on fire, by setting up bonfires, shooting burning arrows into the wooden logs, throwing petards, lit cigarettes, all sorts of things. If a log is burned down, the settlers can put a new one in its place - but they would be even more exposed while doing it. The supply of logs is limited, as is the settlers' ammo… Worse, once one log is on fire, the flame will skip sideways, damaging big chunks of the palisade through a chain reaction, which in this case would be more efficient than a chainsaw… Let me introduce the actors: Unfriendlies: ROS. Settlers: antioxidants. Wood logs: membrane PUFAs. Sending in more armed settlers might give a temporary relief, - but the unfriendlies outnumber them hugely, and procreate fast. Yet all that is needed to fix the problem for good is some cement and sand! Make those logs of concrete – identical shape and size, - and use them instead of wood. Voila! The settlers can now leave a couple of guards to make sure the bad guys would not dig, ram through or climb over, - and go have some long-deserved rest. Meet the saviour: D-PUFA, played by a concrete log.

Admittedly, concrete logs are a bit of a poetic license. D-PUFAs get oxidized, too. It just happens slower. But this delay may give a

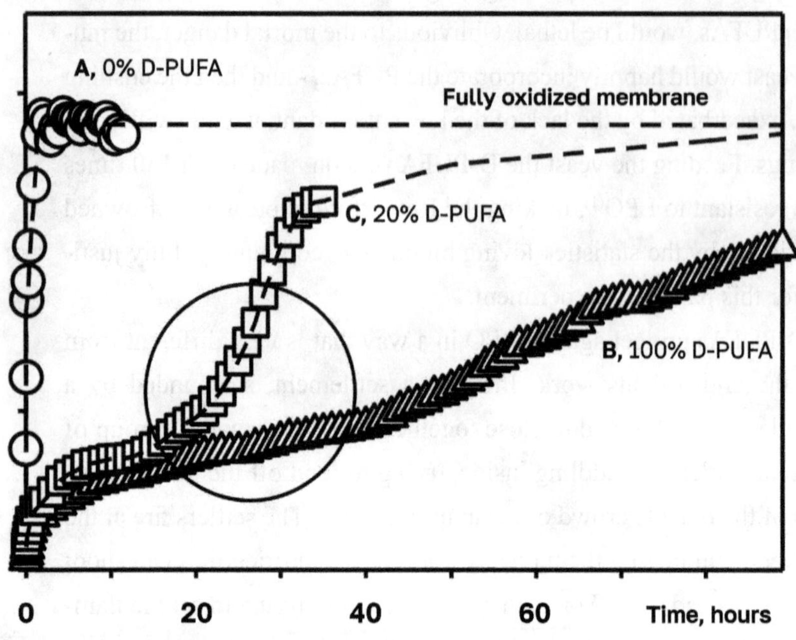

Figure 29. A, instant damage of the H-PUFA membrane, **C,** gradual oxidation of the D-PUFA membrane, and **B,** critical breakdown.

---

chance to antioxidants and other repair mechanisms to get a grip on LPO, catch up with it and keep the problem in check.

With D-PUFAs being that protective, one might be forgiven for contemplating a possibility of using them, in due course, in humans. With one of the first thoughts, naturally, being: with the large stock of H-PUFAs already in the body, and a significant daily dietary intake, what would it take to make sure that most PUFAs in cell membranes are the resilient D-PUFAs? Should people come off their normal diets and take tens of grammes of D-PUFAs daily, for months, waiting for the natural PUFA turnover to deliver enough D-PUFA into all lipid nooks and crannies? This looked like a daunting task – until Cathy Clarke had run her Experiment Number 2. This time, she did not feed the yeast on pure D-PUFAs. Instead, she diluted them with regular H-PUFAs, testing different ratios. The result was unexpected, yet much welcome. Having only about 25% of D-PUFA, on the background of 75% H-PUFA, in the yeast membranes was as protective as having

100% D-PUFA[2].

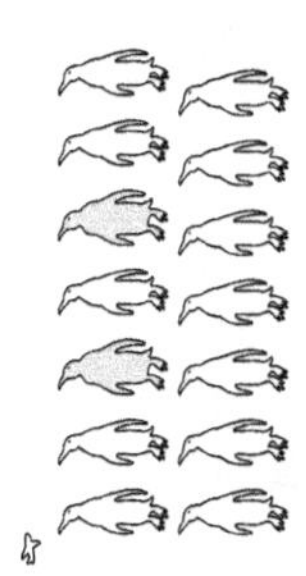

To delve deeper into this fractional D-PUFA protection, liposomes, which model lipid membranes, were employed to reproduce the bilayer structure without any known or unknown living cells-associated biochemical "contamination", like antioxidants & Co. Liposomes with variable quantities of different D-PUFAs in their membranes have been exposed to a cocktail of LPO-initiating chemicals[3]. The graph (*Fig.29*) reveals the favourable feature of fractional protection. 100% H-PUFA membranes (circles) get busted within minutes, while 100% D-PUFA membranes (triangles) slowly smoulder for days, eventually getting oxidized. Membranes with about 20% D-PUFA are holding out for about 20 hours in a way similar to the 100% D-PUFA membranes, - till they can no longer hold out (circle). The ensuing LPO then finishes the membranes off – but still not as quickly as with 100% H-PUFAs. This delay may thus give alive cells a chance, using complementary antioxidant and clean-up defences in tandem with D-PUFAs, to maintain the LPO-free membranes, without the need to have every single PUFA in membranes deuterated, - the level which would be practically unattainable for humans.

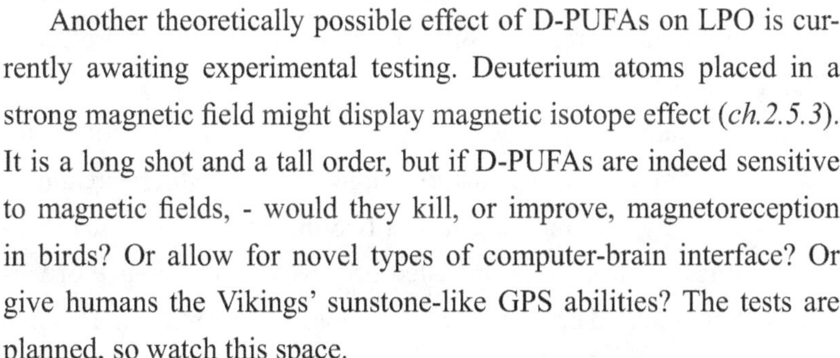

Another theoretically possible effect of D-PUFAs on LPO is currently awaiting experimental testing. Deuterium atoms placed in a strong magnetic field might display magnetic isotope effect (*ch.2.5.3*). It is a long shot and a tall order, but if D-PUFAs are indeed sensitive to magnetic fields, - would they kill, or improve, magnetoreception in birds? Or allow for novel types of computer-brain interface? Or give humans the Vikings' sunstone-like GPS abilities? The tests are planned, so watch this space.

A cartoon may help you visualize the protective effect of D-PUFAs on the chain reaction of PUFA oxidation. To see it, flick through the top page corners with your thumb. The plot is inspired by the Royal Air Force pilots' stories (see *ch.2.6.1*).

6.3. D-PUFAs can break the LPO chains of aging and age-related diseases

"Chains of habit are too light to be felt until they are too heavy to be broken."
— Warren Buffett

"You have nothing to lose but your chains."
— Karl Marx

Significant effort was made trying to understand the protective effect of low D-PUFA levels. Ned Porter, the key authority on LPO from the Vanderbilt University, measured the IE on the chain oxidation of D-LIN and D-LNN in organic solvents[(4)]. In presence of vit E, the IE for D-LIN chain oxidation was found to be 23 (the rate reaction constant was that much slower for D-PUFA related to H-PUFA), compared to 10 in the absence of vit.E. For D-LNN in presence of vit E, the IE was a massive 36. The increase of IE in the presence of vit E is unexpected but useful, as vit E is naturally present in bilayers, likely enhancing the protective effect of D-PUFAs in vivo*.

These IE values, obtained for PUFAs in solution with chaotically distributed molecules, differ from values in structured lipid bilayers, which streamline the chain reaction laterally, by confining it to a two-dimensional space. In cellular membranes things are further complicated because of the presence of various ROS-intercepting antioxidant and repair systems. The mentioned 150-fold increase in yeast survival is difficult to translate into an IE value. For biological applications the increase in survival is a more useful metric compared to IE as it reports on the ultimate protection arising from the combined action of all process-

* IE values in excess of around 7 are believed to involve the quantum tunnelling mechanism, when a particle penetrates through an (energy) barrier which it should not be able to pass. Deuterium, being twice as heavy as hydrogen, would have a much more difficult time crossing that barrier.

es associated with living cells*.

There are other useful but not yet fully understood aspects of D-PUFA (*Box 11*).

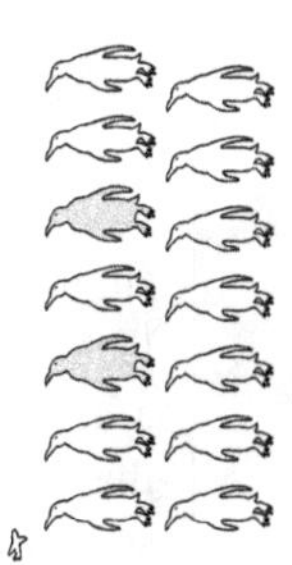

Box 11. THE D-PUFA MAGIC

"It's still magic even if you know how it's done"
— Terry Pratchett

In her Experiment Number 3, Cathy Clarke made another baffling yet welcome observation. LIN partially deuterated at the bis-allylic position, with bis-allylic methylene groups only carrying one D atom instead of two (CHD), would be expected to be as oxidisable (and hence toxic to yeast) as the non-deuterated version, because of the hydrogen atom still available to be abstracted by an ROS species (*ch.2.6*). Yet when yeast incorporated D,H-PUFA into their membranes, the protection they gained was very

Secondary IE | Primary IE

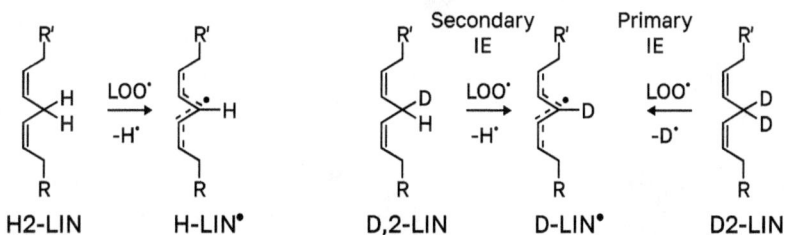

LOO˙ / -H˙ | LOO˙ / -H˙ | LOO˙ / -D˙

H2-LIN | H-LIN• | D,2-LIN | D-LIN• | D2-LIN

similar to that provided by the fully deuterated D2-Lin[2]. The explanation may have to do with the fact that the radical initially formed from LIN upon hydrogen abstraction has a hydrogen atom attached to the carbon-centred radical (H-LIN˙), while radicals formed from both D,H-LIN and D2-LIN are one and the same: D-LIN˙, with a deuterium atom attached to the carbon radical. As the efficiency of hydrogen abstraction from H2-LIN and D,H-LIN is very sim-

* The standard notation is called the kinetic isotope effect (KIE), defined as the ratio of rate constants in solution: KIE = k_H/k_D (rate constants link the rates of chemical reactions with concentrations of reagents). Membrane-confined reactions differ from those in solution, so a different acronym, IE, is used here. The large values observed (IE of 23 in the absence of vit E), compared to a stoichiometric IE (KIE is 6 for a stoichiometric D2-LIN oxidation[5]), are due to the chain reaction format, where the IE may be "amplified" along the chain, resulting in substantially larger overall values. This explains the extent of the D-PUFA protection, as well as the result of Cathy Clarke's Experiment Number 2 (see above). In fact, the concept of "amplifying" the IE along the chain might also eliminate the need to invoke the concept of tunnelling.

ilar, it is unlikely that this step defines the rescuing effect of D,H-LIN on the cell survival. But then the only remaining possibility is the different action of H-LIN˙ and D-LIN˙ on the next step of the process. This is a very unorthodox explanation, as the accepted wisdom suggests that these two radicals should undergo the next, fast step of reacting with molecular oxygen equally avidly. There is a phenomenon known as a secondary IE, whereby a heavy isotope that remains attached to the atom whose other bond gets cleaved affects that bond's strength. However, the secondary IE in D,H-LIN was ruled out in a series of separate experiments(2). Be it as it may, the protective effect of the partially deuterated PUFAs is very useful because D,H-PUFA are present in the synthetically manufactured D-PUFA material.

How can the D-PUFAs be made*? The saga of going through all the trials and tribulations of working out the protocols of D-PUFA synthesis, and the incredible twists of fate pertaining to the relevant patents, while sadly outside the remit of this story, may well be worth a dedicated book. Watch this space.

* Expensive multi-step total synthesis methodology, meticulously developed and scaled up by Andrei Bekish and Vadim Shmanai over ten years(6), has been replaced by the cutting-edge catalytic approach conceived by Dragoslav Vidovic, and reduced to practice by Alexei Smarun. It took years of perseverance and a touch of serendipity(7), finally paving the way for mass-production. John Bower at Liverpool continues pushing the envelope, providing insights for the next generation processes. For many years, Vadim's laboratory has been the purveyor of D-PUFAs for cell, animal and human testing, continuing to develop and supply the materials even when I could not provide any funding for his efforts – a regrettably frequent occurrence. Dragoslav magnanimously assigned his part to me. With so many colorful characters inveigling their way into the reinforced lipid world and scrambling for control using all sorts of veiled tactics, where would D-PUFAs be, without my great friends and superstar collaborators? Thanks a lot, guys!

6.4. PUFA deuteration against age-related degeneration

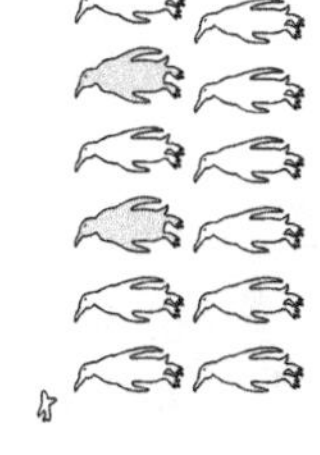

"In testing medicine, we navigate the unknown, driven by hope and guided by science."
— Anonymous

"In the world of science, there is no such thing as too many experiments."
— Carl Sagan

D-PUFAs have been tested in many different systems. Yuri Antonenko's group at Moscow State University has incorporated D-PUFAs into liposomes to find that D-LIN, D-LNN, D-ARA, D-EPA, D-DHA are increasingly resistant to LPO, roughly in line with the total number of the deuterated bis-allylic sites per PUFA molecule[3]. D-PUFAs in liposomes have been found by Mauricio Baptista at Sao Paulo University to inhibit the light irradiation-induced oxidative membrane damage[8], a finding relevant to skin health (see *ch.4.7*).

Murine and human cell models of various diseases known to involve LPO in their aetiology were used to test the D-PUFAs. Compared to cells treated with normal PUFAs, D-PUFAs were found to be protective, reducing cellular stress and extending lifespan. Rob Wilson and co-workers at UPenn have rescued Friedreich's ataxia cells with D-PUFAs by substantially reducing the FA-associated LPO levels[9]. The already mentioned GPX4 is the major hub of the GSH network, responsible for reducing lipid hydroperoxides (LOOH). GPX4 malfunction dramatically increases oxidative stress and inflammation, with relevance to AD and PD. Brent Stockwell's lab at Columbia University has offset the GPX4 deficiency-associated pathology with D-PUFAs[10]. Plamena Angelova at UCL has treated stem cells derived from Progressive Supranuclear Palsy (PSP) patients with D-PUFAs, substantially reducing LPO and oxidative stress. She also reported on the increased levels of GSH and improved mitochondrial morphology, function and membrane potential upon D-PUFA treatment[(11)]. A separate bioenergetics study, carried out by Aleksandr Andreyev at UCSD, has revealed substantial improvements in several mitochondrial parameters, suggesting D-PUFA applications not just in mitochondrial diseases, but also as boosters or normalizers of mitochondrial performance, including in stressful or endurance conditions[(12)]. A cell model of PD suggests that oligomeric aSyn recognizes oxPLs in membranes and squeezes itself into the bilayers, initiating even more LPO in a vicious cycle (see *ch.4.2*). Andrey Abramov's lab at UCL has shown that D-PUFAs can completely stop this aSyn catch-22[(13)]. Useful mechanistic details of ferroptosis such as the involvement of LOX enzymes have been gleaned from cell model studies in the labs of Brent Stockwell and Derek Pratt (Ottawa) using D-PUFAs to stop the LPO, terminating ferroptosis[(14)].

As discussed in *ch.5*, cell culture data, particularly in relation to oxidative stress, can be difficult to interpret if not misleading*, so various animal models have been employed to test D-PUFAs. Many people oppose the use of animals in research, and not just for ethical reasons. Rodents are simply too far away from humans biologically for the findings to be of relevance. If mice were people, we would have conquered cancer a long time ago. Rodents process PUFAs differently, which may partially explain their shorter lifespans. The relevance of neurological, retinal and other disease models to humans is also questionable. How can AMD and AD be modelled in mice if their retinas do not even have maculae, and their brains do not form aBeta-based Lewy bodies**? How can answers to subjective questions be obtained ("How are you feeling today, Mickey? Is your brain clear, or a tad foggy?")? For this reason, people are not talking about say a PD mouse model, but a PD-like phenotype. Still, PUFAs are essential for rodents

---

* On the super high oxygen levels, wrong PUFAs in the medium, and more (see *Box 8*). Suppose measurements reveal high GSH level in cells. Does it mean that oxidative stress is high, and so cells produce more GSH to cope? Or oxidative stress is low, leading to excessive antioxidant levels? Things can only be looked at in context.

** Degu, a cute South-American rodent, does develop the aBeta plaques. More generally, rodents require a dietary choline supply (humans don't), but make their own vit C (humans don't), in addition to the PUFA processing differences, so even extrapolating the general lipid processing data from rodents to humans may not be straightforward.

too, so various insights can be gleaned from the animals, - the rate of D-PUFA incorporation and turnover, some basic disease-relevant mechanisms, distribution of LPO products, memory tests, dose dependence of therapeutic effects (% of D-PUFA in tissues required for disease mitigation), etc. One can give animals 100% D-PUFAs, not diluted by any H-PUFAs, because the dietary composition can be fully controlled – unlike in humans. Experiments would always include a control group with H-PUFAs, so the only difference between all studies would really be the difference between H and D. Positive effects of D-PUFAs have been reported by Jimmy Berbee at Leiden (tragically, Jimmy has passed away, at his prime, during the Covid lockdown) in atherosclerosis mice(15), and Tiffany Thomas at Columbia used D-PUFAs to extend the resilience of red blood cells(16). Mice with Huntington's-like phenotype have been shown by Marie-Francoise Chesselet at UCLA to benefit from the D-PUFA diet(17), and Amy Manning-Bog together with Bill Langston at the SRI reported the same for the PD-like phenotype-bearing mice(18). The late Flint Beal at Cornell, NY demonstrated the strong protective effect of D-PUFAs on the aSyn version of PD-like phenotype in rats(19), corroborating the earlier cell culture findings on the aSyn-LPO vicious cycle (see above). Mark Mattson at the Johns Hopkins treated double-transgenic mice with AD-like phenotype with D-PUFAs, observing the reduction of several aBeta peptide levels(20). Brian Bennet at Kingston, Ontario has run several behavioural tests in ALDH2–deficient mice (which would normally display memory losses consisted with AD, see *ch.2.6.3*) fed on D-PUFAs, reporting the full recovery of their memory capacity back to the wild type level, whereas the H-PUFA control group's memory was badly affected(21). This full protection effect was achieved with the D-PUFA levels making up only about 50% of the total PUFAs in the lipid membranes, attained after several weeks of feeding. The seemingly exceptional degree of protection provided by D-PUFAs in this ALDH2 knockout model has started looking less exceptional when Joshua Dunaief at UPenn used D-DHA to successfully rescue an acute AMD-like phenotype in mice, prevent-

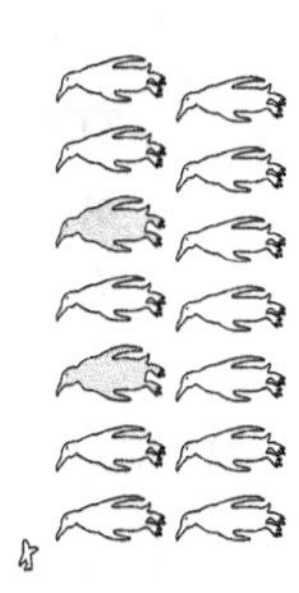

ing the onset of blindness. Two weeks of D-DHA dosing provided a 70% protection; three weeks raised it to 90%; and at four weeks on the D-DHA diet (reaching around 50% of D-DHA in the eyes), the protection has been 100%[22]. LIN, LNN, ARA and to some degree DHA are the main PUFAs in muscle tissue, and age-associated and disease-associated muscle waste conditions such as sarcopenia are linked to LPO, and therapeutic options for preventing muscle atrophy are limited. Diabetes-associated increased oxidative stress accelerates muscle weakening and waste. In a diabetes type 1 mouse model, D-PUFA abolished oxidative stress and prevented muscle atrophy and weakness[23], as has been shown by Hiroaki Eshima in Japan.

D-PUFA are distributed and accumulated in tissues identical to H-PUFA. As a case in point, D-DHA accretion in mice was the fastest in plasma and liver, followed by heart and RBCs. It was slower in retina ($t_{1/2}$ > 20 days)*, while in CNS the build-up was the slowest at $t_{1/2}$ = 29 – 44 days. In retina, the therapeutic target window (20-50% of D-PUFA in membranes) was reached in 2-4 weeks. A robust conversion of D-DHA into retina-specific VLC-PUFA (see *ch.4.1*) has also been confirmed[24].

That the protective effect of D-PUFAs varies from moderate through substantial to strong is not unexpected, as different pathologies may have LPO involved to a different degree. What is surprising is the scope of conditions in which LPO plays a role, - and can be reduced by D-PUFAs. This, and the fact that the protective effect shows up in different animal models of the same disease, speak to the common denominator status of the LPO, affecting every nook and cranny of the body which might have PUFAs in it, - in other words, in every cell.

While inflammation is involved in numerous pathologies and often goes hand in hand with LPO, it relies on enzymatic, stoichiometric oxidation of PUFAs, - that's why it is considered as a separate group in this chapter**. Eicosanoids are

---

* Briefly, $t_{1/2}$ for drugs and for PUFAs is different. Drug absorption is determined by their polarity, while PUFAs are taken up by enzymes. Drugs usually do not accumulate in tissues and have target and off target effects, although none of these evolved to bind drugs. In contrast, PUFAs are substrates for dozens of enzymes. Most drugs are eliminated through a tandem of cytochrome oxidation – oxidation product removal pathways, while PUFAs are burned to CO_2 and water.

** Even though this book is about the chain reaction, the D-PUFAs modulate the acproduced by COX and LOX enzymes with the same rate-limiting step, - hydrogen abstraction at a bis-allylic site of the PUFA. However, it is not a chain reaction (see *ch.4.9*). Most eicosanoids are pro-inflammatory, while notable exceptions, such as lipoxins, resolve (clean up) inflammation. How might D-PUFAs affect this fine balance? Edward Dennis, the world's top authority on lipidomics at UCSD, has used a collection of cells each expressing one particular type of COX or LOX, loaded with one of D-ARA derivatives from a library of seven partially or fully deuterated ARA derivatives (*Fig.28*). He then employed cutting-edge mass-spectrometry to check how each derivative is processed into various eicosanoids by different enzymes. This tour de force study, having looked at many hundreds of eicosanoid ratios, found, expectedly, that deuteration reduced the levels of pro-inflammatory lipid mediators. However, contrary to expectations, deuteration increased the levels of anti-inflammatory lipoxins! While Ed has come up with a plausible explanation[25], an animal test was needed to check if D-ARA would indeed be anti-inflammatory at the level of the entire organism. These plans were delayed as the ravaging Covid pandemic threw a spanner in the works, with lockdowns confining people to their homes, and many a university vivarium being forced to cull their animals as a result. Still, Alla Molchanova at the Institute of Physiology in Minsk managed to dodge the lockdown, successfully testing the anti-inflammatory action of D-ARA in mice, while using H-ARA in a control group. Lung inflammation, triggered by LPS (a bacterial material that kick-starts inflammation in mammals*), was significantly lower in the D-ARA group[26]. D-ARA is a double helping of benefit, a double barrel weapon against various aspects of inflammation and

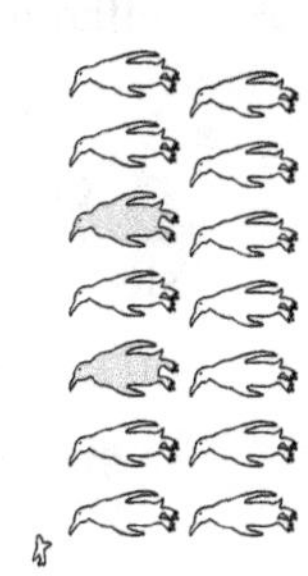

tion of COX and LOX in a very favourable way and accordingly are discussed here as a novel anti-inflammatory approach.

* Mice differ from humans in various aspects of their LOX toolkit. For example, humans have 5-LOX, 12-LOX and 15-LOX, while in mice, in addition to those, there are also 8-LOX, 9-LOX and 13-LOX. For this and many other reasons, medicines intended for humans should be tested in humans. "Render unto Caesar what is Caesar's"

inflammaging, reducing both LPO and pro-inflammatory eicosanoids[27]. The real McCoy of D-PUFA*.

Once in the body, would D-PUFAs become a part of a PUFA turnover, constantly being disconnected and reformatted as TGs and PLs, like in photoreceptor outer segment renewal and Lands cycle (see *ch.5.8*), - or could they get stuck in membranes for good, on account of their increased resistance to oxidation? The experiments with D-LIN, D-ARA and D-DHA have proven that when the dosing is switched from D-PUFAs to H-PUFAs, D-PUFAs do get washed out from organs and tissues, albeit at somewhat slower rate compared to the rate of build-up. For example, mice fed D-DHA as the only n-3 PUFA reached the retinal level of 90% in 10 weeks, while the washout over the following 10 weeks reduced that level to about 20%[22]. Elimination from other tissues was also typically a tad slower than the build-up.

These and many other experiments have all been pointing in the same direction. It seemed that the D-PUFAs have earned their stripes, by efficiently mitigating, or delaying, the horrors of aging and disease described in previous chapters. The time had come to test the compounds in humans. But before we delve into that, let's briefly take a look at one of the most intriguing and enigmatic phenomena in human physiology (*Box 12*).

Box 12. PLACEBO EFFECT: IS IT ALL IN MIND?

"The placebo cures 30% of patients – no matter what they have."
— David Kline

"If your experiment worked – great! If not – that does not mean anything."
— Judy Campisi

Having almost single-handedly begotten the need for clinical trials, the placebo effect is well known and widely exploited – just ask the cosmetics companies. While the extensive literature on the topic allows the reader to delve deeper, below are some cherry-picked facts.

* The large inhibitory effect of D-PUFAs on LPO has to do with the chain reaction format, exponentially amplifying the IE with every step of the chain. Enzymatic PUFA oxidation is linear (stoichiometric), so the effect of D-PUFAs on COX and LOX will be smaller, particularly in presence of H-PUFAs.

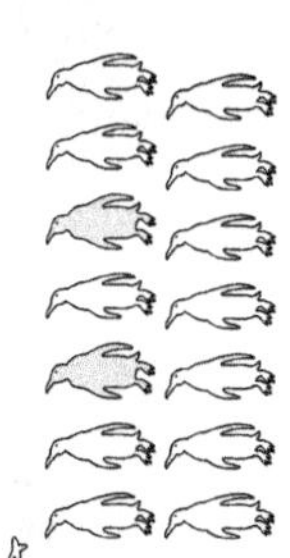

Dopamine, the key feel-good chemical, is implicated in the placebo machinery[28]. This makes clinical trials for diseases that involve the dopamine imbalance (PD, Schizophrenia, ADHD, HD, - note that they all have LPO in their aetiology) challenging, as taking a placebo, while believing it to be a cure, makes one feel better – and that feeling increases dopamine... Placebo's alter ego is nocebo, where harm is expected, and duly felt, from taking a harmless substance labelled as poison.

Excluding trusted caregivers whose presence allegedly makes animals feel less pain, placebo does not affect animals, nor infants. In adults, however, placebo is multifaceted. Even the colour and shape of pills affect the study outcome. Red and yellow placebo pills have been found to be stimulating, while colour blue was depressive[29]. Too bad there was no Ig Nobel prize when that paper was published. Because since then, several placebo varieties were awarded the distinction. Expensive fake medicines[30] and real drugs[31] have been reported by the patients to be more efficacious, and the more expensive, the better (Ig Nobel prize, 2008). Presumably, more expensive drugs are perceived as the real deal, boosting the dopamine, - particularly considering that that study looked at PD. A more recent permutation of the effect was the perception that drugs with (mild) side-effects must be actually doing something, - and hence be real[32] - with duly reported benefits (Ig Nobel prize, 2024). How could one navigate through this minefield? Perhaps making the clinical trial patients aware that the red pill they are ingesting costs $1,500 per dose? Or telling both the drug and the control groups that the green pill they are taking is dirt cheap? I wonder what Ben Goldacre would say.

On average, placebo accounts for 30-50% of the treatment effect in clinical trials. Placebo effect typically peaks at 4-6 months and starts wearing off after a year[33], so long term supplementation with D-PUFAs should not be affected much.

There isn't much in the literature on the influence of placebo and nocebo effects on lifespan. Self-contented optimists often enjoy longer lives, could it be because they (sub)consciously plan to live on forever (see Joseph Heller's epigraph to ch.6)? In contrast, the broader population is so well aware of the global, national and other types of life expectancy metrics, that they may simply grow to (sub)consciously accept the expiry date of say 79 years, - and when they succumb at around this age, could this be a nocebo subtly at play?

D-PUFAs underwent human testing to verify their safety and ensure they "Do no harm". A small cohort of Friedreich's ataxia patients were the first to be exposed, and they tolerated the D-PUFA well. D-PUFAs were given to patients in 1 g gelcaps at up to 9 g per day for 28 days. The treatment was found to be safe and tolerable[34]. Patients with PSP have been treated with D-PUFAs for 26 months, and tolerated the compounds well[11]. Infants and toddlers with INAD also tolerated the D-PUFAs, delivered as an emulsion, well, with no side-effects[35]. According to a meta-study (study of studies), D-PUFAs were evaluated in expanded access clinical studies of several disorders in dozens of subjects[36] with no drug-related adverse effects*. PUFA profiles in RBCs are a good proxy of their body-wide take-up. D-PUFAs have a unique mass-spec signature which allows for their straightforward identification and quantitation in blood. Pharmacokinetic and pharmacodynamic (PK/PD) studies** of blood collected from human subjects dosed on D-PUFAs revealed robust incorporation and turnover of D-PUFAs[37], with no toxicity detected even when large quantities were dosed to animals for extended periods of time***.

---

* One patient with very low BMI on a 9 grammes per day dose developed diarrhoea which cleared once that dose was split in two to be taken twice a day. This was not drug related as a similar side effect would occur if a bolus of olive oil is to be ingested on an empty stomach.

** The low-level detection and, separately, structural confirmation of D-PUFAs posed a significant analytical challenge, but it was brilliantly resolved, through unwavering perseverance, by my good friend and major collaborator, Tom Brenna, who is currently at UT Austin. Remember the Nobel prize-winning DNA and protein sequencing methods? Well, Tom has come up with PUFA sequencing, - using high-end mass-spec techniques! He "walks along" the D-PUFA strand, taking note of where the deuteriums and the double bonds are, - how cool is that!? Biochemistry now has all bases covered, - Edman-Sanger protein sequencing, Sanger DNA sequencing, Brenna PUFA sequencing. ESSB.

*** On the point of D_2O being toxic to mammals at levels above 25% of the total body water. With 42 L of water in a 70 kg human, it would require ingesting about 10 L of heavy water to reach this level. A daily dose of 1g D-DHA contains about 27 mg of D (*Back-of-the-envelope-13*). Assuming all this deuterium gets released into the body as water (which it does not), that would amount to 135 mg of D_2O, or 4-5 orders of magnitude below the toxicity level. The current fad of consuming deuterium-depleted water[38] offers drinking water with D decreased by up to 5-fold (30 ppm), and remembering that the food consumed would still have the normal deuterium abundance, the D-PUFA supplementation should not be making a dent in this deuterium-reducing "drinking diet". Refer to[37] for more detailed calculations.

6.5. An ounce of prevention is worth a pound of cure

"Let's be getting at them before they get at us."
— W.G.Grace

"The time to repair the roof is when the sun is shining."
— John F. Kennedy

At the advanced stages of AD, chunks of brain degrade and disappear, as the liquid-filled, gaping voids become clearly visible on the brain scans. Sadly, it is only in quantum mechanics that the information cannot be lost. The disappeared neurons take with them the memories, and the identity, irreversibly depriving the affected individuals of self. At that stage, it is too late to interfere.

Alive cells enact multiple defence mechanisms in their relentless but futile attempt to curb the LPO and related damage. At first, the damage accumulation is slow and only ever so slightly exceeds the repair efficiency: otherwise, we would not last for as long as we do. Yet gradually but surely, the LPO junk accumulates. You may recall that up to 75% of motor neuron volume may be taken up by lipofuscin. These deposits accelerate the cell damage.

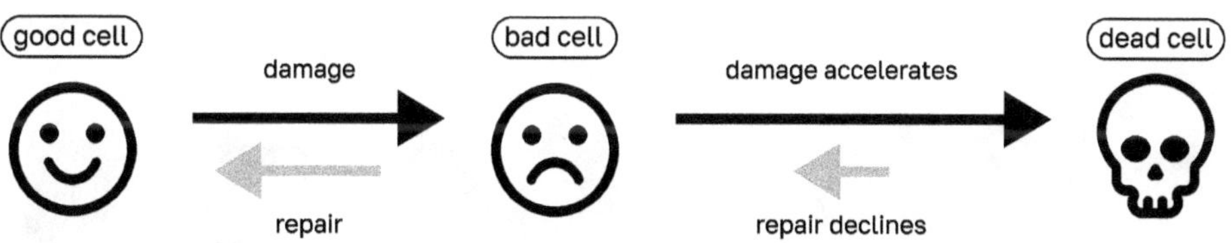

Damage outweighs repair **Overwhelming damage kills**

Sisyphus continues pushing the boulder up a hill, but every time it rolls back down, getting bigger and bigger. Eventually the inevitable demise takes place, with cell death being a point of no return.

However, if the cell is nearing, but not at the tipping point just yet, it may still be rescued. The ever-increasing damage chokes cells even as the defence systems are going full throttle to clean it up, and only gets worse as damage accumulates. So, the requirement to protect these ever-deteriorating cells, and the associated bills, could only be going in one direction.

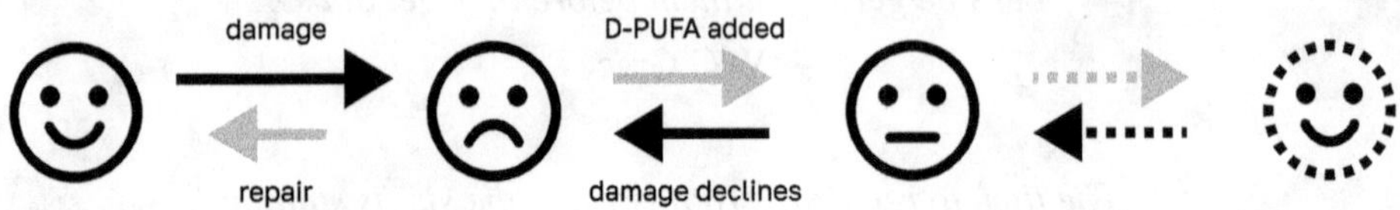

But what if the rate of damage accumulation is curbed to no longer exceed the rate of repair? The cleanup may then start catching up. Postponing cell death or, if the damage is slowed down even more, gradually rescuing cells from the brink back into life, bringing respite. It could be particularly helpful to non-dividing, post-mitotic cells, with implications for aging (you have not forgotten the book title, have you?) and diseases.

In other words, while the best option would be to load up on D-PUFAs as early as possible, starting later in life would also be beneficial*, as long as key cells are still alive, if struggling.

---

* LPO takes years to gradually induce visible detrimental changes to cells and tissues. While the D-PUFA intervention will initiate the repair process and start undoing damage, one has to prepare for a gradual progression, long term supplementation, and a long road ahead.

6.6. D proof of d pudding

"Pudding's good for you if you have it at the right time."
— W.G.Grace

"My dear, here we must run as fast as we can, just to stay in place. And if you wish to go anywhere you must run twice as fast as that."
— Lewis Carroll

Forgive a little pontification. Math reigns supreme in the world of science. Physics looks up to it, itself being looked up to by chemistry, etcetera, all the way down to social sciences. The higher up on the list, the less democracy: mathematician Grigory Perelman's view on the Poincare conjecture cannot be changed by a popular vote. No matter how inarticulate, tongue-tied, and reticent Perelman may be, when facing the other side, charismatic, and eloquent. Particularly when the proof has been accepted, having been verified by a group of top experts*. Here is a scary thought: what if there is a fiendish math problem which only one person on the planet can crack? Scarier still: what if a problem cannot be solved by a human brain at all?

It is a good thing that drug discovery is not math. Even when the mechanism of a drug action is unknown, an observation of changes in living organisms is a good enough proof. You have already seen that **puD$_2$ing** being eaten. Yet as of today, more people travelled into space than took D-PUFAs…

Making lipid membranes more resilient, with all the positive downstream effects, is where the D-PUFAs may help cells just to stay in the same place – that is, in the same age. And for this method of prevention, we don't even need an ounce: a gramme per day may do, to keep the LPO and inflammaging away.

---

* To validate Perelman's proof, the experts had worked their way through a million-dollar grant[39].This is interesting, considering that while toiling for years over the proof of Poincare conjecture, Perelman had zero funding, and lived off his modest savings put aside during short visitor spells at several universities[40]. The world is remarkably fair...

Going back to the fire-proof palisade example. Concrete or granite "logs" would be damage-resistant but expensive and time-consuming to make, forcing the settlers to use inferior wood instead of stone. They would have no doubt opted for more durable building blocks for their fence had they had the budget for it. Likewise, cells just make do with whatever is good enough to last till reproduction. However, if an outsider were to generously invest in stronger materials, to then offer them to cells for free, the cells would happily lap the expensive stuff up, if unconsciously, - and become hardier. This is the way to look at D-PUFAs*. Skilled chemists spend a lot of time and effort lovingly selecting one resilient, premium atom - deuterium - out of thousands of ordinary, "just good enough" feeble hydrogens to hand-craft the superior, damage-resistant molecules. The cells avidly incorporate these D-PUFAs into membranes. Deuterium extends the cells' resilience and expiration date.

Welcome to the Deuterium Imperium.

---

* Perhaps D-PUFAs are a bit more friendly than the concrete palisade logs for the fire-proof purposes. The settlers would need to carry unwieldy posts by hand and dig them in, in the right places, all the while being in the cross-hairs of the unfriendlies. Likewise, drugs face many challenges: delivery is inefficient, particularly across the BBB and BEB, often requiring the use of the pro-drug form; the drug toxicity is an issue (what if a concrete log falls down on the emaciated settlers, mauling them?), and so are the off-target effects; dosing requires precision; et cetera. D-PUFAs are different. They are taken up, and treasured, by the ingestion and delivery systems; they fly through BBB and BEB and incorporate into places where they are needed most (where the LPO ravages); they protect their natural, weak neighbours, and are utterly safe, - undistinguishable from the normal PUFAs. Essentially, they self-deliver, self-insert, and self-protect.

7. Closing remarks

"Watch this space."
— A 19th century formula, placed by magazines in space which was reserved by an advertiser but for which no advertisement was ever submitted

"I believe in no final solutions."
— Stanislaw Lem

Grant reviewers used to find the D-PUFA approach whacky: "This application suggests putting isotopes into humans. What a crazy idea, to poison people with radioactivity!". (For the uninitiated: the Honourable Reviewer was referring to the stable, nonradioactive deuterium, of which we have a couple of grammes in our bodies anyhow, you see). So, if you think D-PUFAs are weird, you are in a good company. But hey, wouldn't desperate diseases call for innovative measures?

The almost* unanimous rejection of the concept in the early 2010s turned out

---

* Thankfully, there are important exceptions to the word "almost". In 2005, I was working on proteomic mass-tags - detectable at very low concentrations - for high-multiplex analysis. The way to introduce mass variability within the same molecular structure seemed to be through stable isotopes, such as D and carbon-13, so I delved into isotope chemistry. Meanwhile, my leisure reading revolved around age-related chemical damage. On Christmas Day 2005, while reading about lysine side-chain oxidation products formed through hydrogen abstraction by ROS at the aminated CH_2 group, it just clicked: this rate-limiting step could be slowed down by deuterating the methylene. Since lysine is an essential amino acid, feeding its deuterated form would ensure its incorporation into all proteins, thereby reducing oxidative damage. Voilà! ROS damages all classes of biomolecules, and hydrogen abstraction is the rate-limiting step in all of them. So deuteration seemed to be also applicable to DNA bases and LPO. What would these deuterated molecules be? Not drugs, - they behave identically to biological building blocks. Yet they are more stable when it comes to damage. Identical except different. Go figure...
From the inception of the idea, one man did more for helping develop the approach than anyone else. Aubrey de Grey was the first person I discussed it with. He published the first, purely theoretical outline in his journal, Rejuvenation research, in 2006[(1)], and has been superhelpful ever since, unwaveringly endorsing the approach. Advice, fund-raising, introductions, you name it. Thank you, Aubrey!

to be a blessing. Lack of funding* forced us to travel the world, a cap in one hand, a vial of D-PUFA in another, looking for labs willing to magnanimously test the concept. Today, the sun never sets on the D-PUFA-testing Deuterium Imperium as there are more than a hundred labs worldwide actively trialling D-PUFAs in their own areas of expertise, spanning way beyond the topics covered in the previous chapters. In retrospect, the feat of simultaneously running that many experiments and models would have been impossible when relying on one's own lab**. Casting the net so wide, with good catches coming from so many areas, enabled connecting the dots into a chain, implicating the LPO as the upstream event behind multiple disorders. The medical profession tends to restrict diseases to narrow silos; yet everything is interconnected,- and the LPO tentacles are always lurking behind. Another challenge with diseases is their sheer number, currently in the tens of thousands. How could these multiple afflictions, particularly at an early stage, be diagnosed, considering that there are only several tens of different symptoms, at most? Would it not make more sense to diagnose, and tackle, LPO as the common denominator?

The graveyard of failed drugs is full of promising ideas that never made it. Rational predictions for a drug's action are very difficult because there are multiple biochemical by-passes, redundancies and compensatory mechanisms, so people rarely respond to a drug in the same way. There is often a bell-shaped distribution, spanning from non-responders to super-responders***. Chemical innovations in drug discovery seldom meet expectations, and even for rare successful examples, a lot of empirical tinkering is required to fine tune the drug molecule. This is not surprising, considering that many approved drugs operate through unknown mechanisms, and even the targets of approved medicines are often unknown[(2)]! SSRI drugs, paracetamol (acetaminophen or Tylenol in the US) and lithium are good examples.

---

* Lack of funding correlated with High Profile Journals rejecting the D-PUFA manuscripts. Likewise, the conference organizers shunned the topic. Or perhaps the invites to attend just went into my spam folder. Will need to check.

** Working in one's own lab, one just may not be agnostic enough to interpret the observations impartially. Another reason unbiased collaborators are a great asset.

*** Basically, all disease models (including in humans) are wrong, but some are useful.

Oxygen is everywhere, busy doing what it does best: oxidizing things. Antioxidants are unable to stop PUFA oxidation, and can even make things worse. Transition metals accumulate with age, yet chelating agents (*ch.2.4*) and antioxidants (*ch.2.8*) alike may make them more active in the Fenton reaction. That there is currently no drug to stop the slow killing inflicted by LPO, the only non-enzymatic chain reaction in biology, in the 3rd Millenium AD, is an affront to civilization…

D-PUFAs are a rare case of physical, not chemical, drug optimisation - thus avoiding the chemistry-versus-biology uncertainty. Their non-enzymatic action is so high upstream that the protective effect will be felt irrespective of the cell or membrane type. Using a bottle of "vegetable" oil as an example (*ch.2.5*): D-PUFAs are an equivalent of storing that bottle closed, in the dark fridge, rather than open, and in the sun. The additional energy for making D-PUFAs, which evolution was too parsimonious to invest while making things just strong enough to reach reproduction, is put in by the outside, third party – us. And the cells gladly take on, and use, this superior, reinforced material…

Remember that foul-smelling Department of Lipids' manhole we nearly fell into, while taking a walk in the Capital of Sciences (*ch.1.1*)? The vertical tunnel with the faint glimmering light at the far end? It was just that, - the light of hope at the end of the tunnel. The light of D-PUFAs. The D-light. The delight.

Conclusion - British

Might be somewhat useful

Acting through a fairly novel mechanism, D-PUFAs seem to be rather useful for allaying LPO. The chain-inhibiting capacity of C-D bonds is not too bad. D-PUFAs appear to muddle through, in some disease models, a tad better than do other approaches. Hallmarks of D-PUFAs, such as little detected toxicity and an unexpected protection at relatively low levels, turn out to be somewhat useful in medicine. With a bit of luck, D-PUFAs could have a role in the long-term mitigation of various age-related conditions, as an incremental step towards healthier aging, a small piece in the "Age-resilient body" LEGO set.

Conclusion – American

Extraordinary effect, exceptional business opportunity: bigger than sliced bread

Boldly challenging conventional wisdom, D-PUFAs are a revolutionary, cutting-edge development, jam-packed with incredible features. Overcoming tough challenges, they have massively outperformed every other therapy in fighting LPO, - the trademark of aging and age-related diseases. Due to their outstanding performance in this task, the potential applications of D-PUFAs are limitless. Their unmatched safety record, in combination with the most surprising ability to stop LPO even when present in membranes at low levels, opens up enormous opportunities for tackling a vast variety of incurable diseases. From inflammatory and metabolic disorders to neurological and retinal pathologies, diseases large and small, as well as the aging process itself (which is, essentially, a collection of various ailments), - have LPO as a common denominator, and so will all benefit from D-PUFAs. Based on a unique principle and therefore entirely orthogonal to other interventions, they can be used in combination with a wide range of existing drugs and supplements. Alternatively, as they have repeatedly proven their mettle by doing a fantastic job of beating all the competition hands down, they can also be used entirely on their own. Boasting a tremendous potential, both as prevention and as cure, the awesome D-PUFAs represent a huge leap forward in the eternal battle to extend American people's dismally short health-spans and lifespans.

The forward-looking statement

And if/when we do get longer health- and lifespans, what's then? Check out *Back-of-the-last-envelope* for one of the pushing-the-envelope options.

But at present, LPO keeps working relentlessly to wreck our bodies. It is good to start resisting it without delay. Finishing up, to maintain consistency, with a couple of stimulating quotes:

"Old age is the most unexpected of all things that happen to a man."
— Leon Trotsky

"The trouble is, you think you have time."
— Gautama Buddha

8. For those who love numbers: Back-of-the-Envelope Estimates

"If something can't be explained on the back of an envelope, it's rubbish."
— Richard Branson

Back-of-the-envelope-1: At sea level, the air pressure is 760 mm Hg, of which the oxygen component, at 21% in dry air, makes up 160 mm Hg. In lungs, air mixes with the previously inhaled, somewhat oxygen-depleted air, so the oxygen level is 104 mm Hg (13.7%). Partial pressure of oxygen (pO2) in arterial blood is 75-100 mm Hg (10-13%), and in oxygen-depleted venous blood it is 30-40 mm Hg (4-5.25%). Haemoglobin in RBCs delivers oxygen from lungs to all tissues and cells, where it diffuses down its concentration gradient. For most tissues, the pO2 is in the range of 23-70 mm Hg (3-9.2%). In cells, oxygen tension is 1-20 mm Hg (0.13-2.6%). In mitochondria, oxygen is down to 0.5 mm Hg (0.06%). These are the average numbers, as oxygen consumption varies greatly between organs[(1)].

Oxygen is 7-10 times more soluble in fat/lipid membranes than in the aqueous media and this is very useful for cells which are essentially lipid membrane bags inside of lipid membrane bags (Fig.1,8). Lipid membranes are basically the relay stations for transferring oxygen to the final destination points. At 37°C the solubility of oxygen in physiological buffer is around 200 μM, or 0.0064 g per litre. This is 1.2×10^{20} molecules, or about 5 ml of oxygen gas per litre. While it is tempting to estimate a number for oxygen molecules per cell, the exercise would be rather futile for volumes of human cells vary by more than five orders of magnitude[(2)].

Back-of-the-envelope-2: Cells vary a lot so this calculation is not gonna be terribly predictive but let's say an enzyme has a 99.99% fidelity, - a very generous assumption, on par with the precision of DNA machinery. Let's say one cycle takes 10^{-3} sec (typically 100 to 10^{-6}), and that there are 10^{9} enzymes per cell (could be 10-fold that). With these assumptions, there will be about $0.01 \times 10^{3} \times 3,600 \times 24 \times 10^{9} = 10^{15}$ errors per cell per day.

Back-of-the-envelope-3: *Ironing out the iron, and other transition metal numbers. There is up to 4 grams of iron in males and up to 3.5 grams in females. Adult humans have around 20-30 trillion red blood cells (RBC) at any given time[(3)], which makes up about 70% of all cells, by number (36 trillion cells in a 70 kg male, 28 trillion in a 60 kg female[(4)], not taking into account 38 trillion microbes and about 380 trillion viruses; for this reason the total, all-inclusive number of cells is sometimes rounded up as 100 trillion, or 1×10^{14}). Each RBC holds 270 million haemoglobin molecules, each of which contains four iron atoms, thus binding four oxygen molecules, or a total of more than a billion oxygen molecules per one RBC. Of course, not all of those RBCs will be loaded as some will be empty, en route to scoop up more oxygen in the lungs, but all of them will have their haemoglobins. 25 trillion RBCs would thus require $25 \times 10^{12} \times 2.7 \times 10^{8} \times 4 = 2.7 \times 10^{22}$ atoms of iron, or 0.045 moles (2.52 g). Some of the remaining 1.5 grams (0.027 moles, or 1.61×10^{22} atoms) of iron are in myoglobin in muscle tissue (0.4-0.5 g of iron in the total of 120-150 g myoglobin). The remaining iron, approximately 1 gram (or 1.1×10^{22} atoms), is stored in cytosol, and released in controlled way, as complexes of ferritin, a protein universal across all phyla of life, and other iron-specific protein carriers. Averaging across all cells in the body minus RBCs (which do not have DNA nor mitochondria, so their haemoglobin bound iron is already accounted for, $3.2 \times 10^{13} - 2.5 \times 10^{13} = 7 \times 10^{12}$) then gives, very roughly, 1.5×10^{9} atoms of iron per cell, a figure very similar to the number of iron atoms in one RBC's haemoglobins. This however would vary greatly depending on the number of mitochondria (from zero to thousands), DNA machinery, etc.*

Back-of-the-envelope-4: *the estimates of brain DHA levels vary. Richard Bazinet suggests a figure of 5 g, while other sources give higher estimates. An average brain, at 1,200 – 1,400 g (Anatole France's brain clocked in at 2,200 g, while Einstein's was 1,230 g, so size may not matter), with similar number of neuronal and non-neuronal cells, and a water content of 73-77 %, has a dry weight of 276 - 378 g (excluding Anatol France's). With lipids at 60% of dry weight, the total lipid content is 165 - 226 g. An alternative estimate: total lipids, at 12 % of the brain's wet weight[(5)], make up 144 – 168 g. Let's average this value to 175 g total lipids in brain. Fatty acids make up about 60 % of the mass of brain-specific lipids (made up of PLs (65%), cholesterol (25%) and plasmalogens and sphingolipids (10%)). Accordingly, brain FAs weigh at $0.6 \times 175 = 105$ g. DHA makes up 12-14 % of the total brain fatty acids[(6)], or about 14 g.*

ARA, at 8.5% of brain FAs[?], would make up around 9 g. These calculations neglect (still largely unknown) difference in FA composition between neurons and other brain cells.

Back-of-the-envelope-5: we breathe in up to 10,000 litres of air per day, of which 21% is oxygen, but we retain only 5% O2, exhaling the rest. This comes to 500 litres of oxygen per day, or around 25 moles. One-fifth is consumed by the brain, which is 5 moles, or 30×10^{23} molecules daily. According to estimates, 10% of that oxygen is converted to ROS in the brain, which would be about 3×10^{23} molecules (1/2 mole) of O2 per day. The brain holds 5-14 g of DHA and a slightly lower quantity of ARA. That's 10-23 g of LCPUFA in total, or, taking the middle value, 15 g LCPUFA per brain, with an average MW (328 + 304)/2 = 316. There is thus about 3×10^{22} molecules of these two LCPUFAs in the brain. So, every day there will be up to 10 ROS per DHA or ARA molecule looking for a hydrogen to abstract. Half-life of brain DHA is 2.5 years (according to one estimate), over which there will be about 10^4 attempts on the life of every molecule. And an ROS species only needs one successful hit on a PUFA to initiate the chain...

Back-of-the-envelope-6: A useful rule of thumb before we delve into PUFAs. A single carbon-carbon bond is 1.54 Å (angstrom, = 10^{-10} metres or 0.1 nm) in length, and an angle between the sp^3 orbitals is 109°5'. The projection of this on a flat surface (a linear length) would amount to (1.54 x cos ((180°-109.5°)/2) = 1.26 Å. So the length of a stearic (C18) acid chain, attached to a glycerol moiety, is about 19 x 1.26 = 23.94 Å. A bilayer should be twice that, or 47.88 Å. This is pretty close to the measured lipid bilayer thickness of about 4.83 nm (48.3 Å). Bonds become shorter the more electrons are involved in their formation, so a carbon-carbon double bond length is 1.33 Å, and the angles also change. Accordingly, there is a curious "homeostasis" of the bilayer thickness, keeping it more or less the same irrespective of the FA components used. ARA, with four double bonds and 20 carbons, and DHA, six double bonds and 22 carbons, would give bilayers of approximately the same thickness as would stearic acid. Change of thickness indicates membrane sickness.

Back-of-the-envelope-7: An estimated rate of the global "vegetable" oil consumption by 2026 is around 260 million tons, or 32 kg per person per year. With the Western population "benefiting" more from this than the rest of the world, the daily consumption *per person in the US and Europe exceeds 100 g/day. This excludes consumption of grain-based foods like bread and grain fed livestock, so the net total would be closer to 150 g/day/person. Not all of it is ingested, as some is used for frying. This represents a recent development, as "vegetable" oils did not exist before late 1880-s. A huge commercial cotton production generated a lot of waste, which was turned into cottonseed oil used for oiling gears until they figured out how to hydrogenate it into a solid "foodstuff" called Chrystal Cottonseed Oil, better known as Crisco. Before 1866, the Western diet consisted mainly of animal fats (tallow, suet and lard) and butter, while coconut and palm oils were mostly used in the East. Gradually the LIN intake rose from 2 g/day (approx. 1 % of the total caloric intake) in 1865, to 5 g/day in 1909, to 18 g/day in 1999, to 29 g/day in 2008 (15 % of the total caloric intake)[8]. Today, the consumption per person exceeds 60 g/day (> 20 % of the daily energy).*

Back-of-the-envelope-8: *There are other factors defining the brain size, too. Increasing the human brain (1.3 kg, see Back-of-the-envelope-4) dimensions by 50 % would increase its weight by 1.5^3 = 3.375 (3.375 x 1.3 = about 5 kg, the size of an elephant's brain). This 3D structure is supported by a cross-sectional 2D area of the neck, which accordingly should scale as $3.375^{2/3}$ = 2.25. Translating into volume, 3.375 x 70 = 240 kg. We would need to be eating accordingly, - pretty much all the time in the wild, with no spare time left for reading, to fill our new Gulliver's brain up with information. That our brain has attained an optimal size also follows from the elephant to human comparison. While we could afford a 5 kg brain with a puny 240 kg body, an elephant must grow to about 4,000 kg to carry his. This costs him 19 h per day of eating time. Clever though they might be, no wonder they do not read much. Cetaceans are the outliers here because the heat exchange and structural support parameters are different in water. An interesting fact with uncertain biological significance: squirrel is the closest to human in terms of brain to body weight ratio.*

Back-of-the-envelope-9: *Human brain makes up 2% of the bodyweight, yet consumes 20% of the entire energy output. A quarter of that, i.e. 5% of the total energy, is expended fixing damaged lipids in the brain(9), mostly through the Lands cycle. The brain surely takes the wellbeing of its fatty acids very seriously indeed.*

Back-of-the-envelope-10: *The electric component of mito membrane potential is 150-200 mV (1.5–2 x* 10^{-1} *V). Thickness of the inner mito membrane is 7 nm (7 x* 10^{-9} *m). The electric field intensity across the inner membrane of normally functioning mitos is thus in the range of 1.5-2 x* 10^{-1} */ 7 x* 10^{-9} *= 2-3 x* 10^{7} *V/m, or 20-30 megavolts per metre. To put things in perspective, the electric field strength around a lightning bolt is on the order of* 10^{4} *V/m, and anything above 2 x* 10^{3} *V/m represents a great lightning threat. A few thousand times less than in mito membranes. Is this* **mitobolic** *miracle only possible because of PUFAs?*

Back-of-the-envelope-11: *Membrane peroxidation index (MPI)= 0 x (% saturated FA) + 0.025 x (% OLE) + 1 x (% LIN) + 2 x (% LNN) + 4 x (% ARA) + 6 x (% EPA) + 8 x (% DHA), and membrane unsaturation index (MUI) = 1 x (% OLE) + 2 x (% LIN) + 3 x (% LNN) + 4 x (% ARA) + 5 x (% EPA) + 6 x (% DHA). The difference between the two is clear from this example. A membrane consisting entirely of OLE has MUI of 100, and MPI of 2.5. However, a membrane made up of 95 % saturated FA and 5 % DHA will have MUI of 30, yet MPI of 40*[(10)]*.*

Back-of-the-envelope-12: *Water molecules dissociate to some degree, yielding* H^{+} *and* HO^{-} *ions. The concentration of protons in pure water is a constant, called "potential of hydrogen" (pH). To avoid using small numbers, a logarithmic scale is employed to express it, as follows:*

$$pH = -\log[H^{+}] = 7.0$$

The equivalent parameter for heavy water, pD, is 7.4, meaning DO-D bonds have a harder time splitting, compared to the HO-H bonds. Quantitatively, the difference in concentrations is:

$$pD - pH = -\log[D^{+}] + \log[H^{+}] = \log([H^{+}]/[D^{+}]) = 0.4, \text{ or } [H^{+}]/[D^{+}] = 2.51.$$

That much more protons is due to the HO-H bond being less stable than the corresponding DO-D bond.

Back-of-the-envelope-13: *Deuterium, discovered by Harold Urey in 1931, is a stable (non-radioactive) isotope of hydrogen. On Earth, it has a natural abundance 0.0156% by atom number, or 0.0312% by mass (MW of D is 2.014, or approx. twice as heavy as H,*

MW 1.008), corresponding to approximately one atom of D per 6500 atoms of H. Human body contains 9.5 mass% (6.65 kg per 70 kg human). Put another way, we have about 6600 moles of hydrogen atoms in our bodies, and therefore about 1.03 moles of deuterium, or 2.07 g. This is an equivalent of 10.3 g heavy water per body, although deuteriums are distributed between the water fraction, estimated at between 50 % (elderly) to 75% (infants) of the body weight, and the remaining molecules. Heavy water is not toxic to mammals at below 15-20% of total water content, shows some toxicity at above 25%, and is lethal to eukaryotes at above 45% (30-40 kg heavy water per body), on account of changing all the enzymatic mechanisms. If, inconceivably, the whole pool of our body DHA was deuterated (D10-DHA) and then, impossibly, all 10 of these deuteriums were released, as D2O, in an instant, that would only add 0.296g of deuterium (or 1.47 g heavy water) to our existing pool of 2.07 g of D (or 10.3 g of heavy water), or 4-5 orders of magnitude below the toxicity level.

Back-of-the-last-envelope, glass of wine in hand, tongue in cheek:

I hope the readers who reached this far would agree: there is nothing more important than trying to extend our absurdly short lives. But what are we gonna do once we do manage to make the membranes impervious to damage (as well as implement lots of other things), and so extend our lifespans? With all the spare years ahead, we have got to keep reaching the new frontiers. "Because it is there", right? So which path(s) should we take? Let's use a telescope and a microscope to demarcate the playing ground first. The size of the observable Universe is 10^{27} metres, and let's for now try to contain our thoughts within. The Planck distance, whatever that is, is 10^{-35} metres. That's all we have, no matter if we travel slower or faster than light... And where exactly are we, on this playing field? Here: at the 10^0 metre scale, we are a mere 27 orders of magnitude away from the edge of the Universe, - but the whole 35 orders of magnitude away from the bottom of the microworld. Applying the "least resistance" rule of thumb, looks like we have to start building a star ship... But wait a minute. What exactly is the Planck length anyway? Bearing in mind that the Fermi distance (10^{-15}) is how big a proton, made up of smaller quarks, is. And that's about where we currently are in the microworld experiments, say 10^{-16}-ish... Now look the other way. One light year is 10^{16} metres, and the nearest star, Proxima Centauri, is 4 x 10^{16} metres away. This way, we are smack in the middle: 10^{16} – 1(us, glass of wine in hand) – 10^{-16}! And can easily go either way, or both, as it is the same number of steps :)

Eagle-eyed readers might have noticed the "one metre - us" discrepancy. We are a bit bigger, but this may be fixed. With the advent of DNA editing technology, we should shrink ourselves down to a one metre height, or less. Smaller folks need less food (including PUFAs), less resources, smaller cars, - and they live longer (what with lower blood pressure, less cells to give rise to cancer, less radioactive potassium in muscles, etc). With global warming, the heat exchange is easier the smaller the body size. And we can cram more of **oursmallerselves** into the future space ships...

There are plenty of abundant atoms around, and unlimited energy (such as the Dyson sphere batteries, conveniently distributed across the space so that we don't even need to tug them along) to make the less abundant atoms, once we figure out the rules of the 10^{-16} world. And atoms and energy are all we need to conquer the 10^{16} world. This may take a bit of time – so the extended lifespan is useful...

So, for now, let's just protect our PUFAs against LPO, and may Lord help us. Cheers.

9. ACKNOWLEDGEMENTS

Figure 30. Piggybacking on the Cinderella story, these 13 pairs of shoes (the 14th is mine) belong to a variously aged crew who kept me going. One soul is conspicuously missing as he prefers to run around barefoot, rather than wear glass slippers. Had he worn them, there would have been four more shoes here.

I am very grateful to my family (*Fig.30*), for putting up with my unavailability and antics for a veritable year, for showing tolerance and understanding, and for providing uninterrupted supplies and unwavering support to the grumpy hermit's confinement. It was really nice to reunite with them a year later, and I am pleased to say the feeling was mutual, as my unkempt and bearded self finally emerged from the voluntary incarceration.

I would not have obtained such wonderful data, nor would I have ever written this book, if it weren't for my wonderful collaborators. A huge thank you to all of you! And particular thanks to Ruben Abagyan, Arina Alaferdova, Gene Andersen, Alexander Andreyev, Yuri Antonenko, Mauricio Baptista, Richard Bazinet, Valery Bochkov, John Bower, Tom Brenna, Yuri Bunimovich, Charles Cantor, Cathy Clarke, Steven Clarke, Michael Crawford, Stephen Cunnane, Josh Dunaief, Thierry Durand, Paul Else, Jean-Marie Galano, Mark Gomelsky, Claudia Gravekamp, Aubrey de Grey, Sergei Gutnikov, Barry Halliwell, Tony Hulbert, Ned Israelsen, Valerian Kagan, Kostya Khrapko, Anatoly Kontush, Vladimir Kovalzon, Gero Miesenböck, Ginger Milne, Nikolai Pestov, Lada Petrovskaya, Ned Porter, Valerie O'Donnell, Malgorzata Rozanowska, Vera Rybko, Olga Sharko, Vadim Shmanai, Alex Smarun, Daniel Tellez, Lex van der Ploeg, Drasko Vidovic, Jan Vijg, Tessa Westbrook, Rob Wilson and Tonya Zenin for their influence on these chapters.

10. REFERENCES

"Expand your references, and you'll immediately expand your life"
— Tony Robbins

"But it's not just a game of findig literary references"
— Dan Simmons

1.1

1. PMID: 28106741
2. ISBN: 9781421445779
3. ISBN: 9781915465122
4. PMID: 37824663
5. PMID: 26451484

1.2

1. Rosella Lorenzi, Discov.News. 19.02.2008

1.3

1. PMID: 38503334
2. ISBN 0-312-36706-6
3. PMID 13332224

2.1

1. 10.1038/s41561-024-01480-8
2. PMID: 31152014

2.3

1. PMID: 13156638
2. N. Shchepinov, EVOLVE OX14 Biology Magazine, V4, July 2023.
3. Judy Campisi, personal communication
4. PMID: 32726907
5. ISBN 9780750308243
6. PMID: 3456598
7. PMID: 17928583
8. ISBN: 0-471-29646-5

2.4

1. PMID 26479784
2. 10.1038/s41561-024-01480-8
3. PMID: 39354222
4. 10.3389/fmars.2016.00134
5. ISBN: 978-0-19-871784-5. Table 3.6
6. PMID: 17009284
7. PMID 10093929
8. E.B. Aleksandrov, public talk

2.5.1

1. PMID: 12323085
2. PMID: 12323090
3. PMID: 34991720
4. ISBN: 9781421445779
5. PMID: 4580982
6. PMID: 32477293
7. PMID: 10823917
8. PMID: 5873387
9. PMID: 28468892
10. PMID: 24704580
11. PMID: 24050531
12. PMID: 19661256
13. PMID: 26901223
14. PMID: 24828044
15. PMID: 27613109
16. ISBN: 9781915465122; ISBN: 0749306688
17. PMID: 6739443
18. ISBN: 9781398720732
19. PMID: 37513547
20. R. Bazinet, personal communication
21. PMID 32643951
22. PMID: 3092868
23. PMID: 36046053
24. PMID: 6348496
25. PMID: 10601692
26. PMID: 12736897
27. PMID: 18343442
28. PMID: 26528354
29. PMID: 7494623
30. PMID: 23631634
31. PMID 31130146
32. PMID: 18973997
33. R. Bazinet, personal communication

34. ISBN: 0192629271
35. PMID 30415628
36. T. Brenna, R. Bazinet, personal communication
37. www.fda.gov/news-events/press-announcements/fda-approves-use-drug-reduce-risk-cardiovascular-events-certain-adult-patient-groups; https://www.ema.europa.eu/en/medicines/human/EPAR/vazkepa]
38. PMID 21562563
39. PMID: 17194275
40. PMID: 27103682
41. PMID: 22855869
42. PMID: 18973997
43. PMID: 36196762
44. PMID: 37357914
45. ISBN: 9780367380748
46. PMID: 36005540
47. PMID: 34400132
48. PMID: 25954194
49. PMID: 16828044
50. PMID: 18292294

2.5.2

2.5.3

1. 10.1111/j.1439-0310.1968.tb00028.x
2. PMID: 18725629
3. A. L. Buchachenko, É. M. Galimov, et al., Dokl. Akad. Nauk SSSR, 228, 379 1976.
4. PMID: 27216936
5. PMID: 2541828; 24070914
6. PMID: 32467161; 12686
7. PMID: 30859366
8. PMID: 34934425; 33737599
9. PMID: 35414166
10. PMID: 4177393
11. PMID: 4949878
12. PMID: 1277808
13. PMID: 4752217
14. PMID: 27833543

2.5.4

1. PMID: 37998212
2. ISBN: 9780691195889
3. PMID: 23206328
4. PMID: 17689877
5. PMID: 28063940
6. PMID: 38920469
7. doi.org/ndqb
8. PMID: 39294965

2.6

1. PMID: 29632885
2. PMID: 21861450
3. PMID: 32071215; PMID: 37124288, Schaich, Karen M. "Lipid oxidation: theoretical aspects." Bailey's industrial oil and fat products 1. Part 7 (2005): 273-303.
4. PMID: 9214578
5. ISBN 978-981-13-6259-0; PMID: 23159885
6. PMID: 2911023
7. PMID: 24262192; 28326165
8. PMID: 21054827; 32371450
9. PMC2646729
10. PMID: 20093384
11. PMC2741612
12. PMC8245805
13. PMID: 34758328
14. PMID: 33147438
15. PMID: 23159885
16. 10.1101/2024.02.25.581768
17. PMID: 5289873
18. PMID: 16802290
19. PMID: 11131034; 10942908; 11481669
20. PMID: 19059309

2.7

1. 10.1002/ejlt.201400114
2. PMID: 25450347

2.8

1. PMID: 18316025; 28357236
2. PMID: 23581571
3. PMID: 2698099
4. PMID: 35150738
5. PMID: 34699937
6. PMID: 1463440
7. PMID: 8827516
8. PMID: 38722242
9. PMID: 16443163
10. ISBN: 978-0-19-871784-5
11. PMID: 11117200 + Paul Else, unpublished
12. PMID: 28785371
13. PMID: 25668303; 25180889
14. PMID: 5787094
15. PMID: 20649545
16. PMID: 18493812
17. PMID: 15557412

18. PMID: 8127329
19. PMID: 32113652
20. PMID:17729110
21. PMID: 11603657
22. PMID: 19500666
23. PMID: 18547875
24. PMID: 17928583
25. PMID: 31740834
26. PMID: 21824100
27. PMID: 13332224

2.9

1. PMID: 26898250
2. www.deingenieur.nl/artikel/racing-cyclist-in-peloton-saves-more-energy-than-previously-thought
3. PMID: 25449649
4. PMID: 1937131
5. PMID: 36789073
6. PMID: 22578669
7. 10.1016/S0924-2244(01)00022-X
8. N. Shchepinov, EVOLVE OX14 Biology Magazine, July 2024, V7
9. PMID: 7945894
10. PMID: 15730188
11. PMID 23767955
12. PMID: 9278044
13. PMID: 28592453
14. PMID: 17928583
15. PMID: 19283520
16. PMID: 16236899
17. PMID: 25380349
18. PMID: 26227873
19. PMID: 18045864
20. PMID: 27346802
21. PMID: 15337754; 31114584
22. PMID: 29706967
23. ISBN: 9781421445779
24. PMID: 37251392; 38001689

3.1

1. Lung Cancer Tied to Vapor From Oil in Stir Frying The NYT November 1, 1987, Section 1, Page 44.
2. PMID: 30485694
3. https://montrealgazette.com/opinion/columnists/the-right-chemistry-no-eating-french-fries-is-not-the-same-as-smoking-cigarettes
4. PMID: 18442969
5. PMID: 25807518
6. https://weekly.chinacdc.cn/article/doi/10.46234/ccdcw2020.166?pageType=en
7. www.scopus.com/inward/record.url?scp=0035716446&partnerID=8YFLogxK
8. ISBN: 978-1922247773
9. As suggested by Tom Brenna

3.2

1. https://omegaquant.com/how-to-tell-if-fish-oil-capsules-are-rancid/
2. 10.5858/2003-127-1603-MOMLIC
3. edepot.wur.nl/121318
4. European Pharmacopoeia 10.0, 2.5.5, 01/2008:20505, p138
5. PMID: 22159321
6. https://www.ncbi.nlm.nih.gov/pmc/articles/PMC3821093/#R13
7. PMID: 25604397
8. PMID: 28011269
9. PMID: 23656645
10. PMID: 38800667
11. PMID: 23843441
12. PMID: 37872251
13. PMID: 38068754
14. PMID: 25856365
15. https://www.acc.org/Latest-in-Cardiology/Clinical-Trials/2018/11/08/22/48/REDUCE-IT
16. PMID: 39900648

3.3

3.4

1. PMID: 22353612
2. PMID: 13307939
3. ISBN: 9781421445779
4. ISBN: 9781398720732
5. PMID: 17125530
6. Tom Brenna, personal communication
7. 10.1038/s41598-023-50119-y
8. PMID: 10817132; 37193692
9. PMID: 27581992
10. PMID: 26607973
11. PMID: 16026335
12. PMID: 34681529
13. PMID: 10477247
14. PMID: 38802606
15. PMID: 36380074
16. 10.3390/agriculture14111889

4.1

1. ISBN: 9780747592860; 9780691195889
2. 10.31857/S0301179824020045
3. PMID: 19951742
4. PMID: 25995483
5. PMID: 23150373
6. PMID: 22405878
7. PMID: 34971765
8. PMID: 3805015
9. PMID: 37513514
10. PMID: 37513514
11. 10.2165/00126839-200405030-00001; PMID: 3205065
12. PMID: 34210002
13. PMID: 2801038
14. PMID: 38092848

4.2

1. ISBN: 1-57059-564-X
2. PMID: 12208348
3. PMID: 1588600; 21782935
4. PMID: 8179604
5. PMID: 8159737
6. PMID: 8978733
7. PMID: 15721985
8. PMID: 32322100
9. PMID: 17243916
10. PMID: 10888369; 26576216
11. PMID: 35884683
12. PMID: 31317824
13. PMID: 26872597
14. PMID: 2875360
15. PMID: 20580911
16. PMID: 25580849
17. PMID: 34730940
18. PMID: 14593171
19. PMID: 24895477
20. PMID: 18625454
21. PMID: 36453394
22. PMID: 33111259
23. PMID: 30347635
24. PMID: 28494957
25. PMID: 25339906
26. PMID: 36361861
27. PMID: 20923426
28. PMID: 11979513
29. PMID: 15363659
30. PMID: 23680468
31. PMID: 38066913

4.3

1. PMID: 9234964
2. ISBN 0199205647; 9781570595646;
3. PMID: 19285551
4. PMID: 10998361
5. PMID: 25945934
6. PMID: 31435505
7. PMID: 34614167
8. PMID: 31435505
9. PMID: 20577992
10. PMID: 9046246
11. PMID: 37074148; 39506167; 10.1101/2023.04.14.536897
12. ISBN: 1-57059-564-X
13. PMID: 20663895
14. PMID: 11408659
15. PMID: 17928583
16. PMID: 26315290
17. PMID: 12912909, 24204965
18. PMID: 14556858
19. PMID: 25843654
20. PMID: 29370159
21. PMID: 38952719
22. PMID: 33707189
23. PMID: 22266017
24. PMID: 25312902
25. PMID: 30737462
26. PMID: 22269163
27. PMID: 22584571
28. PMID: 24117165
29. PMID: 25546574
30. PMID: 33126055
31. PMID: 22566778; 18296342

4.4

1. ISBN: 9780691195889
2. PMID: 24161127
3. PMID: 17684094
4. PMID: 18769139
5. PMID: 31680947
6. PMID: 35149052
7. PMID: 25051888
8. PMID: 22581285
9. PMID: 23271050
10. ISBN: 978-0-19-871784-5, 10.24

4.5

1. PMID: 16251949
2. PMID: 24136970
3. PMID: 38877307
4. PMID: 21621560

5. PMID: 38360946
6. PMID: 30894743; 40108451
7. doi.org/10.5281/zenodo.4544046
8. PMID: 24438530
9. PMID: 20923426
10. PMID: 33467135
11. PMID: 34918026
12. PMID: 32121189
13. PMID: 38916679
14. PMID: 38016470
15. PMID: 32502393

4.6

1. PMID: 14559957
2. PMID: 28572872
3. PMID: 17215962
4. ISBN: 978-1-394-15838-6
5. ISBN: 9781785044540
6. ISBN: 978-0-19-871784-5
7. PMID 11110863
8. PMID 8624775
9. Tom Brenna, personal communication
10. PMID: 14757688
11. ISBN: 978-981-13-6259-0
12. PMID: 27346802
13. PMID: 18045864
14. PMID: 15808418
15. PMID: 39303953
16. PMID: 38397830
17. PMID: 10855528; PMID: 10391905
18. ISBN: 9781421445779
19. PMID: 34764414
20. PMID: 28864332
21. PMID: 32562735
22. PMID: 34595505
23. ISBN: 9781398720732

4.7

1. PMID: 33561467
2. PMID: 7373078
3. PMID: 21558561
4. Frankel, E.N. Hydroperoxide formation. In Lipid Oxidation, 2nd ed.; Oily Press Lipid Library Series; University of California: Davis, CA, USA, 2005; Chapter 2; pp. 25–50.
5. PMID: 11826181
6. PMID: 23000938
7. Tom Brenna, personal communication
8. PMID: 11333111
9. PMID: 17376148
10. PMID: 33561467
11. PMID: 17009284
12. PMID: 34797713
13. PMID: 18637798
14. PMID: 33670907
15. PMID: 20215107
16. PMID: 25653189
17. ISBN: 978-0-19-871784-5;
18. PMID: 37459506
19. PMID: 28084040
20. PMID: 28315451
21. PMID: 9277149
22. 10.1016/j.jece.2020.104211
23. PMID: 34563935
24. PMID: 32119999
    R.R.Anderson, Photochem Photobiol 2004,80:155 Invited Editorial

4.8

1.
2. PMID: 18285327
3. PMID: 1937131
4. PMID: 11895126
5. PMID: 22632970
6. PMID: 34436298
7. PMID: 27506793
8. PMID: 29632885
9. PMID: 29053969
10. PMID: 29731704
    PMID: 35354936

4.9

1. ISBN: 9781780723723
2. PMID: 25838312
3. PMID: 9597150
4. PMID: 25316652
5. PMID: 32808658
6. PMID: 26139350
7. PMID: 39500419
8. PMID: 26561701
9. ISBN: 9781398720732
10. PMID: 30665457; 30971478; 16203828
11. PMID: 35126121
12. PMID: 21838667
13. PMID: 29882120
14. PMID: 8251267
15. PMID: 24674584
16. PMID: 23644052
17. PMID: 18706086
18. PMID: 29889594
19. PMID: 23258410
20. PMID 6089195
21. PMID: 25139562; 29757195

22. PMID: 16151530
23. PMID: 35308198
24. PMID: 33087142

5

1. PMID: 17928583
2. PMID: 7842141, 10337442
3. PMID: 34362885
4. PMID: 9005875
5. PMID: 16677102
6. PMID: 742290
7. PMID: 15855390
8. ISBN: 9780367380748
9. PMID: 26942670
10. PMID: 17928583
11. PMID: 11128446
12. 10.1093/icb/38.2.341
13. PMID: 17077193
14. PMID: 16620917
15. PMID: 23105667
16. PMID: 12091096
17. PMID: 1277808
18. PMID: 17157356
19. PMID: 11413492
20. PMID: 7568133
21. PMID: 12681474
22. PMID: 24923566
23. PMID: 31809755
24. PMID: 35048548
25. PMID: 32839400
26. PMID: 27453792; 27607581
27. PMID: 29090099
28. Personal communication
29. PMID: 35370799
30. PMID: 26748744
31. PMID: 35883714
32. PMID: 12974942
33. PMID: 30616998
34. ISBN: 3-7186-4983-7
35. PMID: 13054692
36. PMID: 39485277
37. PMID: 22915358
38. PMID: 12610228
39. PMID: 31366722
40. ISBN: 978-0-19-871784-5, p658
41. PMID: 2058418
42. PMID: 10742195
43. PMID: 12595829
44. Tom Brenna, personal communication.
45. PMID: 15937020
46. PMID: 28540446, 23494666
47. PMID: 16781458
48. PMID: 17964285
49. PMID: 35620545
50. PMID: 24962132
51. PMID: 17054663
52. PMID: 30881592
53. PMID: 18547875
54. PMID: 10973932
55. PMID: 25906193
56. 10.1093/gerona/60.7.847; PMID: 29424490
57. PMID: 15689603
58. PMID: 24783202
59. PMID: 24241129
60. PMID: 22499901
61. PMID: 25035131
62. PMID: 23840953
63. PMID: 38710468

5.7

1. PMID: 24780861
2. PMID: 37074683
3. 10.1007/978-3-030-26780-3; 10.1007/978-3-031-39171-2
4. PMID: 9307931
5. PMID: 17313320; 12788931
6. PMID: 28735096
7. PMID: 19622391
8. PMID: 10934055
9. PMID: 36009187
10. PMID: 20629847
11. PMID: 15307867
12. PMID: 27999257
13. PMID: 22396448
14. PMID: 25837702
15. PMID: 3319229

6

1. www.newscientist.com/article/mg20827844-000-heavy-hydrogen-keeps-yeast-looking-good/
2. PMID: 22705367
3. PMID: 30851224
4. PMID: 24380377
5. PMID: 16491182
6. 10.1002/slct.201600955
7. PMID: 29131956
8. PMID: 35276579
9. PMID: 25499576
10. PMID: 34931062
11. PMID: 34202031

12. PMID: 25578654
13. PMID: 25580849
14. PMID: 27506793, 29632885
15. PMID: 28655430
16. PMID: 35557972
17. PMID: 29933522
18. PMID: 21906664
19. PMID: 33308320
20. PMID: 29579687
21. PMID: 29024570
22. PMID: 35257475
23. PMID: 39956472
24. PMID: 35870486
25. PMID: 29206462
26. PMID: 35453366
27. PMID: 37897398
28. PMID: 20083013
29. PMID: 514753
30. PMID: 18319411
31. PMID: 25632091
32. PMID: 38701224
33. PMID: 8936626
34. PMID: 29624723
35. PMID: 32685351
36. J. Clin. Tox. 2021; V11; Issue5; 1000490.
37. PMID: 32871154
38. PMID: 38732643
39. newyorker.com/magazine/2006/08/28/manifold-destiny?currentPage=all/
40. arxiv.org/pdf/math.DG/0211159

7

1. PMID: 17378752
2. PMID: 22711801

8

1. PMID: 37406413
2. PMID: 8660387
3. ISBN: 978-0-7352-1361-6
4. PMID 37722043
5. PMID: 20329590; 3916238
6. PMID: 11724467; 22254110
7. PMID: 5865382; 4302302
8. ISBN: 9781398720732
9. PMID: 24780861
10. ISBN: 9781421445779

www.ingramcontent.com/pod-product-compliance
Lightning Source LLC
Chambersburg PA
CBHW061138220326
41599CB00025B/4277